W9-AKD-676

WIRING DIAGRAMS
for
LIGHT and POWER

Revised by Roland E. Palmquist

THEODORE AUDEL & CO.
a division of

HOWARD W. SAMS & CO., INC.
4300 West 62nd Street
Indianapolis, Indiana 46268

THIRD EDITION

1978 PRINTING

International Standard Book Number: 0-672-23232-4
Library of Congress Catalog Card Number: 75-7017

FOREWORD

The importance of wiring diagrams of electrical machinery and associated relays and instruments is well known to all electrical workers. This volume has been completely revised and expanded to provide up-to-date reference to the accepted standard wiring diagrams for a wide variety of applications. While it is impossible to cover all possible power system arrangements and operating conditions, attention has been given to the conditions most often encountered in standard practice.

Numerous illustrative, diagrammatic examples are given, especially in the parts dealing with power transformers and synchronizing connection of alternating-current generators, due to the importance of this subject whenever electric energy is generated and transmitted.

Because other symbols and methods of wiring may be possible in electrically equivalent circuits, great care should be observed when connecting electrical apparatus. The diagrams furnished by the manufacturer of equipment to be installed should be followed in each individual case.

ROLAND E. PALMQUIST

Contents

WIRING SYMBOLS

Throughout this course and all through your career as an electrician, you will have need for "electrical symbols." They will be used in schematics, drawings, and numerous other places. This chapter is inserted at this point so that you might familiarize yourself with the meaning of these symbols and know where to look to find the meaning when you come on symbols.

The total coverage of electrical and electronic symbols may be obtained from American National Standards Institute, Inc., 1430 Broadway, New York, N.Y. 10018.

Try to become familiar with the following symbols, so that you will recognize them, and thus many explanations in this text may be avoided.

GRAPHIC ELECTRICAL WIRING SYMBOLS

Compiled by American Standard Graphic electrical wiring symbols for architectural and electrical layout drawings.

0.1 DRAFTING PRACTICES *applicable to graphic electrical wiring symbols.*

a. Electrical layouts should be drawn to an appropriate scale or figure-dimensions noted. They should be made on drawing sheets separate from the architectural or structural drawings or the drawing sheets for mechanical or other facilities.

Clearness of drawings is often reduced when all different electric systems to be installed in the same building area are laid out on the same drawing sheet. Clearness is further reduced when an extremely small drawing scale is used. Under these circumstances, each or certain of the different systems should be laid out on separate drawing sheets. For example, it may be better to show signal system outlets and circuits on drawings separate from the lighting and power branch circuit wiring.

b. Outlet and equipment locations with respect to the building should be shown as accurately as possible on the electrical drawing sheets to reduce reference to architectural drawings. Where extremely accurate final location of outlets and equipment is required, figure dimensions should be noted on the drawings. Circuit and feeder run lines should be so drawn as to show their installed location in relation to the building insofar as it is practical to do so. The number and size of conductors in the runs should be identified by notation when the circuit run symbol does not identfy them.

c. *All branch circuits, control circuits and signal system circuits should be laid out in complete detail on the electrical drawings including identification of the number, size and type of all conductors.*

d. *Electric wiring required in conjunction with such mechanical facilities as heating, ventilating and air conditioning equipment, machinery and processing equipment should be included in detail in the electrical layout insofar as possible when its installation will be required under the electrical contract. This is desirable to make reference to mechanical drawings unnecessary and to avoid confusion as to responsibility for the installation of the work.*

e. *A complete electrical layout should include at least the following on one or more drawings:*

   1. *Floor plan layout to scale of all outlet and equipment locations and wiring runs.*
   2. *A complete schedule of all of the symbols used with appropriate description of the requirements.*
   3. *Riser diagram showing the physical relationship of the service, feeder and major power runs, units substations, isolated power transformers, switchboards, panel boards, pull boxes, terminal cabinets and other systems and equipment.*
   4. *Where necessary for clearness, a single line diagram showing the electrical relationship of the component items and sections of the wiring system.*
   5. *Where necessary to provide adequate information elevations, sections and details of equipment and special installations and details of special lighting fixtures and devices.*
   6. *Sections of the building or elevation of the structure showing floor to floor, outlet and equipment heights, relation to the established grade, general type of building construction, etc. Where practicable, suspended ceiling heights indicated by figure dimensions on either the electrical floor plan layout drawings or on the electrical building section or elevation drawings.*
   7. *Where necessary to provide adequate information plot plan to scale, showing the relation of the building or*

structure to other buildings or structures, service poles, service manholes, exterior area lighting, exterior wiring runs, etc.

8. *In the case of exterior wiring systems for street and highway lighting, area drawings showing the complete system.*
9. *Any changes to the electrical layout should be clearly identified on the drawings when such changes are made after the original drawings have been completed and identified on the drawing by a revision symbol.*Δ

0.2 EXPLANATION SUPPLEMENTING THE SCHEDULE OF SYMBOLS

a. *GENERAL*

1. TYPE OF WIRING METHOD OR MATERIAL REQUIREMENT: *When the general wiring method and material requirements for the entire installation are described in the specifications or specification notations on drawings, no special notation need be made in relation to symbols on the drawing layout, e.g., if an entire installation is required by the specifications and general reference on the drawings to be explosion proof, the outlet symbols do not need to have special identification.*

   When certain different wiring methods or special materials will be required in different areas of the building or for certain sections of the wiring system or certain outlets, such requirements should be clearly identified on the drawing layout by special identification of outlet symbols rather than only by reference in the specifications.
2. SPECIAL IDENTIFICATION OF OUTLETS: *Weather proof, vapor tight, water tight, rain tight, dust tight, explosion proof, grounded or recessed outlets or other special identification may be indicated by the use of uppercase letter abbreviations at the standard outlet symbol, e.g.:*

Weather proof	*WP*
Vapor tight	*VT*
Water tight	*WT*
Rain tight	*RT*
Dust tight	*DT*
Explosion proof	*EP*

Grounded	***G***
Recessed	***R***

The grade, rating and function of wiring devices used at special outlets should be indicated by abbreviated notation at the outlet location.

When the standard Special Purpose Outlet symbol is used to denote the location of special equipment or outlets or points of connection for such equipment, the specific usage will be identified by the use of a subscript numeral or letter alongside the symbol. The usage indicated by different subscripts will be noted on the drawing schedule of symbols.

b. LIGHTING OUTLETS

1. IDENTIFICATION OF TYPE OF INSTALLATION: ***A major variation in the type of outlet box, outlet supporting means, wiring system arrangement and outlet connection and need of special items such as plaster rings or roughing-in cans, often depends upon whether a lighting fixture is to be recessed or surface mounted. A means of readily differentiating between such situations on drawings has been deemed to be necessary. In the case of a recessed fixture installation the standard adopted consists of a capital letter R drawn within the outlet symbol.***
2. FIXTURE IDENTIFICATION: ***Lighting fixtures are identified as to type and size by the use of an upper-case letter, placed alongside each outlet symbol, together with a notation of the lamp size and number of lamps per fixture unit when two or more lamps per unit are required. A description of the fixture identified by the letter will be given either in the drawing schedule of symbols, separate fixture schedule on the drawing or in the electrical specifications.***
3. SWITCHING OF OUTLETS: ***When different lighting outlets within a given local area are to be controlled by separately located wall switches, the related switching will be indicated by the use of lower-case letters at the lighting and switch outlet locations.***

c. SIGNALLING SYSTEMS

1. BASIC SYMBOLS: ***Each different basic category of signalling system shall be represented by a distinguishing***

Basic Symbol. Every item of equipment or outlet comprising that category of system shall be identified by that basic symbol.

2. IDENTIFICATION OF INDIVIDUAL ITEMS: *Different types of individual items of equipment or outlets indicated by a basic system symbol will be further identified by a numeral placed within the open system basic symbol. All such individual symbols used on the drawings shall be included on the drawing schedule of symbols.*

3. USE OF SYMBOLS: *Only the basic signalling system outlet symbols are included in this Standard. The system or schedule of numbers referred to in (2) above will be developed by the designer.*

4. RESIDENTIAL SYMBOLS: *Signalling system symbols for use in identifying certain specific standardized residential type signal system items on residential drawings are included in this Standard. The reason for this specific group of symbols is that a descriptive symbol list such as is necessary for the above group of basic system symbols is often not included on residential drawings.*

d. POWER EQUIPMENT

1. ROTATING EQUIPMENT: *At motor and generator locations, note on the drawing adjacent to the symbol the horsepower of each motor, or the capacity of each generator. When motors and generators of more than one type or system characteristic, i.e., voltage and phase, are required on a given installation, the specific types and system characteristics should be noted at the outlet symbol.*

2. SWITCHBOARDS, POWER CONTROL CENTERS, UNIT SUBSTATIONS AND TRANSFORMER VAULTS: *The exact location of such equipment on the electrical layout floor plan drawing should be shown.*
   A detailed layout including plan, elevation and sectional views should be shown when needed for clearness showing the relationship of such equipment to the building structure or other sections of the electric system.
   A single line diagram, using American Standard Graphic Symbols for Electrical Diagrams—Y32.2.

should be included to show the electrical relationship of the components of the equipment to each other and to the other sections of the electric system.

e. SYMBOLS NOT INCLUDED IN THIS STANDARD

1. *Certain electrical symbols which are commonly used in making electrical system layouts on drawings are not included as part of this Standard for the reason that they have previously been included in American* ***Standard Graphic Symbols for Electrical Diagrams,*** *W32.2.*
   ASA policy requires that the same symbol not be included in two or more Standards. The reason for this is that if the same symbol was included in two or more Standards, when a symbol included in one Standard was revised, it might not be so revised in the other Standard at the same time leading to confusion as to which was the proper symbol to use.
2. *Symbols falling into the above category include, but are not limited to, those shown below. The reference numbers are the American Standard Y32.2 item numbers.*

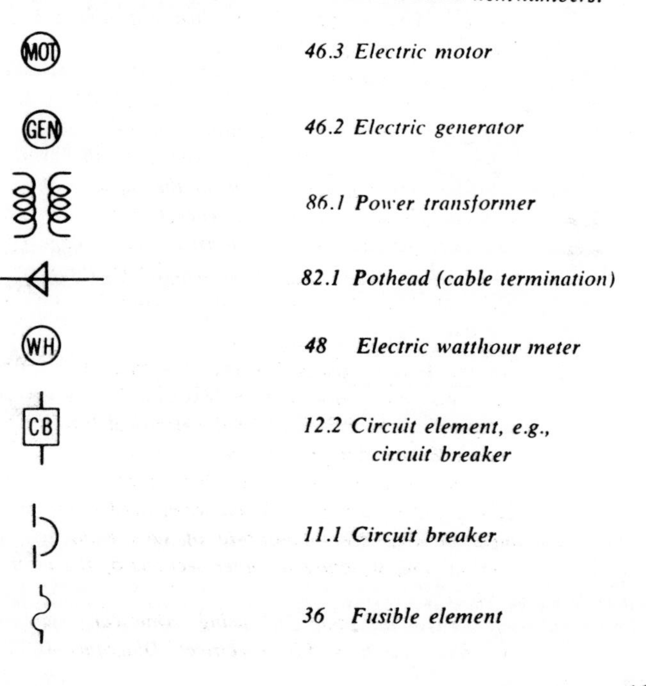

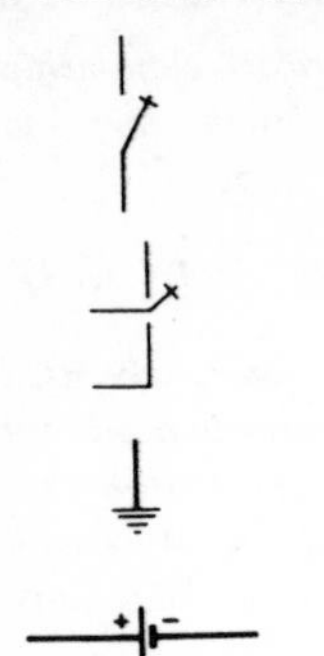

76.3 Single-throw knife switch

76.2 Double-throw knife switch

13.1 Ground

7 Battery

LIST OF SYMBOLS

1.0 Lighting Outlets

Ceiling **Wall**

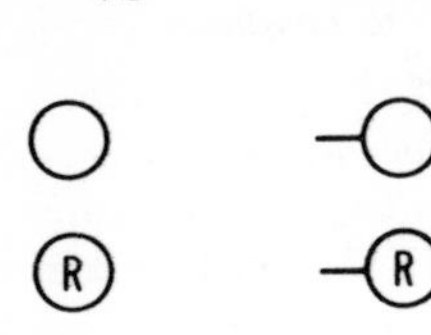

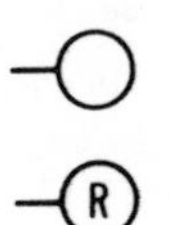

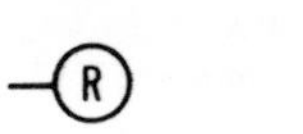

1.1 Surface or pendant incandescent mercury vapor or similar lamp fixture

1.2 Recessed incandescent mercury vapor or similar lamp fixture

1.3 Surface or pendant individual fluorescent fixture

1.4 Recessed individual fluorescent fixture

1.5 Surface or pendant continuous-row fluorescent fixture

*1.6 *Recessed continuous-row fluorescent fixture*

*1.7 **Bare-lamp fluorescent strip*

*In the case of combination continuous-row fluorescent and incandescent spotlights, use combinations of the above standard symbols.

**In the case of continuous-row bare-lamp flourescent strip above an area-wide diffusing means, show each fixture run, using the standard symbol; indicate area of diffusing means and type by light shading and/or drawing notation.

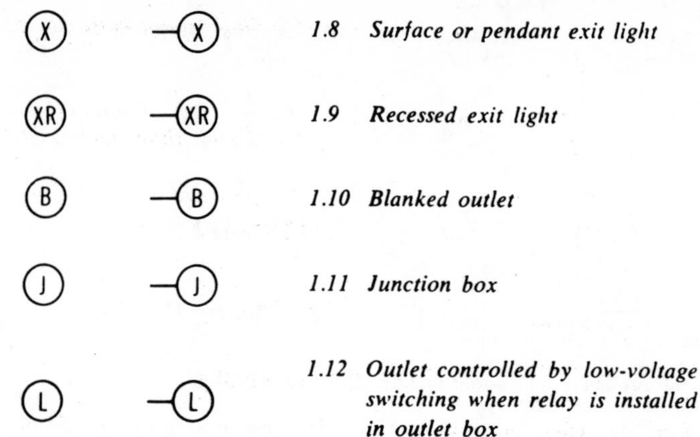

1.8 Surface or pendant exit light

1.9 Recessed exit light

1.10 Blanked outlet

1.11 Junction box

1.12 Outlet controlled by low-voltage switching when relay is installed in outlet box

2.0 Receptacle Outlets

Where all or a majority of receptables in an installation are to be of the grounding type, the upper-case letter abbreviated notation may be omitted and the types of receptacles required noted in the drawing list of symbols and/or in the specifications. When this is done, any non-grounding receptacles may be so identified by notation at the outlet location.

Where weather proof, explosion proof or other specific types of devices are to be required, use the type of upper-case subscript letters referred to under Section 0.2 item a-2 of this Standard. For example, weather proof single or duplex receptacles would have the upper-case subscript letters noted alongside of the symbol.

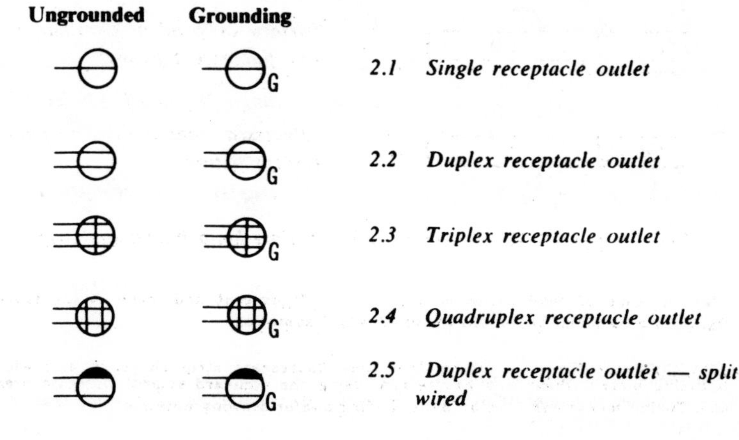

2.1 Single receptacle outlet

2.2 Duplex receptacle outlet

2.3 Triplex receptacle outlet

2.4 Quadruplex receptacle outlet

2.5 Duplex receptacle outlet — split wired

ELECTRICAL SYMBOLS

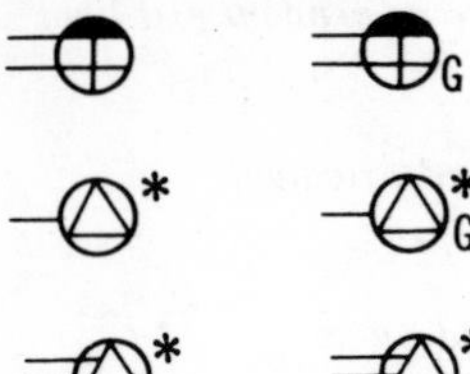

2.6 Triplex receptacle outlet — split wired

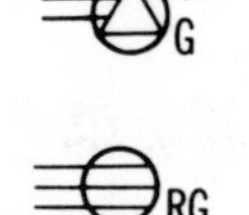

*2.7 *Single special-purpose receptacle outlet*

*2.8 *Duplex special-purpose receptacle outlet*

2.9 Range outlet

Ungrounded **Grounding**

2.10 Special-purpose connection or provision for connection. Use subscript letters to indicate function (DW — dishwasher; CD — clothes dryer, etc.)

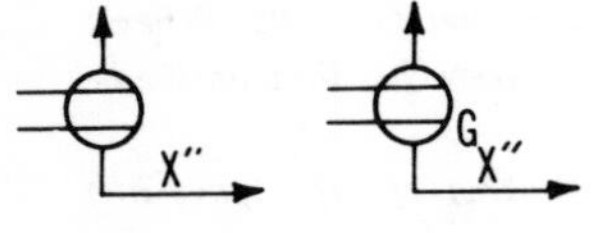

2.11 Multi-outlet assembly. (Extend arrows to limit of installation. Use appropriate symbol to indicate type of outlet. Also indicate spacing of outlets as x inches.)

2.12 Clock Hanger Receptacle

2.13 Fan Hanger Receptacle

2.14 Floor Single Receptacle Outlet

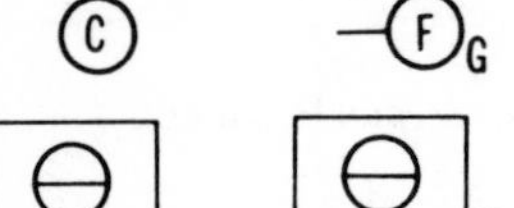

2.15 Floor Duplex Receptacle Outlet

*2.16 *Floor Special-Purpose Outlet*

---

*Use numeral or letter either within the symbol or as a subscript alongside the symbol keyed to explanation in the drawing list of symbols to indicate type of receptacle or usage.

2.17 Floor Telephone Outlet—Public

2.18 Floor Telephone Outlet—Private

Not a part of the Standard: Example of the use of several floor outlet symbols to identify a 2, 3, or more gang floor outlet

4.0 Institutional, Commercial, and Industrial Occupancies (Cont'd)

Basic Symbol	Examples of Individual Item Identification (Not a part of the standard)	
	1	*Nurses' Annunciator (can add a number after it as* 1 *24 to indicate number of lamps)*
	2	*Call station, single cord, pilot light*
	3	*Call station, double cord, microphone-speaker*
	4	*Corridor dome light, 1 lamp*
	5	*Transformer*
	6	*Any other item on same system—use numbers as required.*
		4.2 II. Paging System Devices (any type)

Keyboard

Flush annunciator

Ungrounded **Grounding**

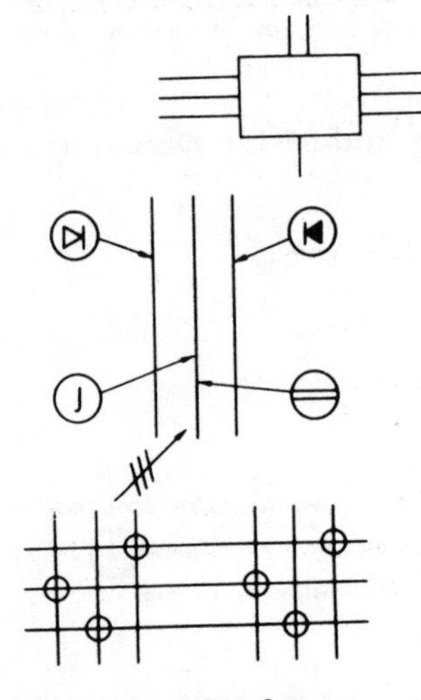

2.19 Underfloor Duct and Junction Box for Triple, Double or Single Duct System as indicated by the number of parallel lines

Not a part of the Standard: Example of use of various symbols to identify location of different types of outlets or connections for underfloor duct or cellular floor systems

2.20 Cellular Floor Header Duct

3.0 Switch Outlets

S	*3.1*	*Single-pole switch*
S_2	*3.2*	*Double-pole switch*
S_3	*3.3*	*Three-way switch*
S_4	*3.4*	*Four-way switch*
S_K	*3.5*	*Key-operated switch*
S_P	*3.6*	*Switch and pilot lamp*
S_L	*3.7*	*Switch for low-voltage switching system*

S_{LM} *3.8 Master switch for low-voltage switching system*

—⊖S *3.9 Switch and single receptacle*

═⊖S *3.10 Switch and double receptacle*

S_D *3.11 Door switch*

S_T *3.12 Time switch*

S_{CB} *3.13 Circuit breaker switch*

S_{MC} *3.14 Momentary contact switch or pushbutton for other than signalling system*

Ⓢ *3.15 Ceiling pull switch*

SIGNALLING SYSTEM OUTLETS

4.0 Institutional, Commercial, and Industrial Occupancies

Basic Symbol	Examples of Individual Item Identifiation (Not a part of the standard)		
+○		4.1	**I. Nurse Call System Devices (any type)**

4.0 Institutional, Commercial, and Industrial Occupancies (Cont'd)

Basic Symbol	Examples of Individual Item Identification (Not a part of the standard)	
	3	*2-Face annunciator*
	4	*Any other item on same system—use numbers as required.*
		4.3 III. Fire Alarm System Devices (any type) including Smoke and Sprinkler Alarm Devices
	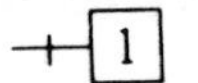	*Control panel*
	2	*Station*
	3	*10" Gong*
	4	*Pre-signal chime*
	5	*Any other item on same system—use numbers as required.*

4.0 Institutional, Commercial, and Industrial Occupancies (Cont'd)

Basic Symbol	Examples of Individual Item Identification (Not a part of the standard)		
		4.4	**IV. Staff Register System Devices (any type)**
			Phone operators' register
			*Entrance register—flush*
	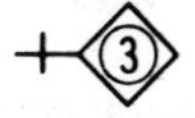		*Staff room register*
			Transformer
			Any other item on same system—use numbers as required.
		4.5	**V. Electric Clock System Devices (any type)**
			Master clock
			12" Secondary—flush

4.0 Institutional, Commercial, and Industrial Occupancies (Cont'd)

Basic Symbol	Examples of Individual Item Identification (Not a part of the Standard)	
		12" Double dial—wall mounted
		18" Skeleton dial
	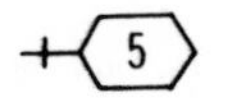	*Any other item on same system—use numbers as required.*
		4.6 VI. Public Telephone System Devices
	 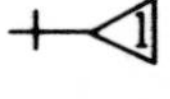	*Switchboard*
	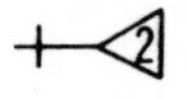	*Desk phone*
		Any other item on same system—use numbers as required.
		4.7 VII. Private Telephone System Devices (any type)
		Switchboard
		Wall phone
		Any other item on same system—use numbers as required.

5.0 Residential Occupancies (Cont'd)

5.5 Chime

5.6 Annunicator

5.7 Electric door opener

5.8 Maid's signal plug

5.9 Interconnection box

5.10 Bell-ringing transformer

5.11 Outside telephone

5.12 Interconnecting telephone

5.13 Radio outlet

5.14 Television outlet

6.0 Panelboards, Switchboards and Related Equipment

*6.1 Flush mounted panelboard and cabinet**

*6.2 Surface mounted panelboard and cabinet**

6.3 Switchboard, power control center, unit substations—should be drawn to scale*

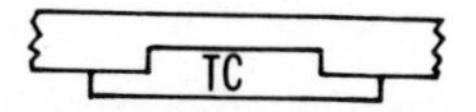

6.4 Flush mounted terminal cabinet (In small-scale drawings the TC may be indicated alongside the symbol)*

6.0 Panelboards, Switchboards and Related Equipment (Cont'd)

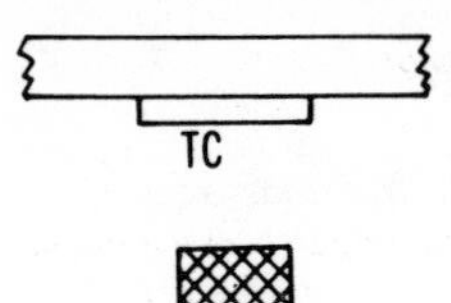

6.5 *Surface mounted terminal cabinet* (In small-scale drawings the TC may be indicated alongside the symbol)*

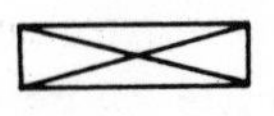

6.6 *Pull box (Identify in relation to wiring section and sizes)*

6.7 *Motor or other power controller**

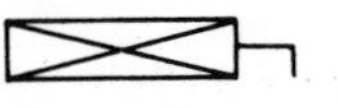

6.8 *Externally operated disconnection switch**

6.9 *Combination controller and disconnection means**

7.0 Bus Ducts and Wireways

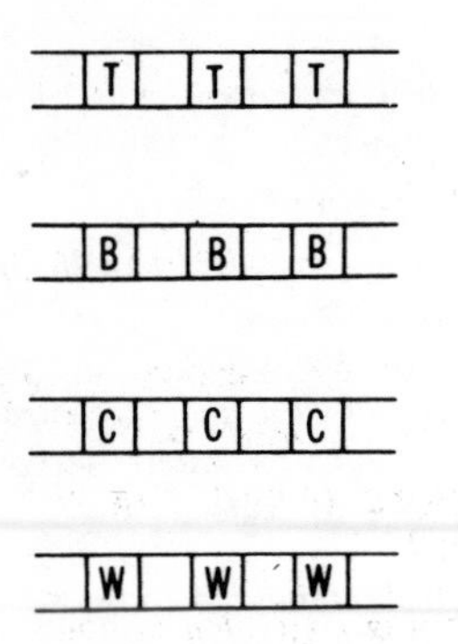

7.1 *Trolley duct**

7.2 *Busway* (Service, feeder, or plug-in)**

7.3 *Cable trough ladder or channel**

7.4 *Wireway**

*Identify By Notation or Schedule.

8.0 Remote Control Stations for Motors or other Equipment*

8.1 Pushbutton station

F *8.2 Float switch—mechanical*

L *8.3 Limit switch—mechanical*

P *8.4 Pneumatic switch—mechanical*

8.5 Electric eye—beam source

8.6 Electric eye—relay

T *8.7 Thermostat*

9.0 Circuiting

Wiring method identification by notation on drawing or in specification

9.1 Wiring concealed in ceiling or wall

9.2 Wiring concealed in floor

9.3 Wiring exposed

9.0 Circuiting

Wiring method identification by notation on drawing or in specification

Note: Use heavy-weight line to identify service and feeders. Indicate empty conduit by notation CO (conduit only)

2 1

9.4 Branch circuit home run to panel board. Number of arrows indicates number of circuits. (A numeral at each arrow may be used to identify circuit number.) Note: Any circuit without further identification indicates two-wire circuit. For a greater number of wires, indicate with cross lines, e.g.:

3 wires;

4 wires, etc.

Unless indicated otherwise, the wire size of the circuit is the minimum size required by the specification.
Identify different functions of wiring system, e.g., signalling system by notation or other means.

9.5 Wiring turned up

9.6 Wiring turned down

10.0 Electric Distribution or Lighting System, Underground

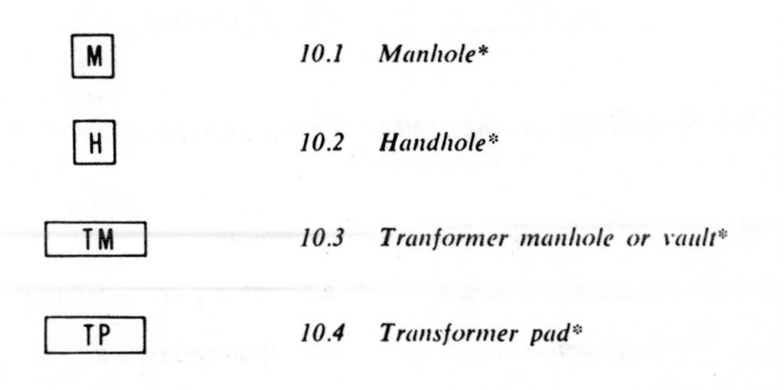

*10.1 Manhole**

*10.2 Handhole**

*10.3 Tranformer manhole or vault**

*10.4 Transformer pad**

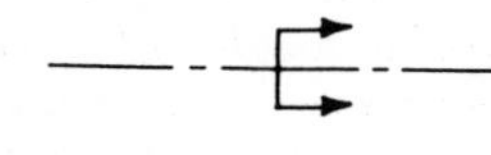

10.5 *Underground direct burial cable (Indicate type, size and number of conductors by notation or schedule)*

10.6 *Underground duct line (Indicate type, size, and number of ducts by cross-section identification of each run by notation or schedule. Indicate type, size, and number of conductors by notation or schedule)*

10.7 *Street light standard feed from underground circuit**

11.0 Electric Distribution or Lighting System Aerial

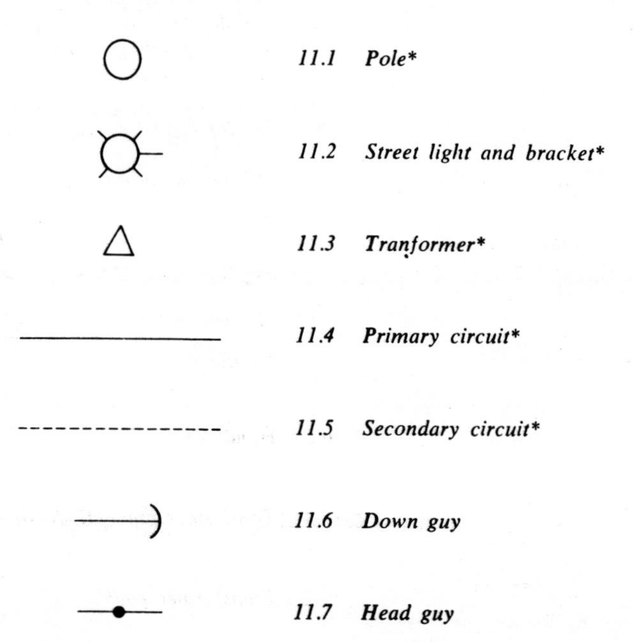

11.1 *Pole**

11.2 *Street light and bracket**

11.3 *Tranformer**

11.4 *Primary circuit**

11.5 *Secondary circuit**

11.6 *Down guy*

11.7 *Head guy*

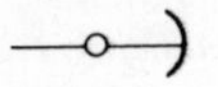

11.8 Sidewalk guy

*11.9 Service weather head**

4 Arrester, Lightning
Arrester (Electric surge, etc.)
Gap

4.1 General

4.2 Carbon block

Block, telephone protector

The sides of the rectangle are to be approximately in the ratio of 1 to 2 and the space between rectangles shall be approximately equal to the width of a rectangle.

4.3 Electrolytic or aluminum cell
This symbol is not composed of arrowheads.

4.4 Horn gap

4.5 Protective gap
These triangles shall not be filled.

4.6 Sphere gap

4.7 Valve or film element

*Identify By Notation or Schedule.

4.8 Multigap, general

4.9 Application: gap plus valve plus ground, 2 pole

7 Battery

The long line is always positive, but polarity may be indicated in addition. Example:

IEC

7.1 Generalized direct-current source

7.2 One cell

IEC

7.3 Multicell

7.3.1 Multicell battery with 3 taps

7.3.2 Multicell battery with adjustable tap

11 Circuit Breakers

If it is desired to show the condition causing the breaker to trip, the relay-protective-function symbols in item 66.6 may be used alongside the break symbol.

11.1 General

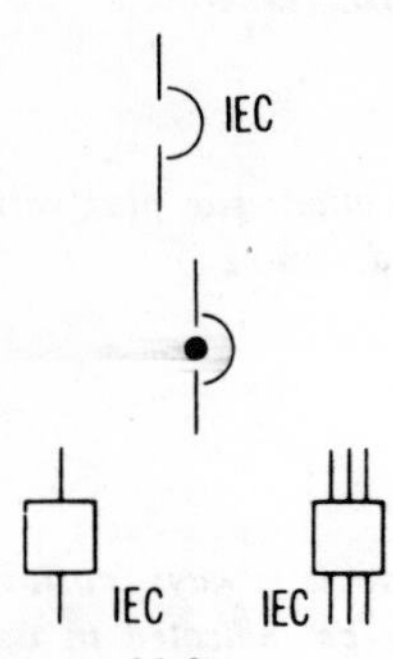

11.2 Air circuit breaker, if distinction is needed; for alternating-current breakers rated at 1,500 volts or less and for all direct-current circuit breakers

11.2.1 Network protector

11.3 Circuit breaker, other than covered by item 11.2. The symbol in the right column is for a 3-pole breaker.

See note 11.3A

11.3.1 On a connection or wiring diagram, a 3-pole single-throw circuit breaker (with terminals shown) may be drawn as shown.

See note 11.3A

11.4 Applications

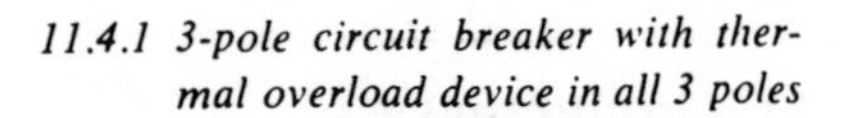

11.4.1 3-pole circuit breaker with thermal overload device in all 3 poles

Note 11.3A—On a power diagram, the symbol may be used without other identification. On a composite drawing where confusion with the general circuit element symbol (item 12) may result, add the identifying letters CB inside or adjacent to the square.

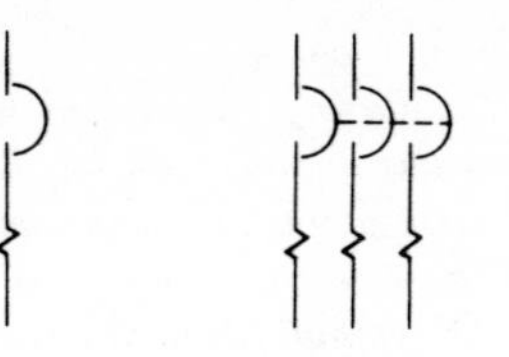

11.4.2 3-pole circuit breaker with magnetic overload device in all 3 poles

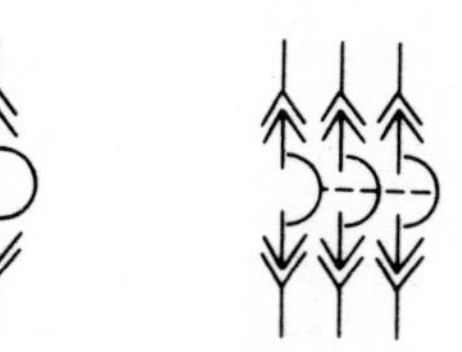

11.4.3 3-pole circuit breaker, drawout type

13 Circuit Return

13.1 Ground

IEC

(A) A direct conducting connection to the earth or body of water that is a part thereof.

(B) A conducting connection to a structure that serves a function similar to that of an earth ground (that is, a structure such as a frame of an air, space, or land vehicle that is not conductively connected to earth).

13.2 Chassis or frame connection

IEC

A conducting connection to a chassis or frame of a unit. The chassis or frame may be at a substantial potential with respect to the earth or structure in which this chassis or frame is mounted.

13.3 Common connections

Conducting connections made to one another. All like-designated points are connected.

**The asterisk is not a part of the symbol. Identifying values, letters, numbers, or marks shall replace the asterisk.*

15 Coil, Magnetic Blowout ‡

23 Contact, Electrical

For build-ups or forms using electrical contacts, see applications under CONNECTOR (item 18), RELAY (item 66), SWITCH (item 76). See DRAFTING PRACTICES (item 0.4.6).

23.1 Fixed contact

23.1.1 Fixed contact for jack, key, relay, etc.

23.1.2 Fixed contact for switch

23.1.3 Fixed contact for momentary switch

See SWITCH (item 76.8 and 76.10).

23.1.4 Sleeve

‡The broken line - —— indicates where line connection to a symbol is made and is not a part of the symbol.

23 Contact, Electrical (Cont'd)

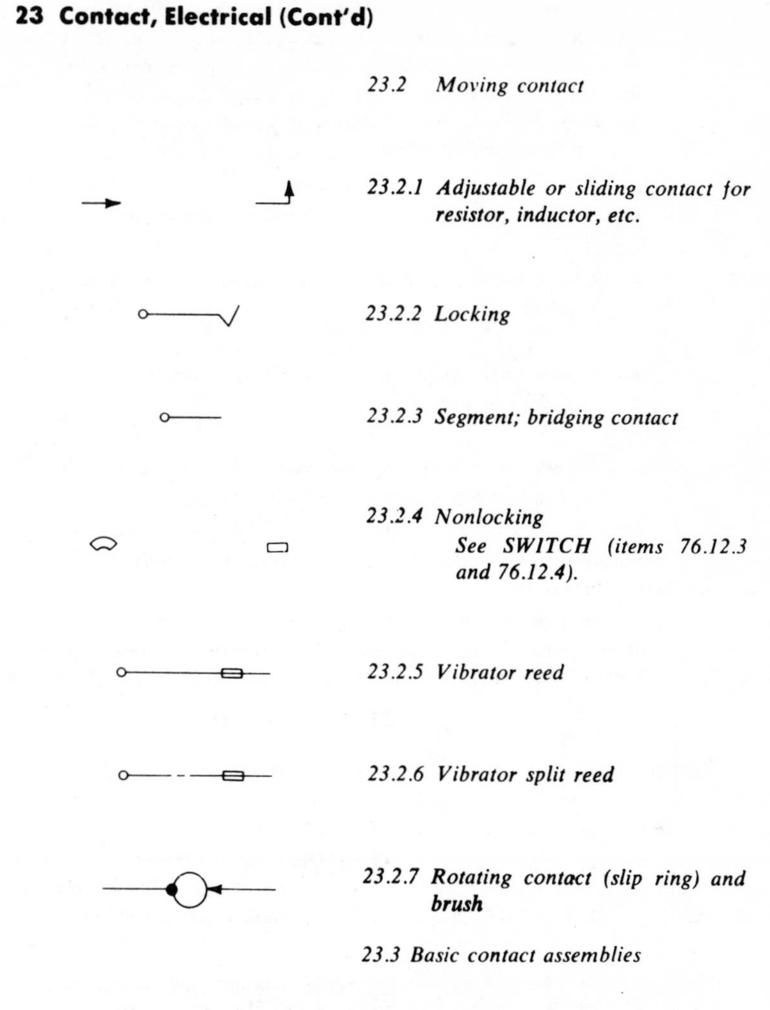

23.2 Moving contact

23.2.1 Adjustable or sliding contact for resistor, inductor, etc.

23.2.2 Locking

23.2.3 Segment; bridging contact

23.2.4 Nonlocking
See SWITCH (items 76.12.3 and 76.12.4).

23.2.5 Vibrator reed

23.2.6 Vibrator split reed

23.2.7 Rotating contact (slip ring) and brush

23.3 Basic contact assemblies

The standard method of showing a contact is by a symbol indicating the circuit condition it produces when the actuating device is in the deenergized or nonoperated position. The actuating device may be of a mechanical, electrical, or other nature, and a clarifying note may be necessary with the symbol to explain the proper point at which the contact functions, for example, the point where a contact closes or opens as a function of changing pressure, level, flow, voltage, current, etc. In cases where it is desirable to show contacts in the energized or operated condition and where confusion may result, a clarifying note shall be added to the drawing. Auxiliary switches or contacts for circuit breakers, etc., may be designated as follows:

(a) Closed when device is in energized or operated position,

(b) Closed when device is in de-energized or nonoperated position,

(aa) Closed when operating mechanism of main device is in energized or operated position,

(bb) Closed when operating mechanism of main device is in de-energized or nonoperated position.

See American Standard C37.2-1962 for further details.

In the parallel-line contact symbols showing the length of the parallel lines shall be approximately 1¼ times the width of the gap (except for item 23.6)

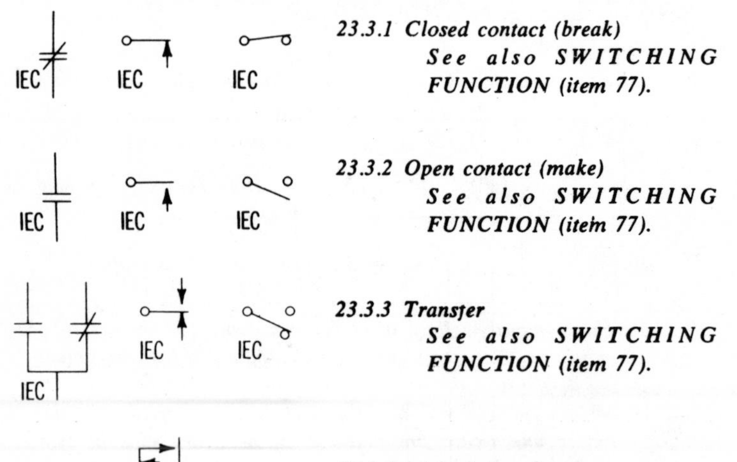

23.3.1 Closed contact (break)
See also SWITCHING FUNCTION (item 77).

23.3.2 Open contact (make)
See also SWITCHING FUNCTION (item 77).

23.3.3 Transfer
See also SWITCHING FUNCTION (item 77).

23.3.4 Make-before-break

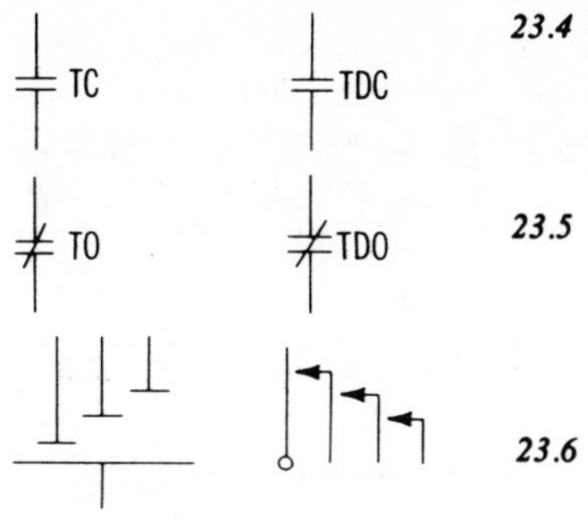

23.4 *Application: open contact with time closing (TC) or time delay closing (TDC) feature*

23.5 *Application: closed contact with time opening (TO) or time delay opening (TDO) feature*

23.6 *Time sequential closing*

24 Contactor

See also RELAY (item 66).

Fundamental symbols for contacts, coils, mechanical connections, etc., are the basis of contactor symbols and should be used to represent contactors on complete diagrams. Complete diagrams of contactors consist of combinations of fundamental symbols for control coils, mechanical connections, etc., in such configurations as to represent the actual device.

Mechanical interlocking should be indicated by notes.

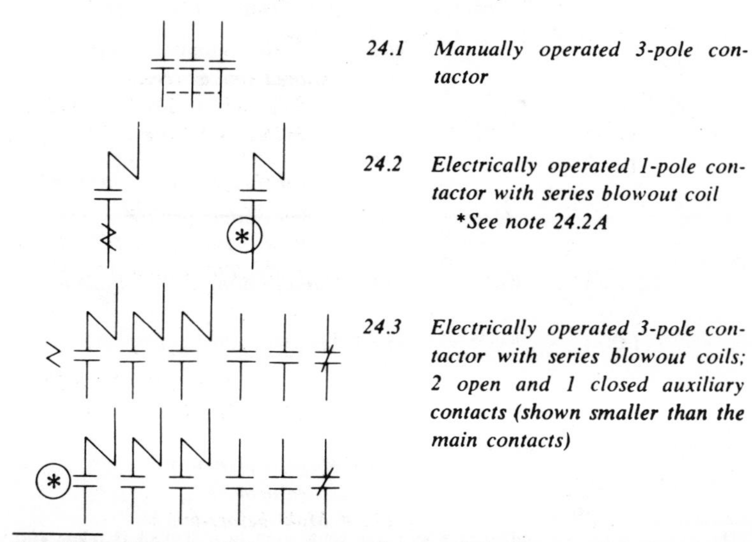

24.1 *Manually operated 3-pole contactor*

24.2 *Electrically operated 1-pole contactor with series blowout coil*
**See note 24.2A*

24.3 *Electrically operated 3-pole contactor with series blowout coils; 2 open and 1 closed auxiliary contacts (shown smaller than the main contacts)*

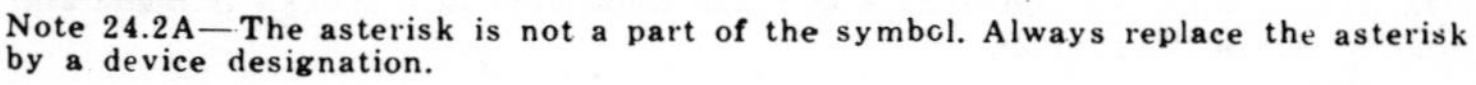
Note 24.2A—The asterisk is not a part of the symbol. Always replace the asterisk by a device designation.

24 Contactor (Cont'd)

24.4 Electrically operated 1-pole contactor with shunt blowout coil

46 Machine, Rotating

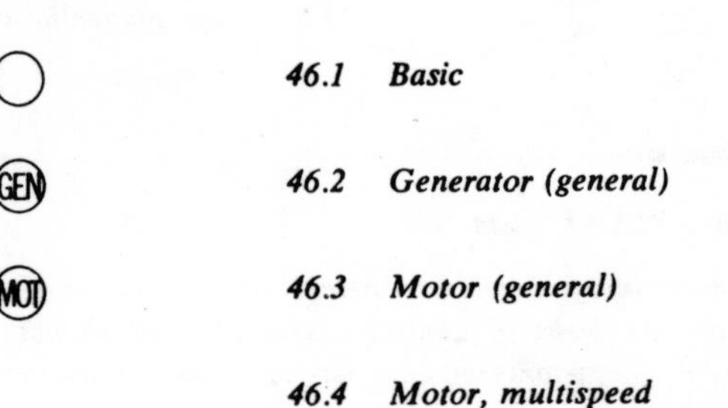

46.1 Basic

46.2 Generator (general)

46.3 Motor (general)

46.4 Motor, multispeed

USE BASIC MOTOR SYMBOL AND NOTE SPEEDS

46.5 ‡Rotating armature with commutator and brushes

46.6 Field, generator or motor
Either symbol of item 42.1 may be used in the following items.

IEC *46.6.1 Compensating or commutating*

IEC *46.6.2 Series*

IEC *46.6.3 Shunt, or separately excited*

46.6.4 Magnet, permanent
See item 47.

‡The broken line - - - indicates where line connection to a symbol is made and is not a part of the symbol.

46.7 *Winding symbols*
Motor and generator winding symbols may be shown in the basic circle using the following representation.

46.7.1 *1-phase*

46.7.2 *2-phase*

46.7.3 *3-phase wye (ungrounded)*

46.7.4 *3-phase wye (grounded)*

46.7.5 *3-phase delta*

46.7.6 *6-phase diametrical*

46.7.7 *6-phase double-delta*

46.8 *Direct-current machines; applications*

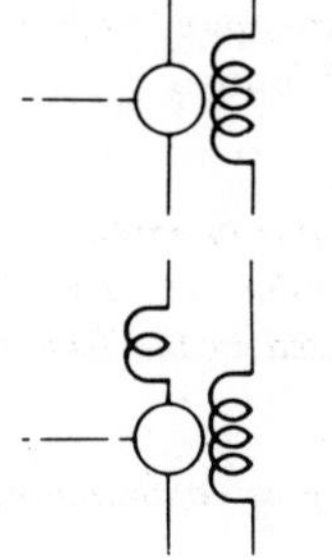
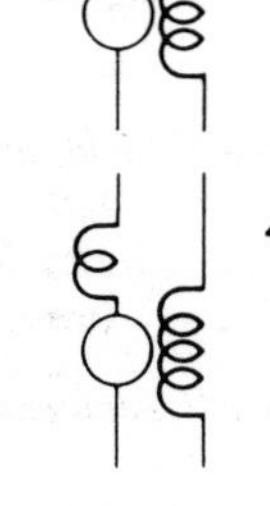

46.8.1 ‡*Separately excited direct-current generator or motor*

46.8.2 ‡*Separately excited direct-current generator or motor; with commutating or compensating field winding or both*

46.8.3 ‡*Compositely excited direct-current generator or motor; with commutating or compensating field winding or both*

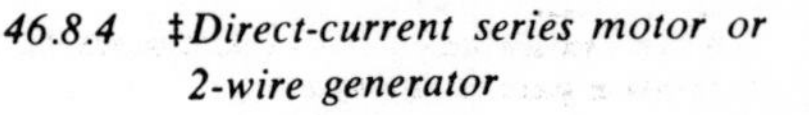

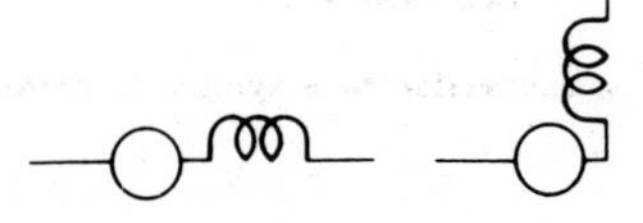

46.8.4 ‡*Direct-current series motor or 2-wire generator*

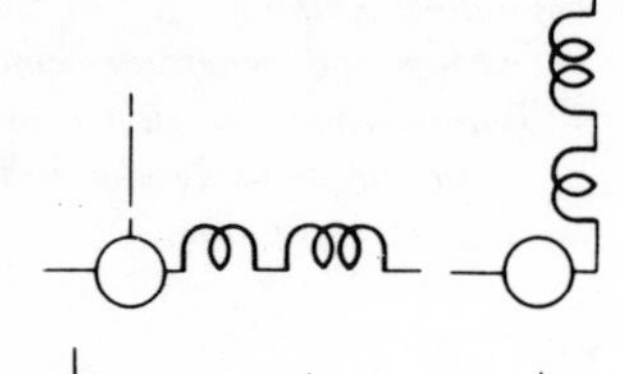

46.8.5 ‡*Direct-current series motor or 2-wire generator; with commutating or compensating field winding or both*

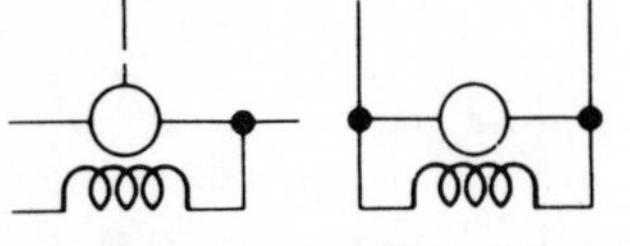

46.8.6 ‡*Direct-current shunt motor or 2-wire generator*

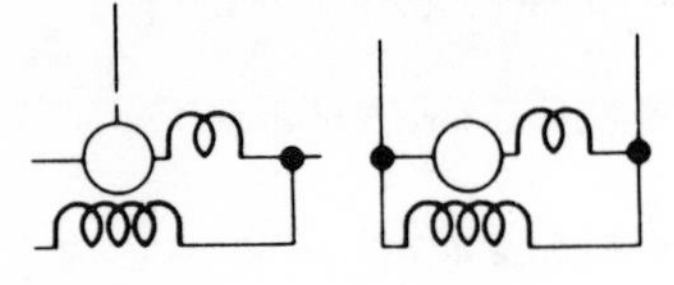

46.8.7 ‡*Direct-current shunt motor or 2-wire generator; with commutating or compensating field winding or both*

46.8.8 ‡*Direct-current permanent-magnet-field generator or motor*

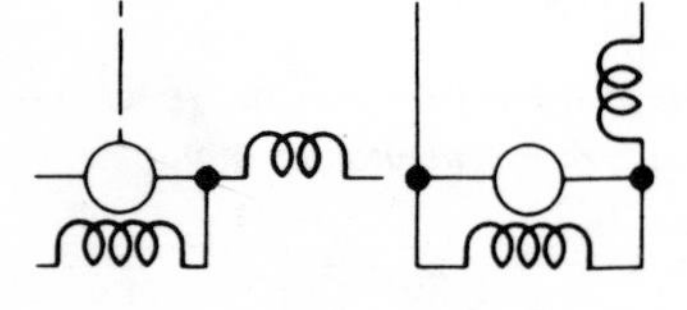

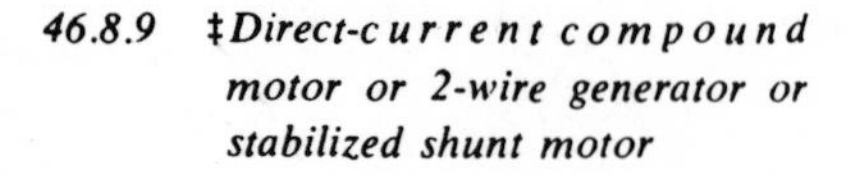

46.8.9 ‡*Direct-current compound motor or 2-wire generator or stabilized shunt motor*

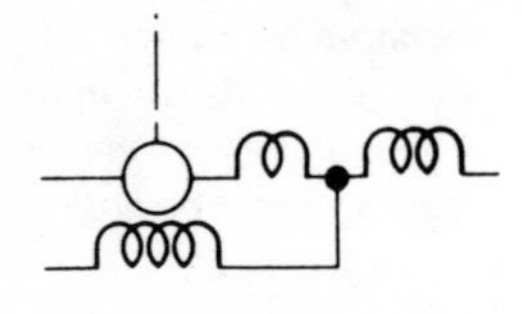

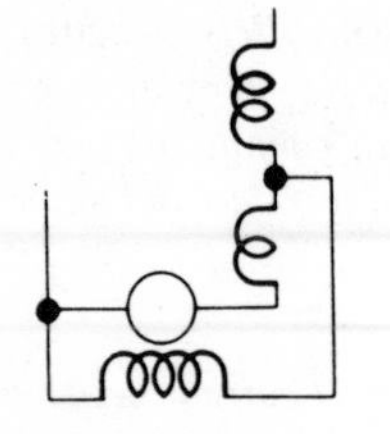

46.8.10 ‡*Direct-current compound motor or 2-wire generator or stabilized shunt motor; with commutating or compensating field winding or both*

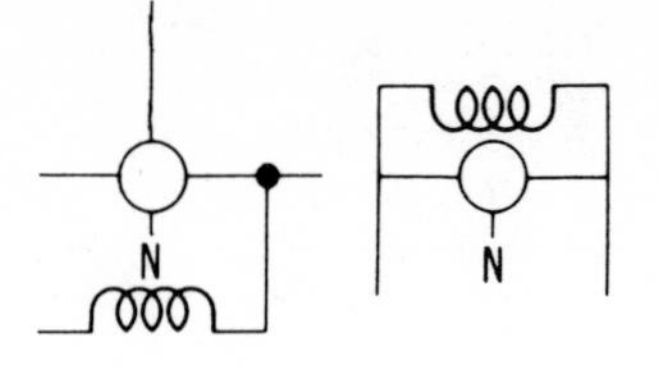

46.8.11 ‡*Direct-current 3-wire shunt generator*

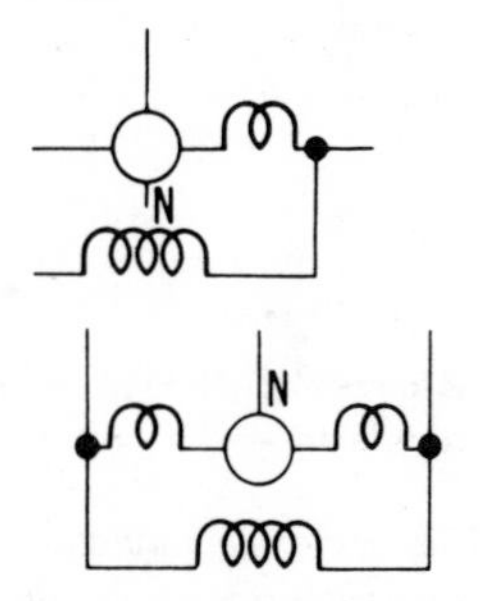

46.8.12 ‡*Direct-current 3-wire shunt generator; with commutating or compensating field winding or both*

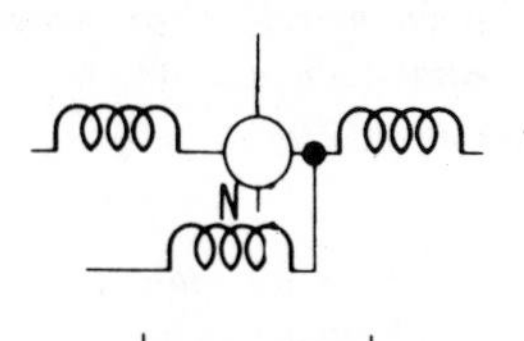

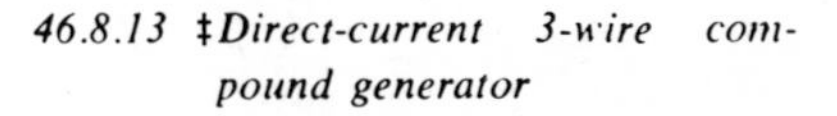

46.8.13 ‡*Direct-current 3-wire compound generator*

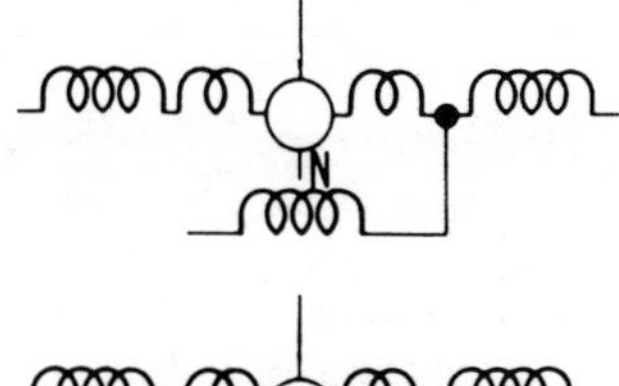

46.8.14 ‡*Direct-current 3-wire compound generator; with commutating or compensating field winding or both*

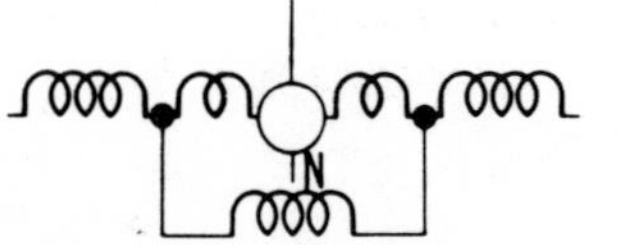

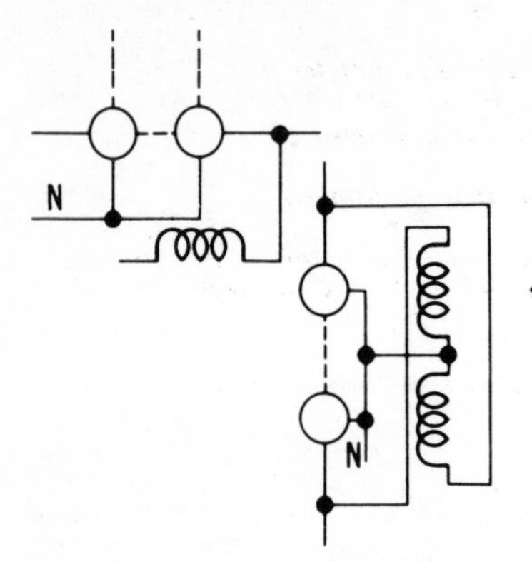

46.8.15 ‡*Direct-current balancer, shunt wound*

46.9 *Alternating-current machines; applications*

46.9.1 ‡*Squirrel-cage induction motor or generator, split-phase induction motor or generator, rotary phase converter, or repulsion motor*

46.9.2 ‡*Wound-rotor induction motor, synchronous induction motor, induction generator, or induction frequency converter*

46.9.3 ‡*Alternating-current series motor*

48 Meter Instrument

Note 48A—The asterisk is not a part of the symbol. Always replace the asterisk by one of the following letter combinations, depending on the function of the meter or instrument, unless some other identification is provided in the circle and explained on the diagram.

A *Ammeter IEC*

---

‡The broken line - ——indicates where line connection to a symbol is made and is not a part of the symbol.

AH	*Ampere-hour*
CMA	*Contact-making (or breaking) ammeter*
CMC	*Contact-making (or breaking) Clock*
CMV	*Contact-making (or breaking) voltmeter*
CRO	*Oscilloscope or cathode-ray oscilograph*
DB	*DB (decibel) meter*
DBM	*DBM (decibels referred to 1 milliwatt) meter*
DM	*Demand meter*
DTR	*Demand-totalizing relay*
F	*Frequency meter*
G	*Galvanometer*
GD	*Ground detector*
I	*Indicating*
INT	*Integrating*
μA or UA	*Microammeter*
MA	*Milliammeter*
NM	*Noise meter*
OHM	*Ohmmeter*
OP	*Oil pressure*
OSCG	*Oscillograph string*
PH	*Phasemeter*
PI	*Position indicator*
PF	*Power factor*
RD	*Recording demand meter*
REC	*Recording*
RF	*Reaction factor*
SY	*Synchroscope*
TLM	*Telemeter*
T	*Temperature meter*
THC	*Thermal converter*
TT	*Total time*

V	*Voltmeter*
VA	*Volt-ammeter*
VAR	*Varmeter*
VARH	*Varhour meter*
VI	*Volume indicator; meter, audio level*
VU	*Standard volume indicator; meter, audio lever*
W	*Wattmeter*
WH	*Watthour meter*

58 Path, Transmission, Conductor, Cable, Wiring

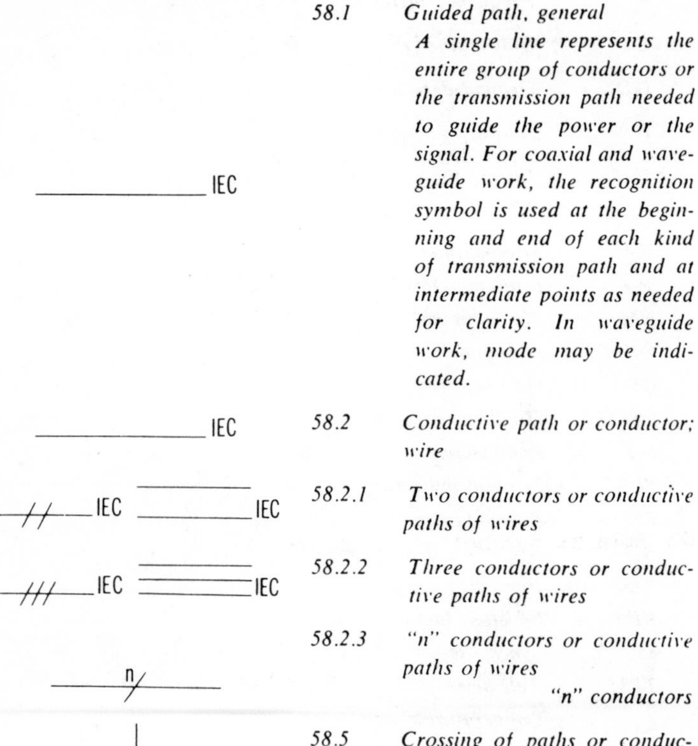

58.1 *Guided path, general*
A single line represents the entire group of conductors or the transmission path needed to guide the power or the signal. For coaxial and waveguide work, the recognition symbol is used at the beginning and end of each kind of transmission path and at intermediate points as needed for clarity. In waveguide work, mode may be indicated.

58.2 *Conductive path or conductor; wire*

58.2.1 *Two conductors or conductive paths of wires*

58.2.2 *Three conductors or conductive paths of wires*

58.2.3 *"n" conductors or conductive paths of wires*
"n" conductors

58.5 *Crossing of paths or conductors not connected*
The crossing is not necessarily at a 90-degree angle.

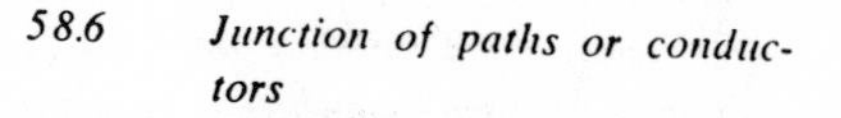

58.6 Junction of paths or conductors

58.6.1 Junction (if desired)

IEC

58.6.1.1 Application: junction of paths, conductor, or cable. If desired indicate path type, or size

IEC

58.6.1.2 Application: splice (if desired) of same size cables. Junction of conductors of same size or different size cables. If desired indicate sizes of conductors

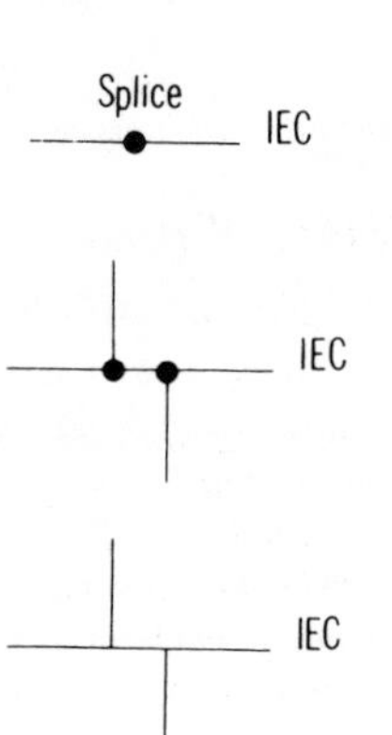

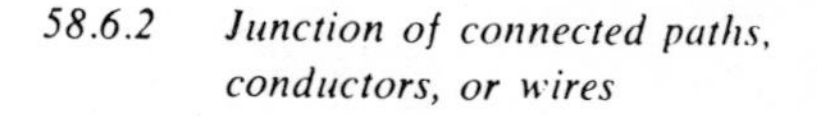

58.6.2 Junction of connected paths, conductors, or wires

OR ONLY IF REQUIRED BY SPACE LIMITATION

IEC

63 Polarity Symbol

\+ IEC *63.1 Positive*

− IEC *63.2 Negative*

76 Switch

See also FUSE (item 36); CONTACT, ELECTRIC (item 23); and DRAFTING PRACTICES (items 0.4.6 and 0.4.7).

Electrical Symbols

Fundamental symbols for contacts, mechanical connections, etc., may be used for switch symbols.

The standard method of showing switches is in a position with no operating force applied. For switches that may be in any one of two or more positions with no operating force applied and for switches actuated by some mechanical device (as in air-pressure, liquid-level, rate-of-flow, etc., switches), a clarifying note may be necessary to explain the point at which the switch functions.

When the basic switch symbols in items 76.1 through 76.4 are shown on a diagram in the closed position, terminals must be added for clarity.

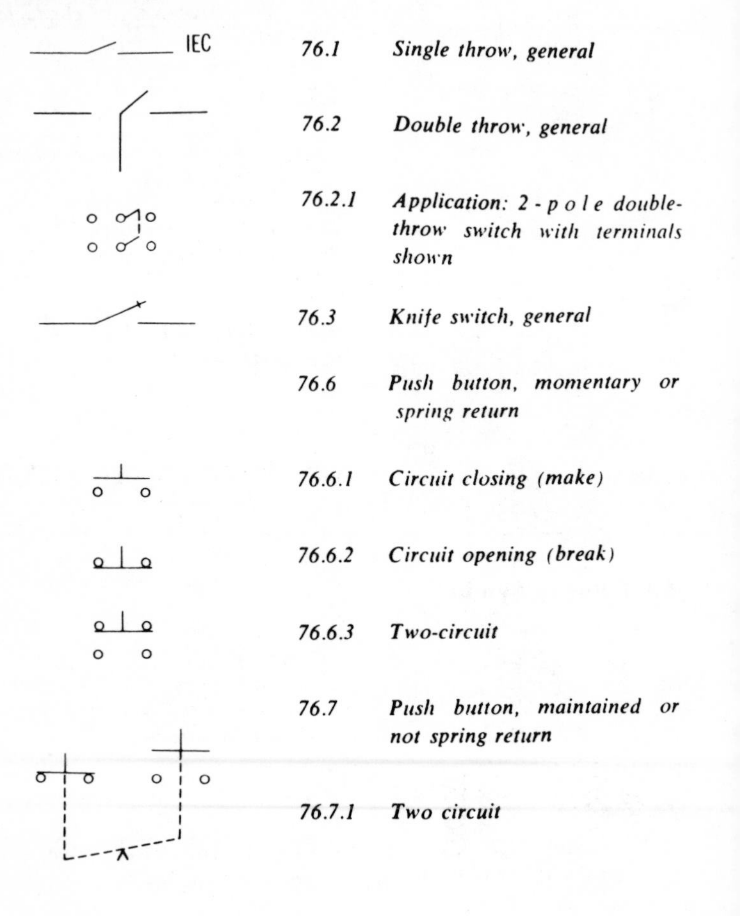

76.1 *Single throw, general*

76.2 *Double throw, general*

76.2.1 *Application: 2-pole double-throw switch with terminals shown*

76.3 *Knife switch, general*

76.6 *Push button, momentary or spring return*

76.6.1 *Circuit closing (make)*

76.6.2 *Circuit opening (break)*

76.6.3 *Two-circuit*

76.7 *Push button, maintained or not spring return*

76.7.1 *Two circuit*

86 Transformer

86.1 General

IEC

Either winding symbol may be used. In the following items, the left symbol is used. Additional windings may be shown or indicated by a note. For power transformers use polarity marking H_1, X_1, etc., from American Standard C6.1-1956. For polarity markings on current and potential transformers, see items 86.16.1 and 86.17.1. In coaxial and waveguide circuits, this symbol will represent a taper or step transformer without mode change.

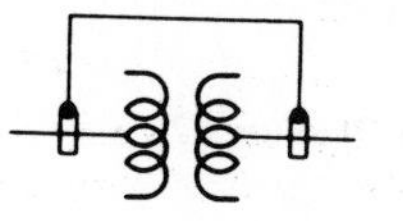

86.1.1 Application: transformer with direct-current connections and mode suppression between two rectangular waveguides

86.2 If it is desired especially to distinguish a magnetic-core transformer

86.2.1 Application: shielded transformer with magnetic core shown

86.2.2 Application: transformer with magnetic core shown and with a shield between windings. The shield is shown connected to the frame

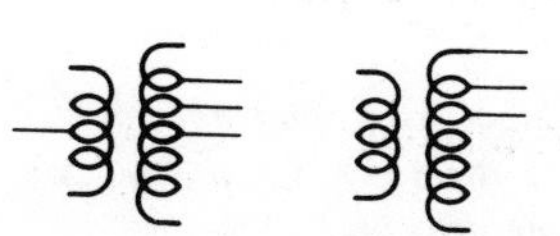

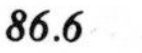

86.6 With taps, 1-phase

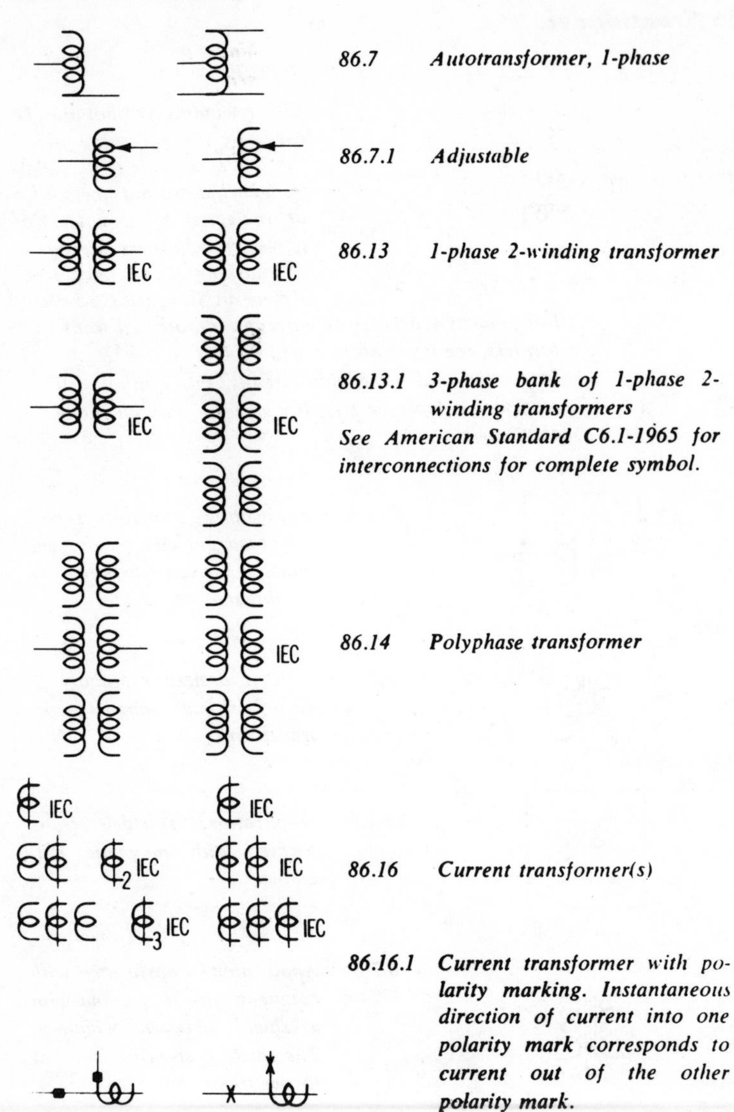

86.7 *Autotransformer, 1-phase*

86.7.1 *Adjustable*

86.13 *1-phase 2-winding transformer*

86.13.1 *3-phase bank of 1-phase 2-winding transformers*
See American Standard C6.1-1965 for interconnections for complete symbol.

86.14 *Polyphase transformer*

86.16 *Current transformer(s)*

86.16.1 *Current transformer with polarity marking. Instantaneous direction of current into one polarity mark corresponds to current out of the other polarity mark.*

Symbol used shall not conflict with item 77 when used on same drawing.

86.16.2‡ *Bushing-type current transformer*

86.17 *Potential transformer(s)*

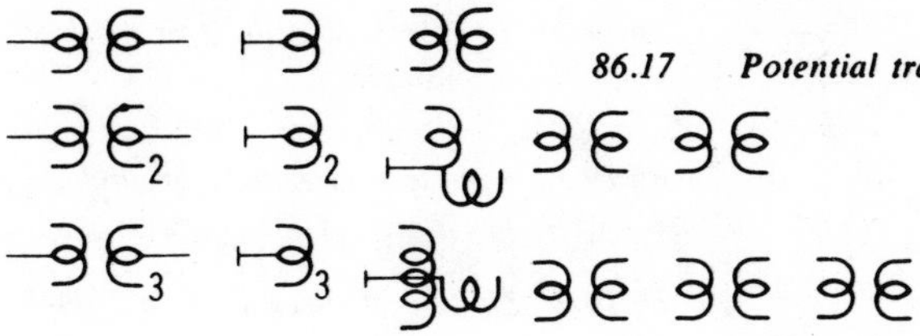

2 2

3 3

86.17.1 *Potential transformer with polarity mark. Instantaneous direction of current into one polarity mark corresponds to current out of the other polarity mark.*

Symbol used shall not conflict with item 77 when used on same drawing.

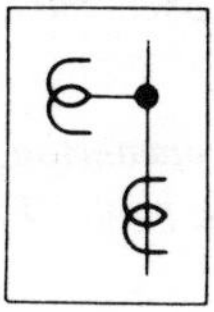

86.18 *Outdoor metering device*

86.19 *Transformer winding connection symbols*
For use adjacent to the symbols for the transformer windings.

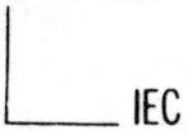

IEC

86.19.1 *2-phase 3-wire, ungrounded*

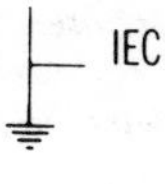

IEC

86.19.1.1 *2-phase 3-wire, grounded*

86.19.2 *2-phase 4-wire*

‡The broken line - — - indicates where line connection to a symbol is made and is not a part of the symbol.

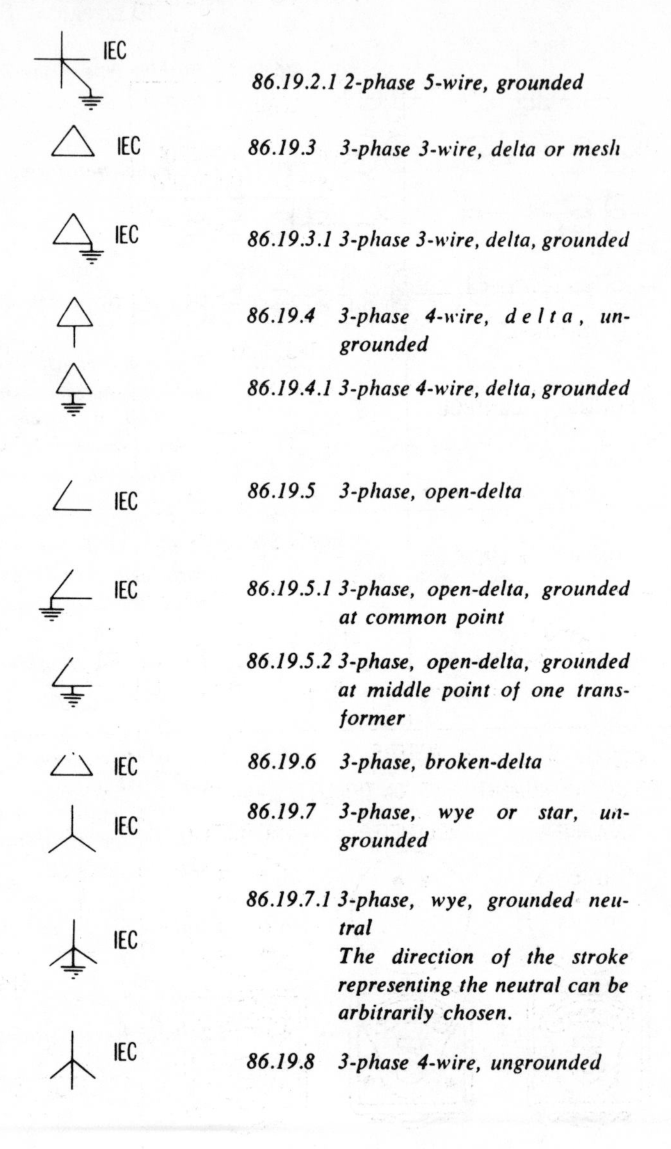

86.19.2.1 2-phase 5-wire, grounded

86.19.3 3-phase 3-wire, delta or mesh

86.19.3.1 3-phase 3-wire, delta, grounded

86.19.4 3-phase 4-wire, delta, ungrounded

86.19.4.1 3-phase 4-wire, delta, grounded

86.19.5 3-phase, open-delta

86.19.5.1 3-phase, open-delta, grounded at common point

86.19.5.2 3-phase, open-delta, grounded at middle point of one transformer

86.19.6 3-phase, broken-delta

86.19.7 3-phase, wye or star, ungrounded

86.19.7.1 3-phase, wye, grounded neutral
The direction of the stroke representing the neutral can be arbitrarily chosen.

86.19.8 3-phase 4-wire, ungrounded

WIRING SYMBOLS

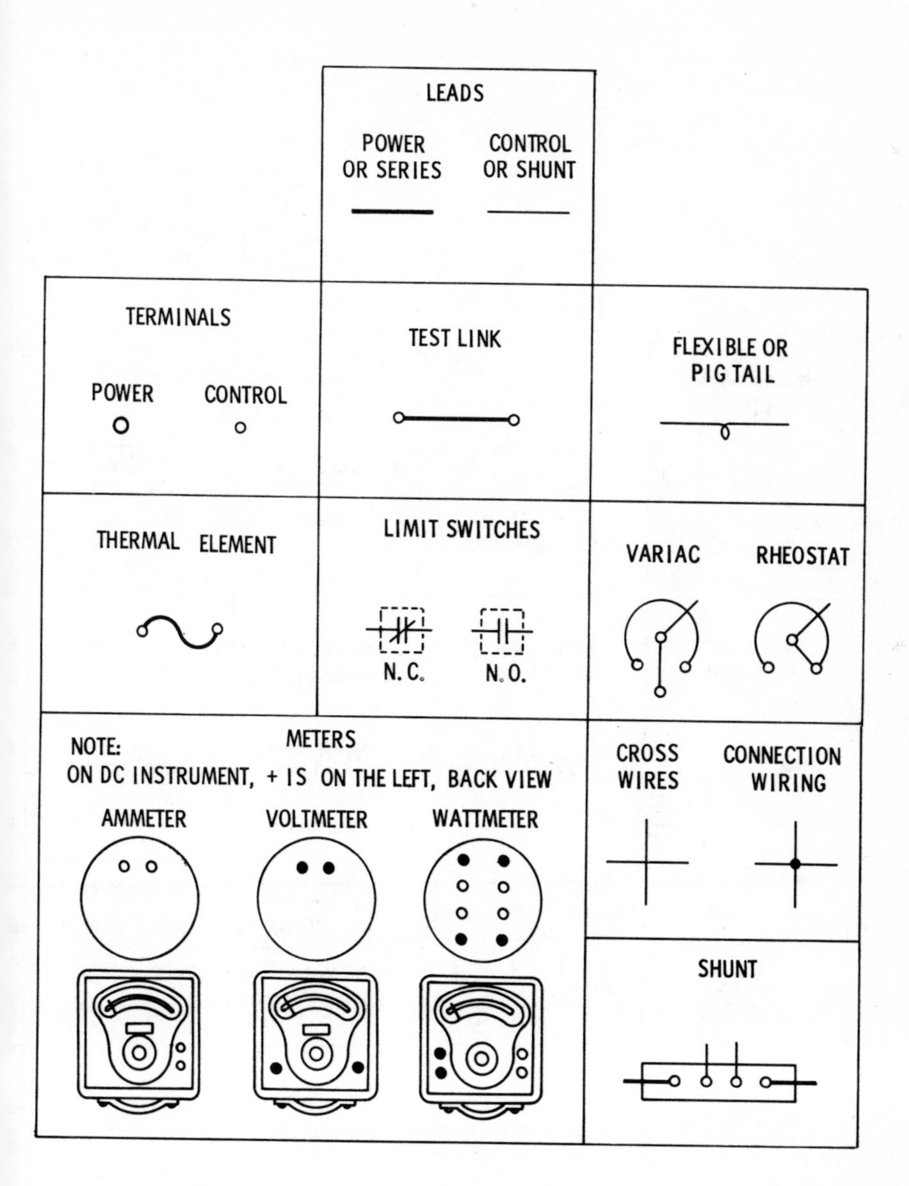

WIRING SYMBOLS

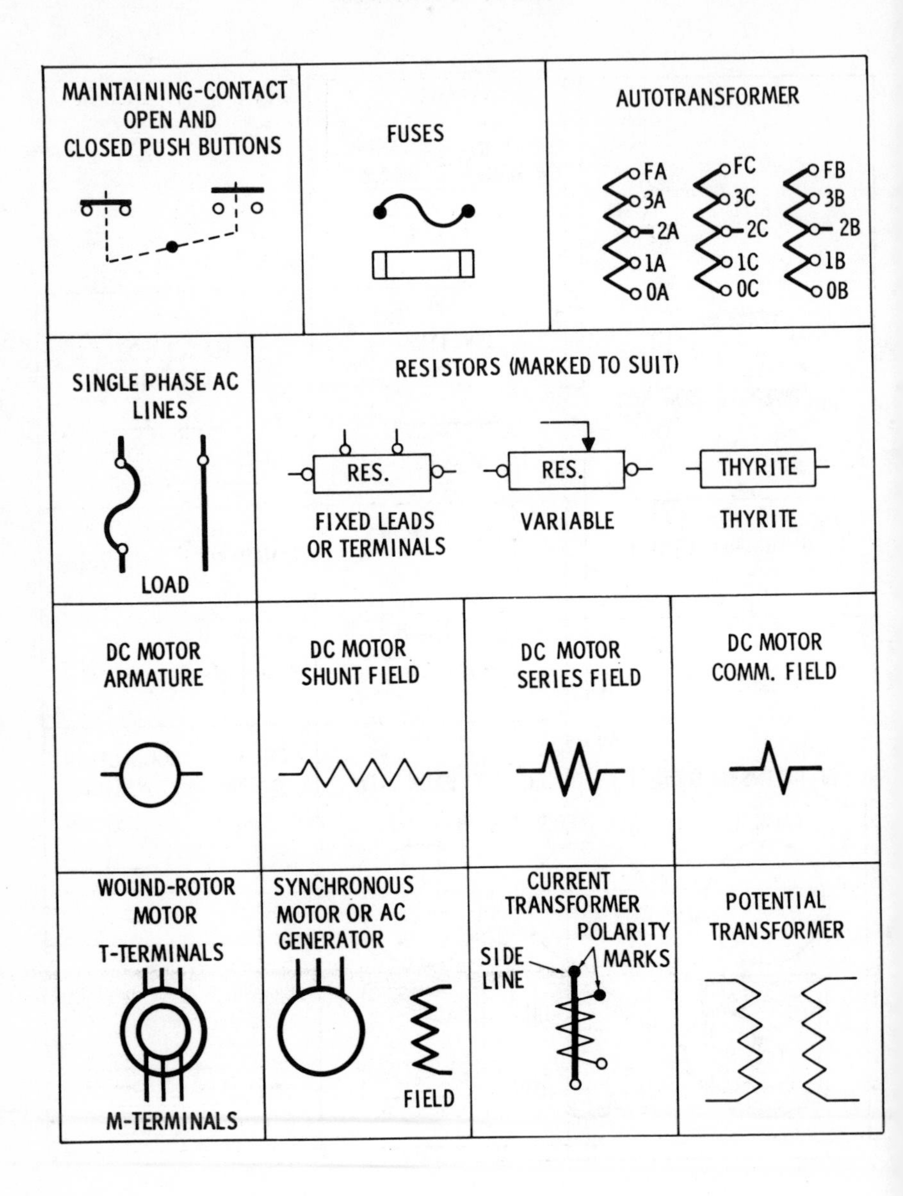

WIRING SYMBOLS

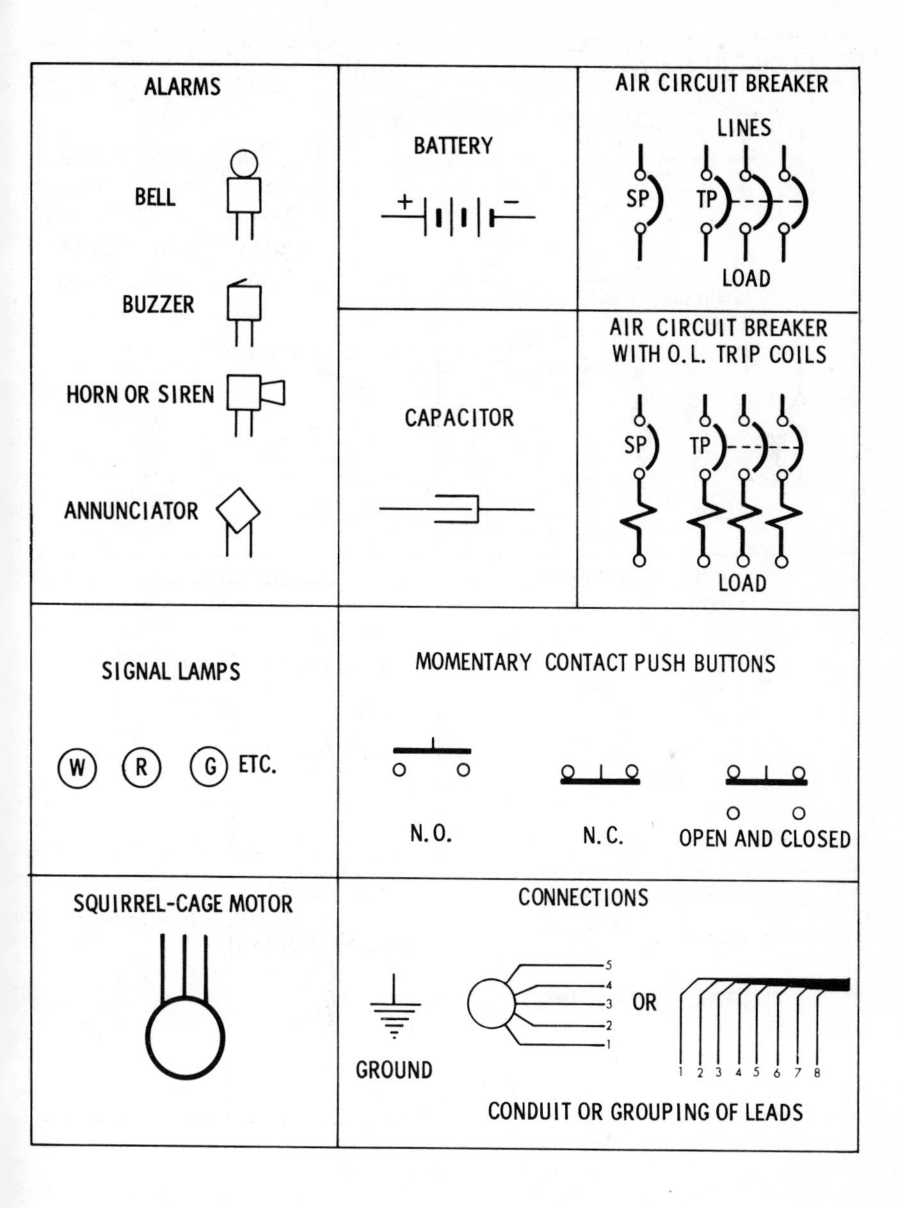

WIRING SYMBOLS

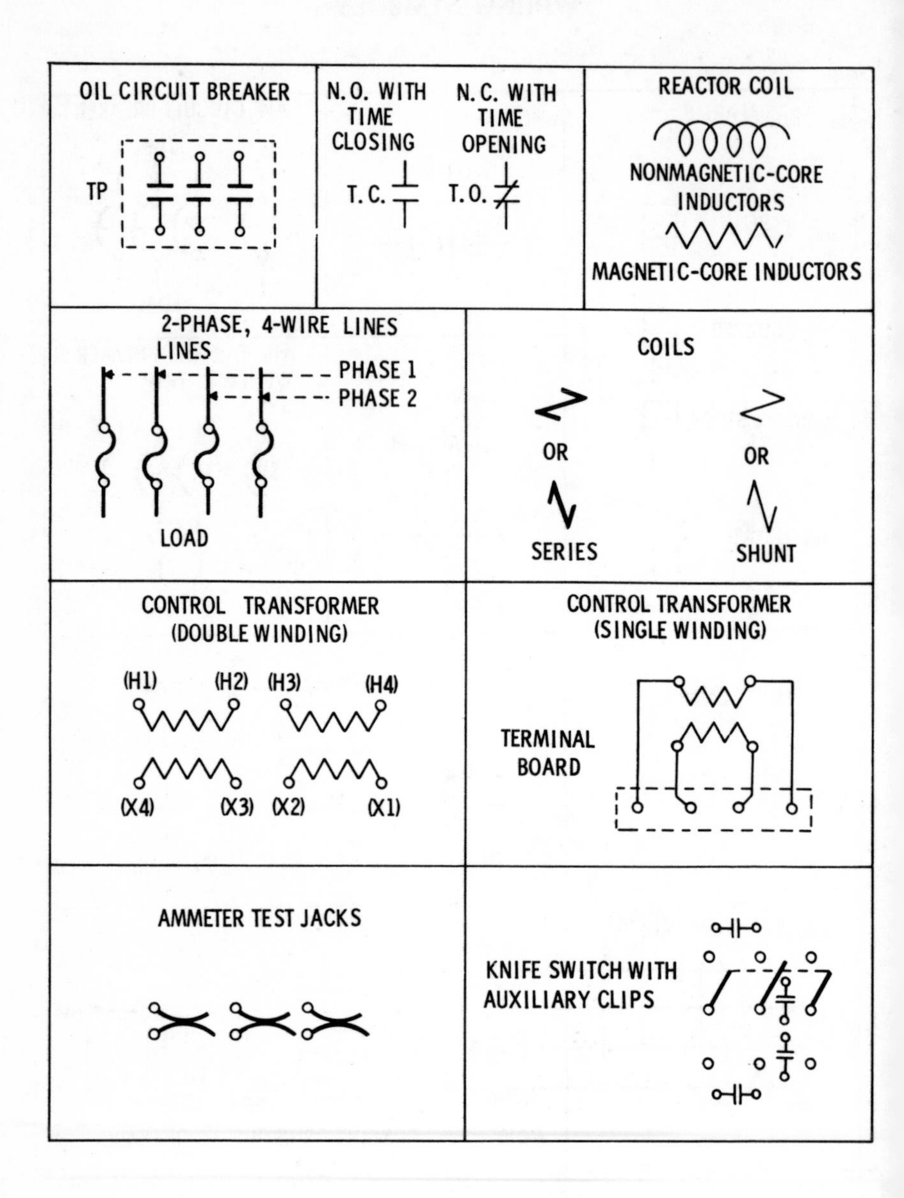

WIRING SYMBOLS

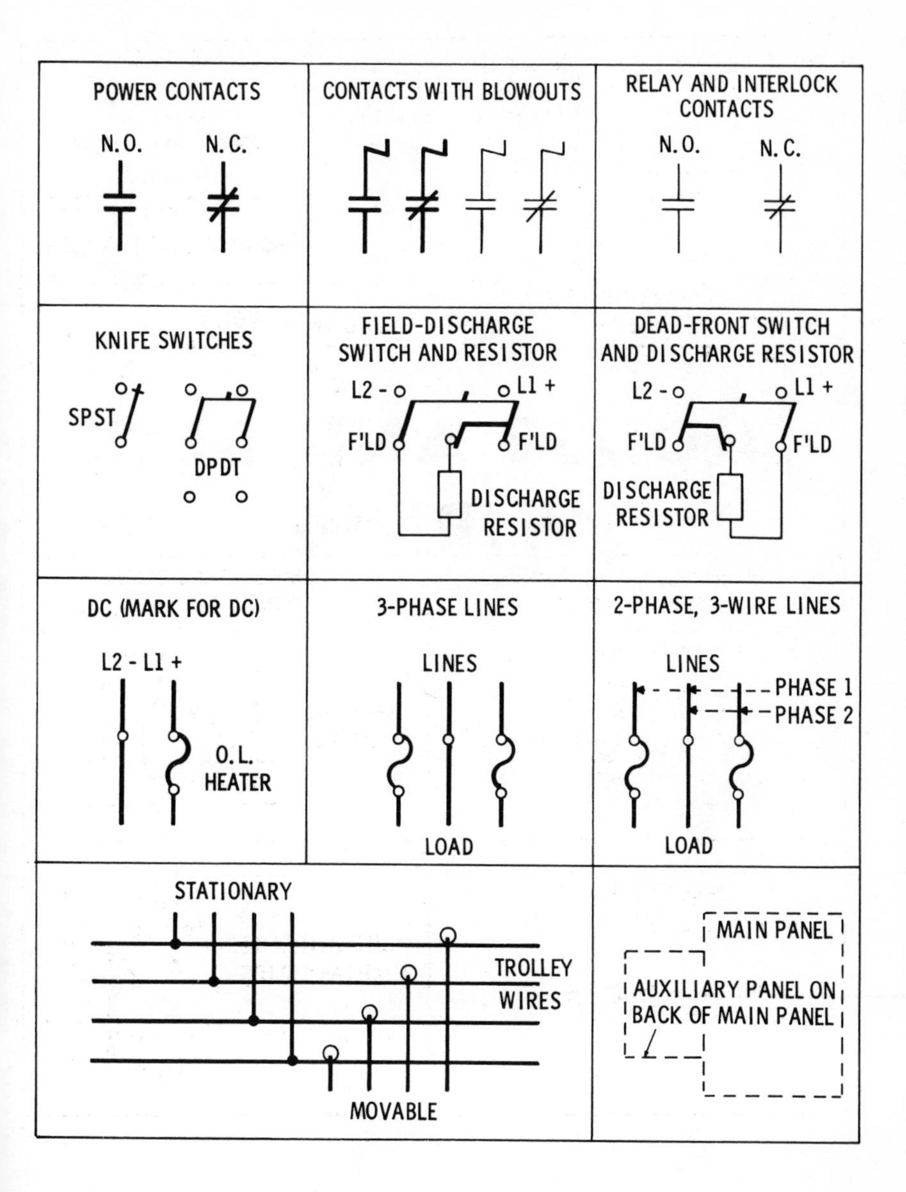

HOUSE AND BELL WIRING

VARIOUS LAMP-CONTROL SCHEMES

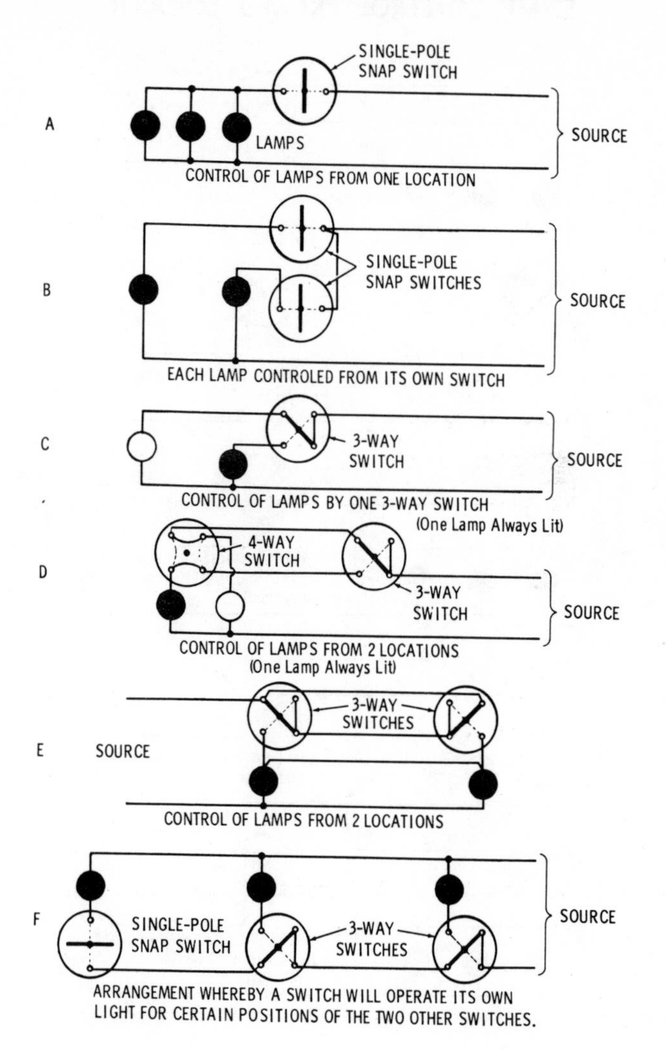

In the lamp-control diagrams represented above, Fig. A illustrates the connection when one single-pole snap switch is used.

Fig. B shows how two lights (or two groups of lights) can be controlled individually from a set of two single-pole switches.

Figs. C to F illustrate a series of special types of lamp control used in test circuits or in any location where particular control schemes are desirable.

LAMP CONTROL FROM 2 LOCATIONS

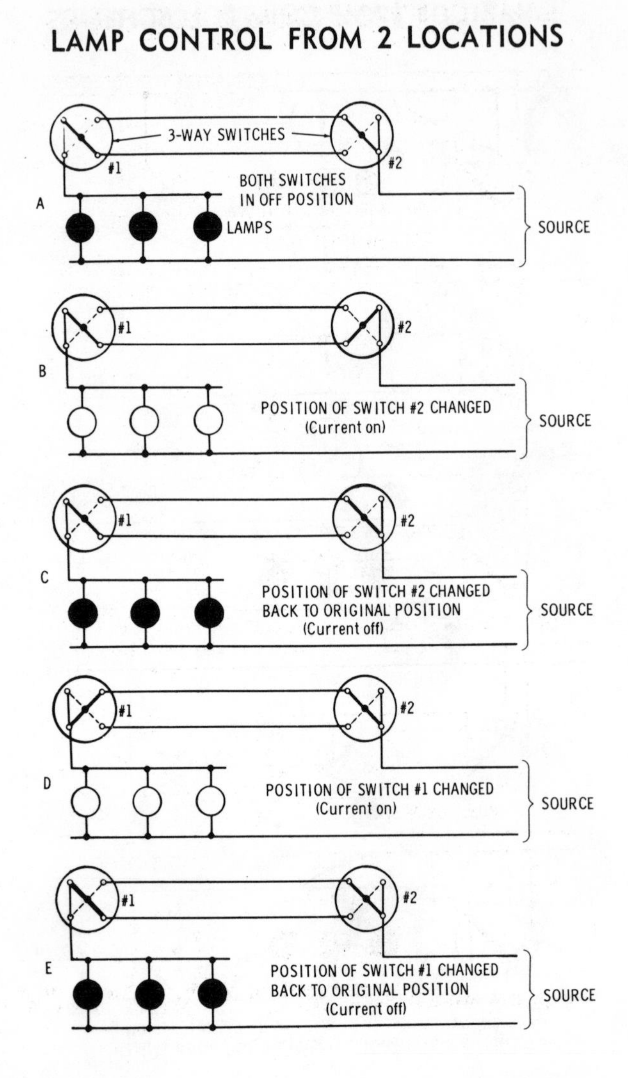

A convenient and often used method for control of a lamp or a group of lamp from two points by means of 3-way switches is shown in the diagrams. The lamps ma be extinguished or lighted from either switch regardless of the position of the othe When both switches are in the positions shown in Fig. A, the lamps are extinguishe and can be illuminated by the operation of switch No. 1 or 2. If as shown in di gram, No. 2 switch is operated the lamps will be illuminated, and can now be e tinguished from either switch. A typical sequence of operation is shown diagrar matically in Figs. A to E.

LAMP CONTROL FROM 2 LOCATIONS

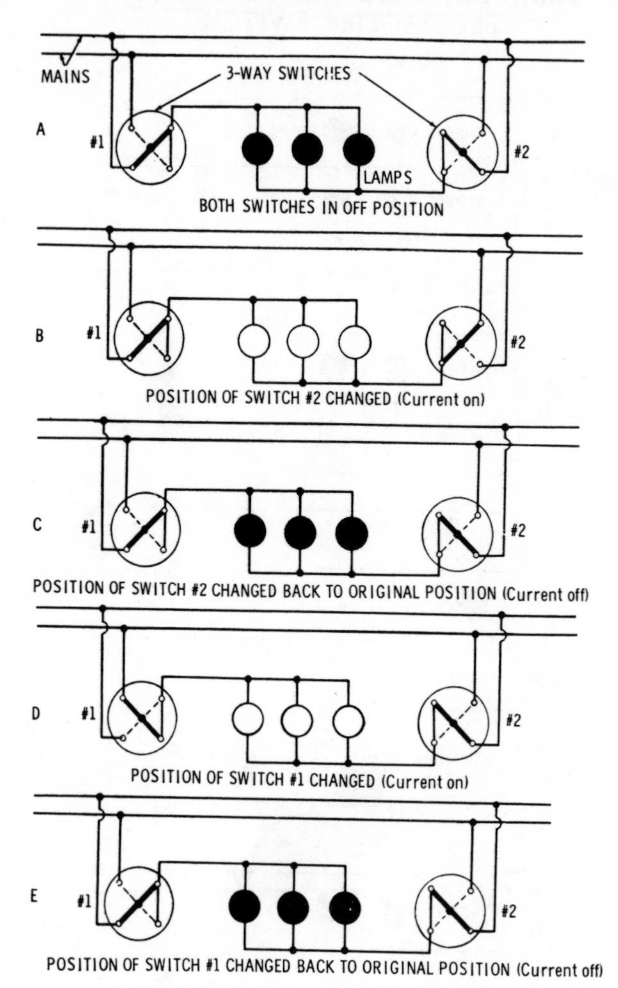

This connection provides an economical means of lamp control from two loca-ions. Although not permissible under the National Electric Code, it is shown only s an electrically possible circuit. As in the previous connections shown, both switches re in off position in Fig. A, the lamps are extinguished and can be lit by operating ither switch. If switch No. 2, Fig. B is operated to position "S" the lamps will be luminated and can be extinguished again from any one of the two switches. Figs. to E inclusive show the lamps lighted or extinguished, depending on position of witch No. 1, relative to the position of switch No. 2.

LAMP CONTROL FROM 2-CIRCUIT ELECTROLIER SWITCH

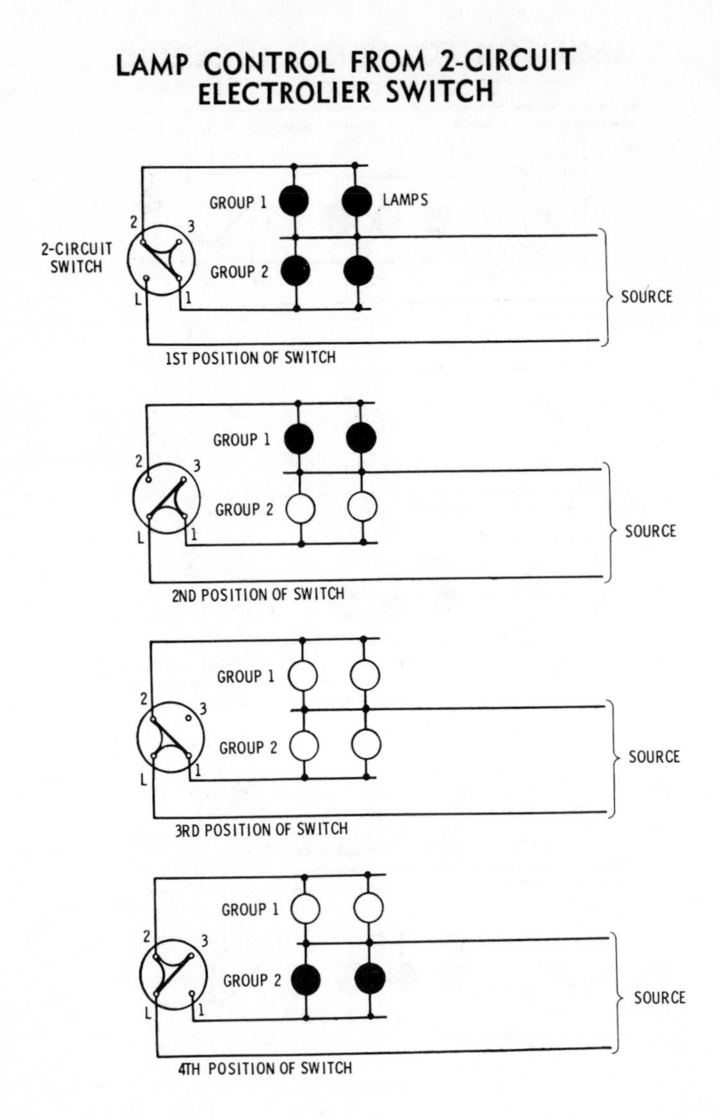

Large fixtures or electroliers are often wired so that lights can be controlled in two or more independent groups. As shown in the diagram the two groups of lamps are extinguished in the first position of the switch. When the switch is moved to the second position, group No. 2 will be illuminated. In the third position the maximum amount of brightness is obtained as both groups of lamps are illuminated and finally in the fourth position, group No. 1 only is lit. This switch may not be considered as standard, it is only one of several arrangements.

LAMP CONTROL FROM 3-CIRCUIT ELECTROLIER SWITCH

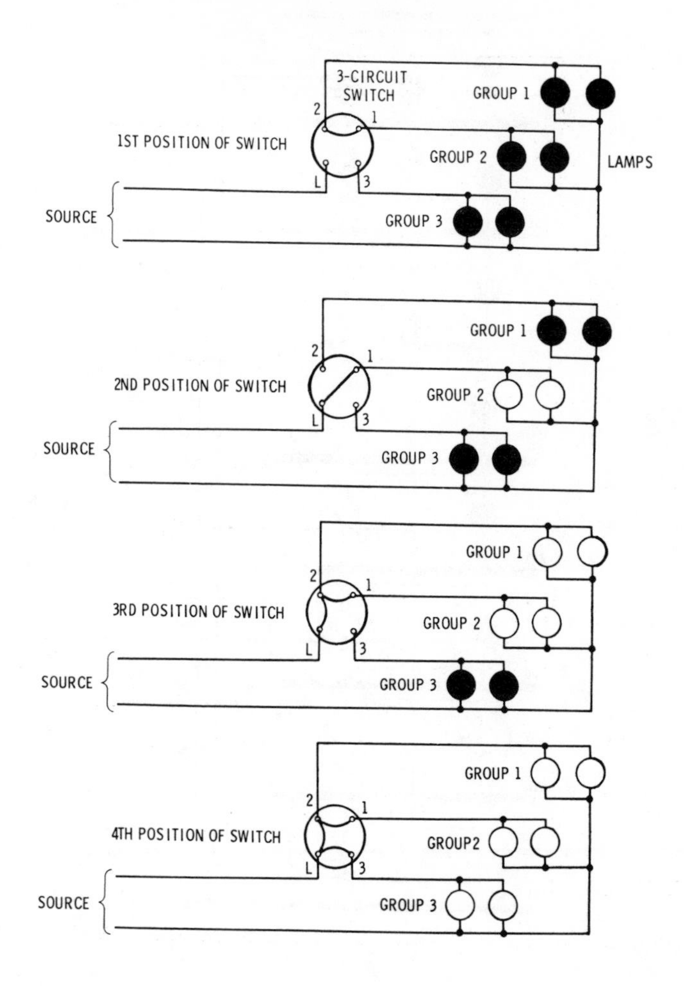

A 3-circuit electrolier switch from which three groups of lamps are controlled is shown above. The sequence of operation is depicted diagrammatically and is principally the same as shown in the previous 2-circuit switch. In the 4th position maximum illumination is obtained, with all lamps lighted. The switch shown is typical only among a great variety of switches manufactured for electrolier or dome-lamp control. The current carrying capacity of the switch as well as potential of the source to be connected should be considered for each individual application.

STAIRWAY LAMP-CONTROL WIRING

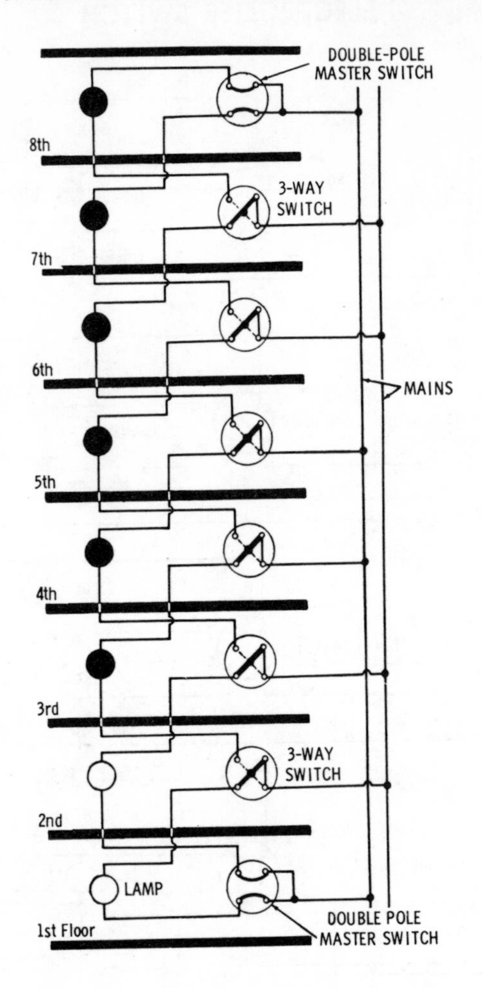

As shown the switches used in this type of light control consist of two double-pole switches, interconnected on the first and last floor, and one 3-way switch for each floor. The sequence of operation is as follows: Closing switch on the first floor lights lamp on first and second floor. Operating the switch on the second floor extinguishes the light on the first floor and lights the lamp on the third floor, etc. This operation is continued until the top floor is reached. In other words the switch on each floor should be operated in passing. It can be readily seen that this light control arrangement lends itself to operation of lamps irrespective of number of floors encountered.

CONTROL OF LIGHTS FROM THREE LOCATIONS BY USE OF TWO, 3-WAY SWITCHES AND ONE 4-WAY SWITCH.

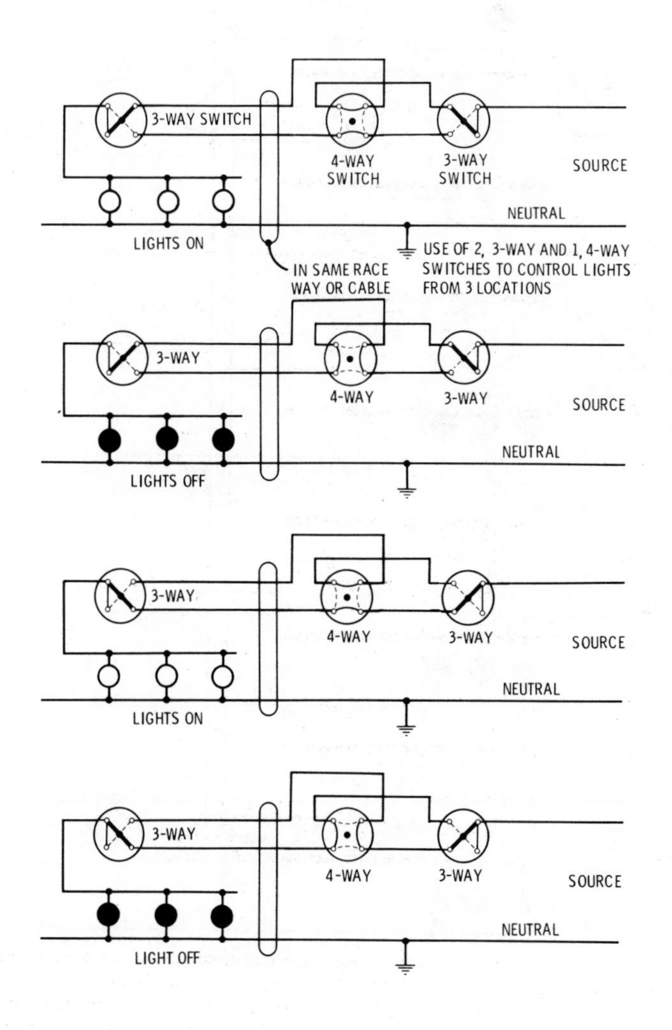

By following through the above diagrams, one may readily see that the lights may be turned **on** or **off** from any of the three locations at any time.

CONTROL OF LAMPS FROM MORE THAN ONE LOCATION BY MEANS OF 3-WAY AND 4-WAY SWITCHES

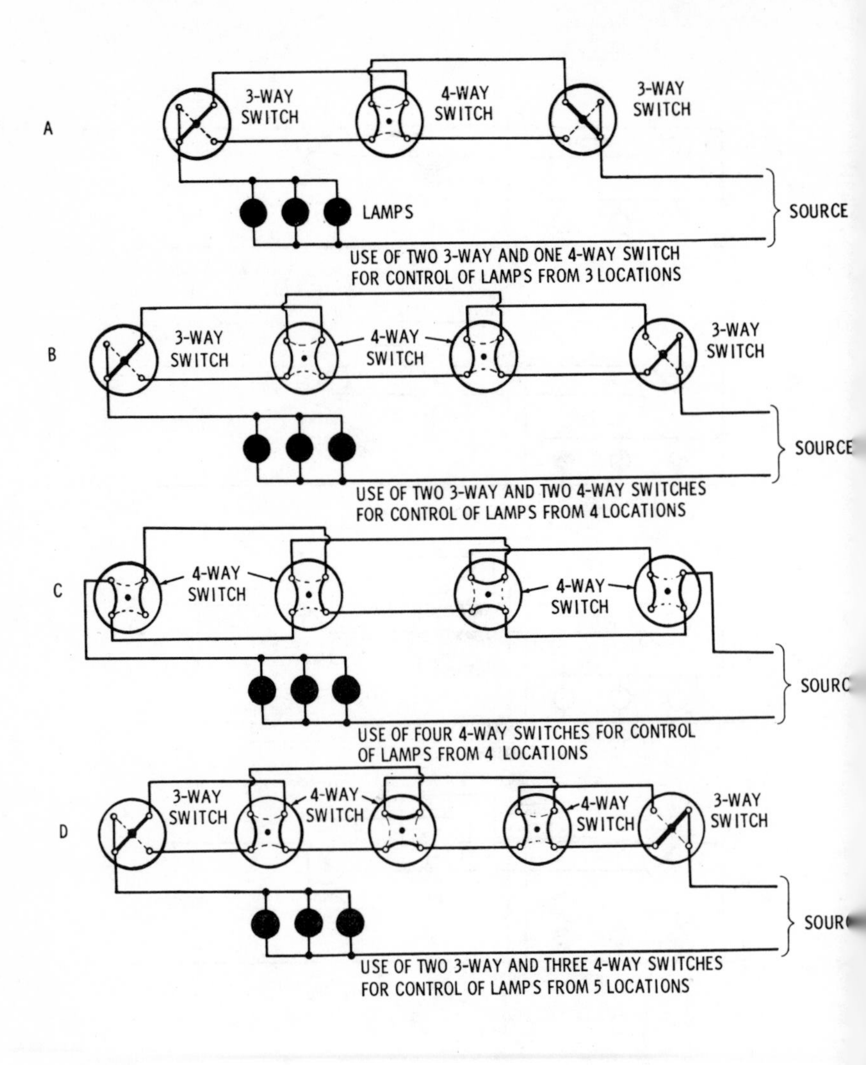

The connection diagrams shown in Figs. A to D, illustrate the conventional metho of lamp control when using 3- and 4-way switches. With reference to Fig. A, it obvious that for any additional point of control desired a 4-way switch connect the same as the middle switch must be used. See Figs. B to D.

Two 3-way switches (one in a garage and one in the house) control the garage lights from either location, plus a duplex receptacle that remains energized The wiring may be overhead or underground. An equipment grounding conductor must also be run from the service entrance equipment in the house to the garage, for grounding the noncurrent carrying parts, such as fixture enclosures and wall boxes for receptacles. The equipment grounding conductor was left out of the illustration to simplify the sketch.

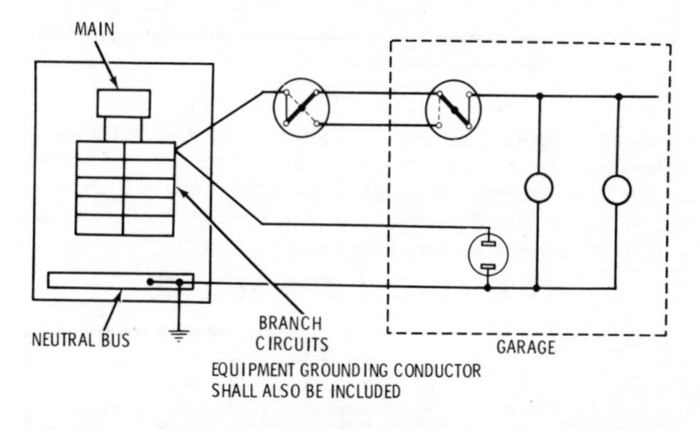

There have also been other methods using only three conductors, but this is illegal since the National Electrical Code states that the neutral is not to be switched unless all conductors are simultaneously disconnected.

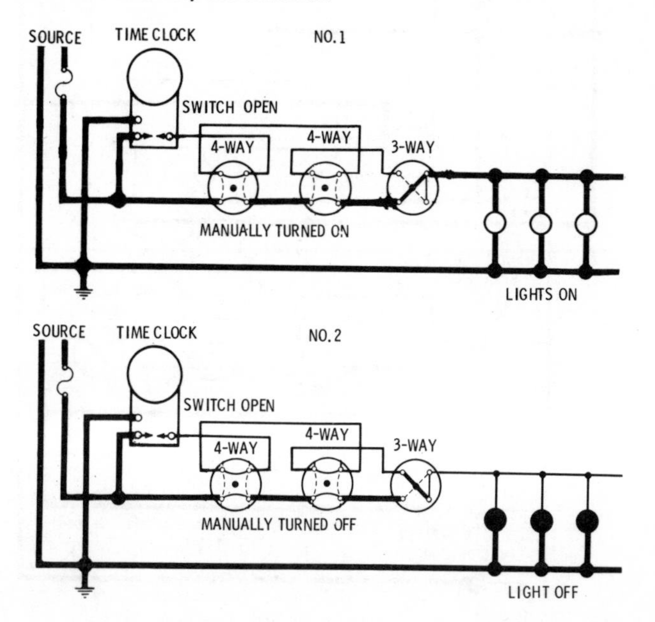

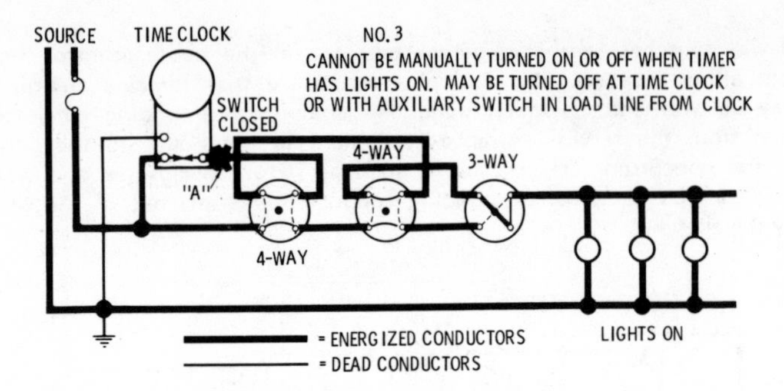

Wiring for outside lighting around a residence may be controlled with switches from several locations in the residence. A clock control may be added to turn the lights on and off automatically. A single pole-single throw switch may be installed at point "A", in Sketch No. 3 so that you may control the lights manually instead of automatically should you wish to do so at any time.

TWO CONDUCTOR SERVICE

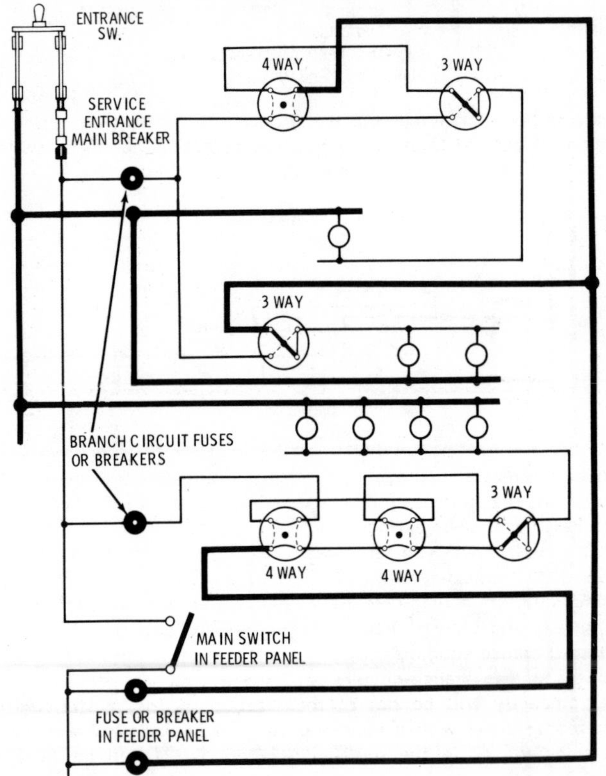

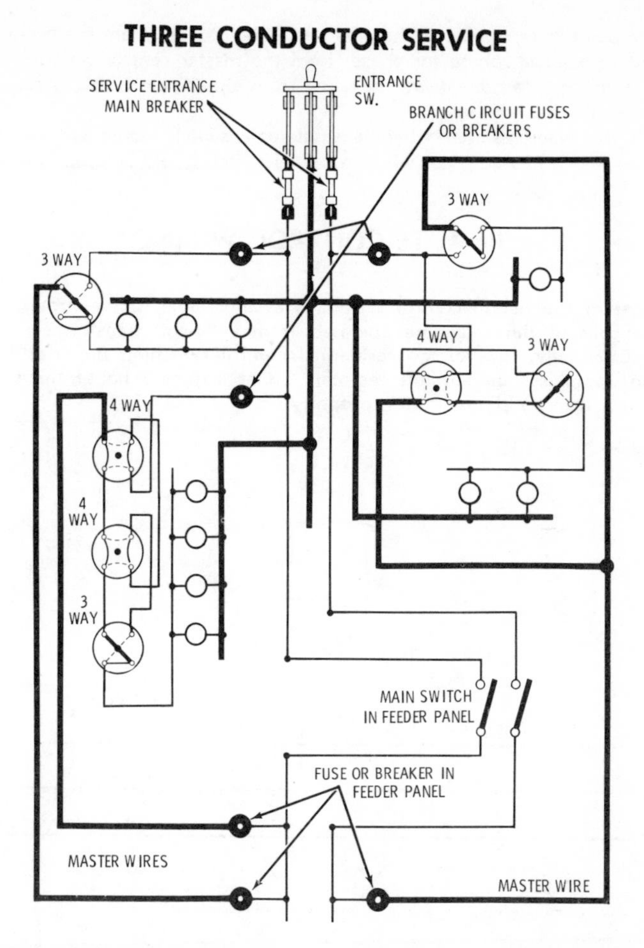

Master control for turning on lights in the rooms of a home from a point or location such as the master bedroom is a security measure. When the master control switch is turned on, the switches in the rooms controlled will not turn off the lights. The only way that they may be extinguished is by unscrewing bulbs.

The diagram showing two-wire service is rare at the present time, as 3-wire service is generally used. In these diagrams, an entrance switch and branch-circuit fuses are shown for simplicity. In actual cases a main breaker will be used in the service-entrance equipment and branch circuit breakers will be used for the various circuits.

In the service-entrance equipment, a 2-pole breaker of ample capacity to handle the loads on the sum of the branch-circuits to be used will be installed and conductors of the proper ampacity will be run to the location of the master-control. At this point another feeder panel with a main breaker will be installed and branch circuits of the right amperage as in the service-equipment panel will be installed for the various circuits.

The one point to remember is that the circuit being used from the service-entrance panel **must correspond,** phase for phase, from the master control panel (feeder panel) so that the voltages to ground are the same from either point. This is no problem as it may readily be checked out.

Note in the diagrams that where a single-pole switch would ordinarily be used, a three-way switch is used Also where a 3-way switch would be used, a 4-way switch is used.

REMOTE-CONTROL WIRING

The heart of the remote-control system is a single-pole, single-throw, double-coil relay. The coils of this relay are operated from a 24-volt transformer and remote switches. Essentially, low-voltage residential control switching provides for turning lights and appliances **on** and **off** remotely either within a house, its surrounding grounds, or other buildings on the property.

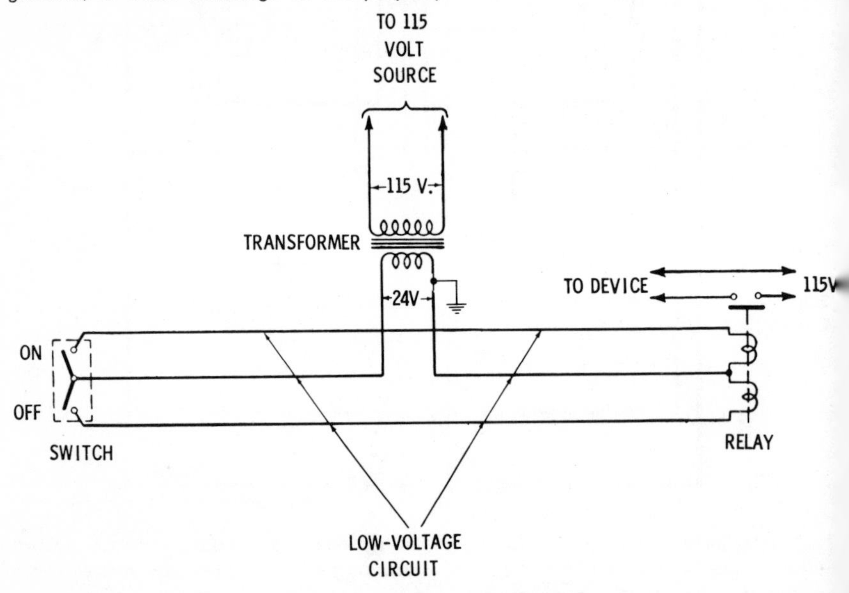

Schematic Diagram of Remote-Control Wiring Circuit Showing One Relay Controlled from One Switch—The relay contacts are operated by momentarily energizing either the opening or the closing coils. The contacts remain latched in either the opening or closed position with no further application of control power.

The relay is mounted by inserting the metal barrel through a one-half inch knockout in a standard outlet box or other metal enclosure. Spring actuated dogs hold the relay in position when installed.

When installed the necessary physical separation of the power circuit and the control circuits as required by the National Electrical Code is automatically established with the power circuits confined in the metal enclosure of the raceway, and the control circuit isolated by being on the outside.

REMOTE-CONTROL WIRING

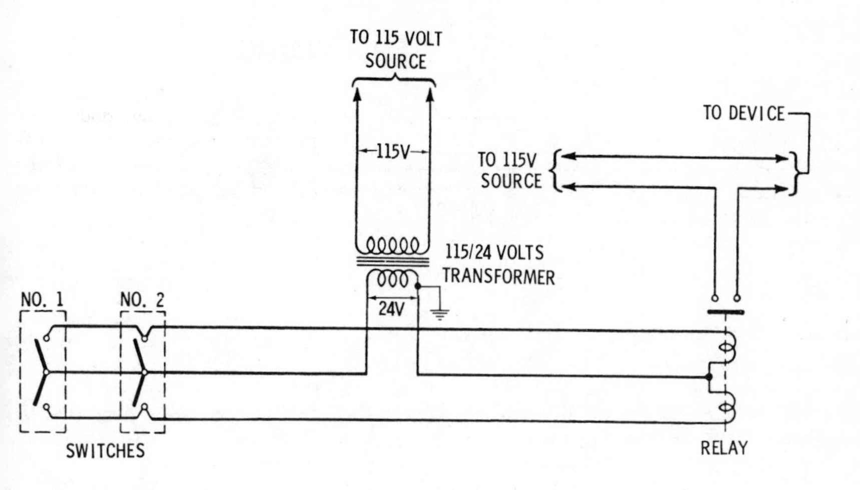

Schematic Diagram of Remote-Control Wiring Circuit Showing One Relay Controlled from Either One of Two Push Button Stations.

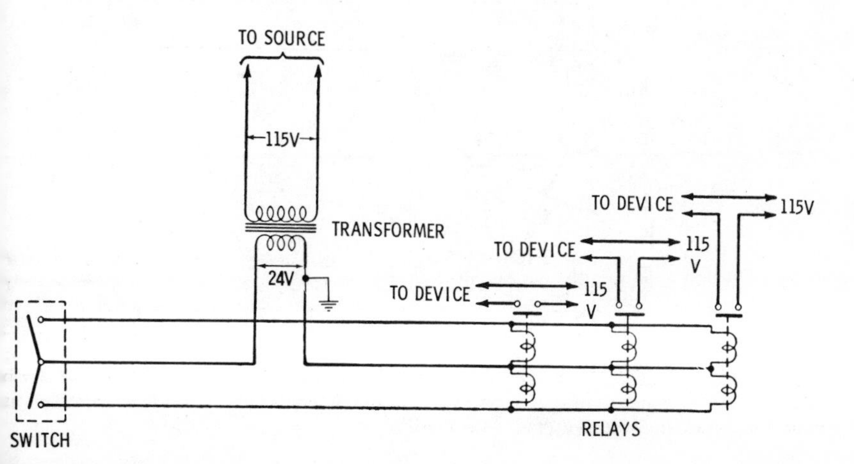

Schematic Diagram Showing a Number of Relays Controlled from a Single Push Button Station.

REMOTE-CONTROL WIRING

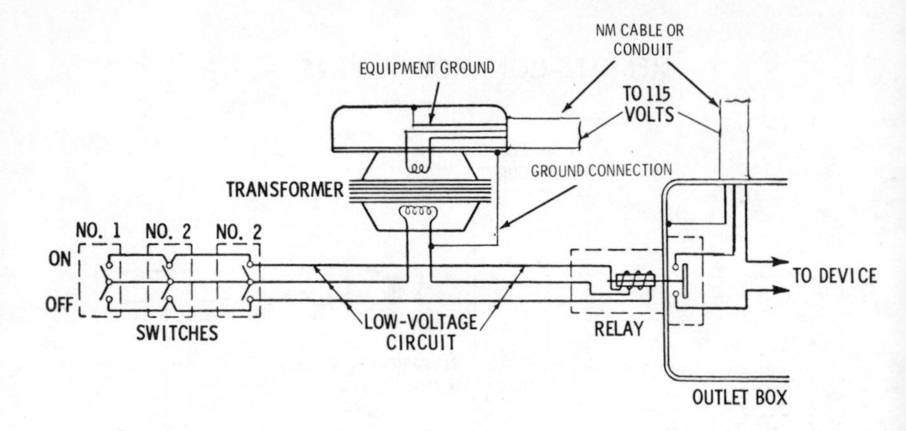

Pictorial Wiring Diagram Illustrating One Relay Controlled from Any One of Three Push Button Stations.

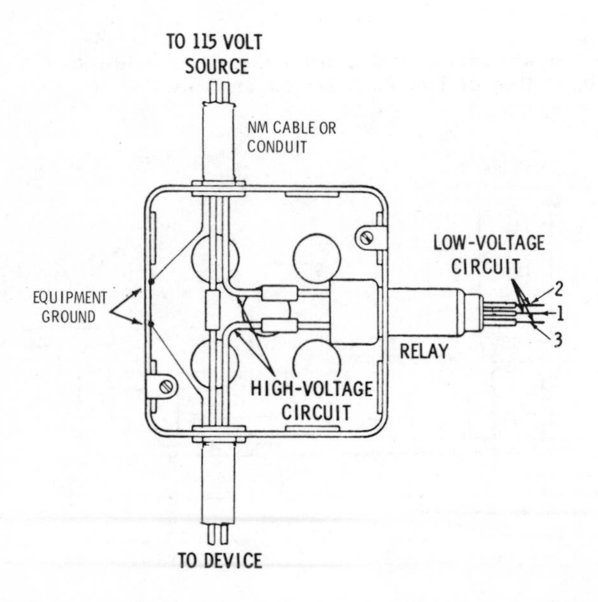

Remote-Control Relay Mounted in Outlet Box.

REMOTE-CONTROL WIRING

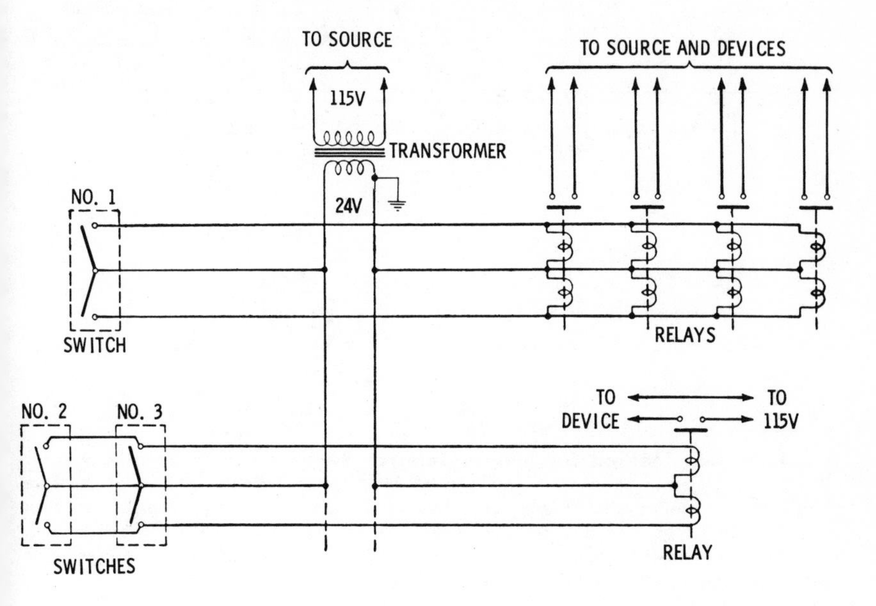

Method of Transformer Connection for Two or More Branch Circuits.

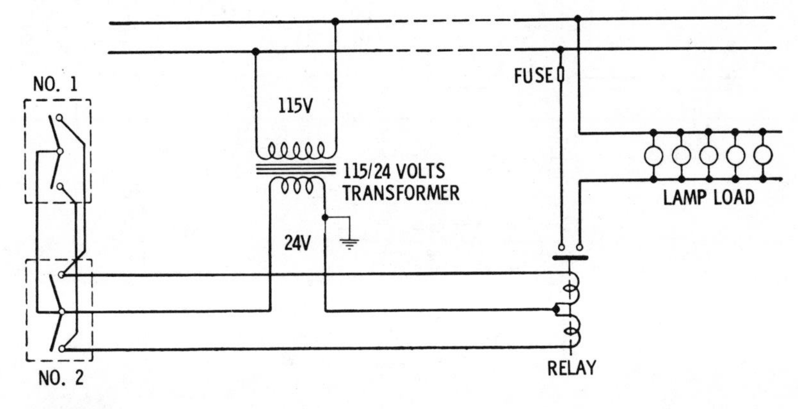

Group of Lamps Controlled from Either One of Two Push Button Stations.

REMOTE-CONTROL WIRING

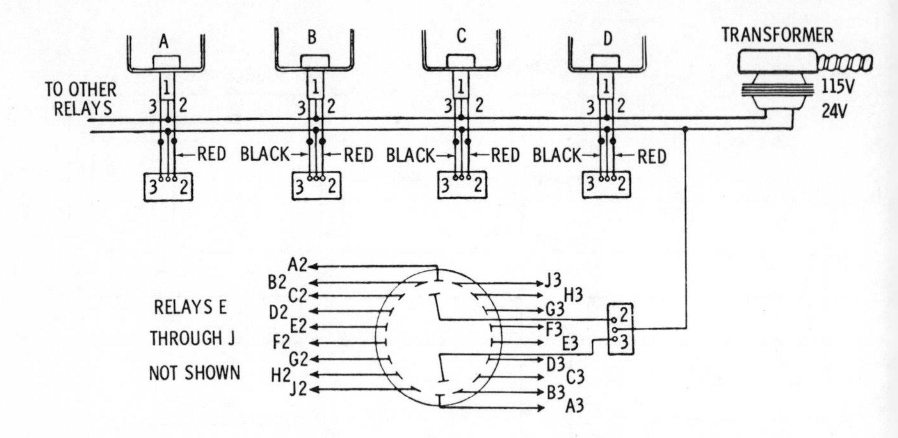

Typical Wiring Method for Master Selector Switch—With the master selector switch it is possible to control any one of nine circuits independently or operate all circuits simultaneously. To control individual circuits, it is only necessary to turn selector switch to circuit desired, then press control switch for on or off as desired. To control all circuits, press control switch for on or off, while turning selector switch through full sweep of all nine positions.

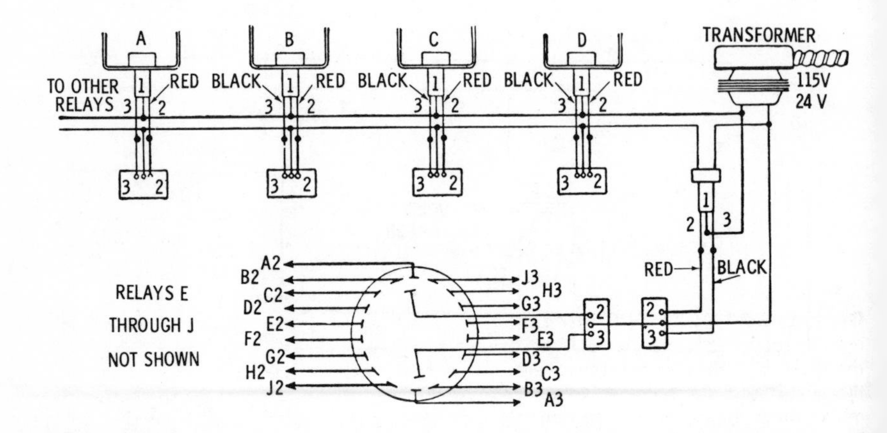

Remote-Control Wiring Hook-Up Showing Connections to Master Selector Switch with Lockout Circuit.

REMOTE CONTROL WIRING

A master switch may turn on or off any number of light circuits at one time. A large diode (D1) is common to all relays (S 1, S 2, and S 3) from the ungrounded side of the transformer. The other side of the transformer is common to the center terminal of all of the switches. There are diodes in each conductor from the master switch, two outside terminals and the outside terminals of the lighting circuit switches. Diode (D1) is reversed from the other six diodes. Thus, switches 1, 2 and 3, may be operated independently or the master switch can take over. The light circuits may however, be turned off independently. The diode (D1) must be large enough to handle the current to all relays involved.

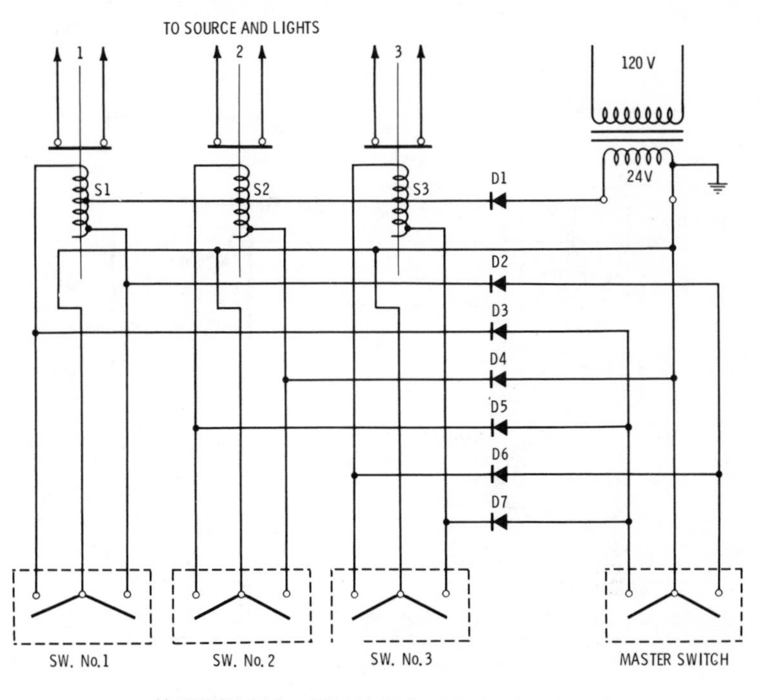

INTERNAL CONNECTION OF BELLS

Operation of Series-Vibrating Bell—When the push button is operated, the current energizes the magnet and attracts the armature, causing the hammer to strike the bell, but before it reaches the end of the stroke the contact breaker breaks the circuit, and the hammer, influenced by the tension of the armature spring rapidly moves back to its initial position, thus completing the cycle.

Operation of Single-Stroke Bell—When the push button is operated the current energizes the magnet and attracts the armature, causing the hammer to strike the bell. The armature remains in the attracted position so long as the current flows

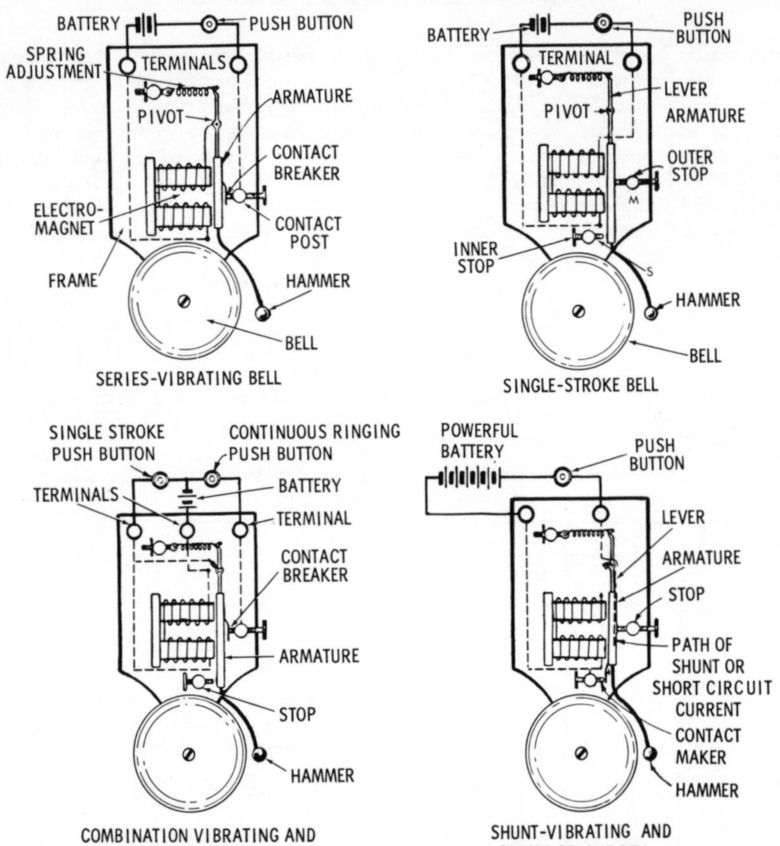

through the magnet. When connection with the battery is broken, the hammer spring pulls the armature back against **M.** A stop **S,** averts the motion of the armature, momentum springing the lever and causing the hammer to strike the bell.

Operation of Combination Vibrating and Single-Stroke Bell—This bell is essentially a vibrating bell with the addition of a third terminal and a stop to prevent continued contact of the hammer with the bell when working single stroke.

Operation of Shunt-Vibrating and Single-Stroke Bell—This is simply an ordinary shunt bell with a switch arranged so that the short circuit through the contact maker, armature, and lever may be cut out, thus restricting the current to the magnet winding.

Operation of Differentially-Wound Vibrating Bell—When the battery circuit is closed, current flows through the magnetizing winding and energizes the magnets which in turn attract the armature. The contact maker closes the circuit through

INTERNAL CONNECTION OF BELLS

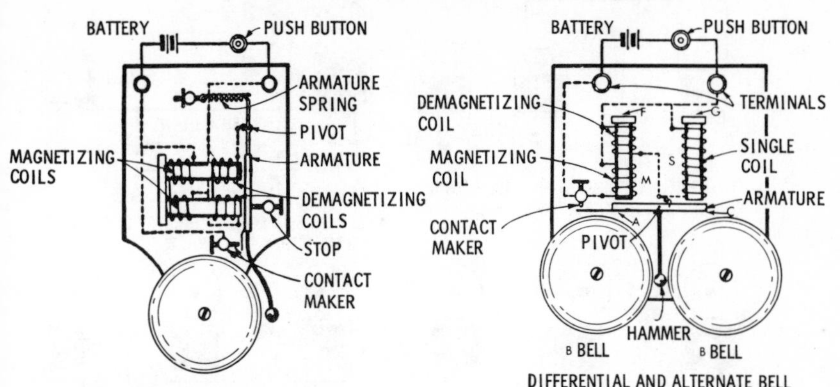

DIFFERENTIALLY-WOUND VIBRATING BELL

DIFFERENTIAL AND ALTERNATE BELL

the demagnetizing coils, which demagnetize the magnets. The armature spring pulls the armature back against the stop, while the contact maker breaks the circuit through the demagnetizing coils.

Operation of Differential and Alternate Bell—When the battery circuit is closed by means of the push button, current flows through the magnetizing winding **M** and energizes **F,** which attracts end **A** of the armature. The contact maker closes circuit through magnetizing coil, and the single coil **S,** of magnet **G.** Then the demagnetizing coil demagnetizes **F,** and as a result, magnet **G** attracts end **C** of the armature. The contact maker breaks the circuit through demagnetizing coil **D,** and single coil **S** of magnet **G,** completing the operation cycle.

VARIOUS BELL CIRCUITS

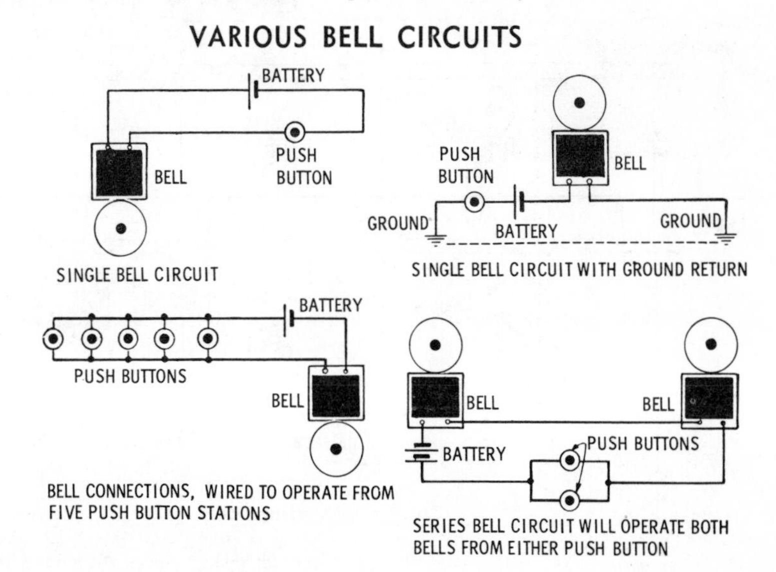

SINGLE BELL CIRCUIT

SINGLE BELL CIRCUIT WITH GROUND RETURN

BELL CONNECTIONS, WIRED TO OPERATE FROM FIVE PUSH BUTTON STATIONS

SERIES BELL CIRCUIT WILL OPERATE BOTH BELLS FROM EITHER PUSH BUTTON

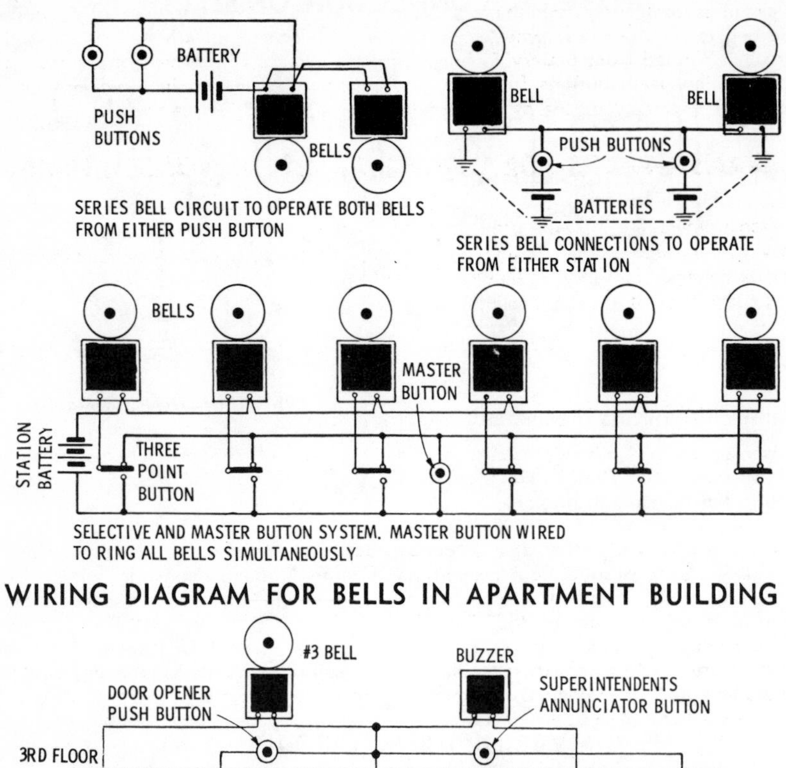

SERIES BELL CIRCUIT TO OPERATE BOTH BELLS FROM EITHER PUSH BUTTON

SERIES BELL CONNECTIONS TO OPERATE FROM EITHER STATION

SELECTIVE AND MASTER BUTTON SYSTEM. MASTER BUTTON WIRED TO RING ALL BELLS SIMULTANEOUSLY

WIRING DIAGRAM FOR BELLS IN APARTMENT BUILDING

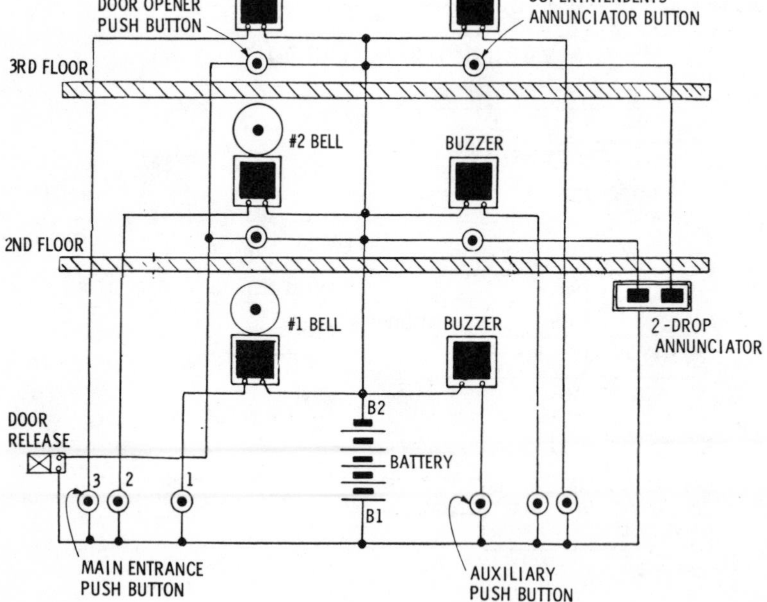

Operation—When for example, push button to apartment on **3rd** floor is operated, a circuit is completed from battery B_1 through bell # **3** and to battery B_2 causing the bell to ring. Similarly when door opener push button on 4th floor is pressed, a circuit is formed from battery B_2 energizing the release coil, which opens the door. The auxiliary push buttons from the service entrance function in a similar manner, notifying tenant by means of buzzer of the presence of service man.

TRANSFORMERS FOR OPERATING BELLS AND/OR CHIMES

Transformers are usually used for the operation of door bells and/or chimes. They may have a single voltage on the low-voltage output or may have taps for various voltages. Batteries as illustrated in the preceding illustrations are very seldom used but in these illustrations they show the voltage source for bell operation.

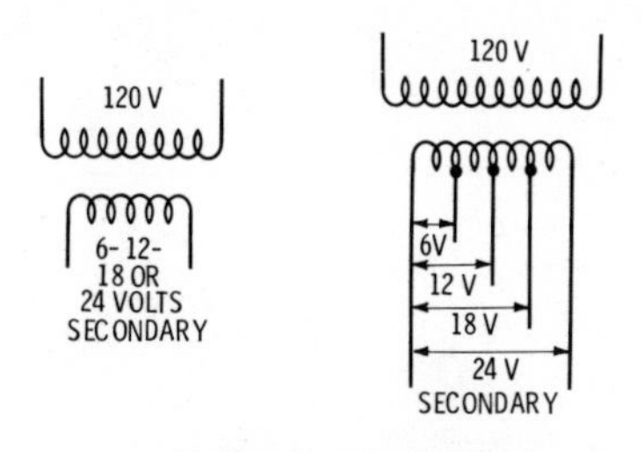

DOOR CHIMES

Door chimes have solenoids, the plungers of which hit tempered metal for gong sounds. The most common chimes give a double chime for the front door and a single chime for the back door. The push buttons often have neon bulbs across the

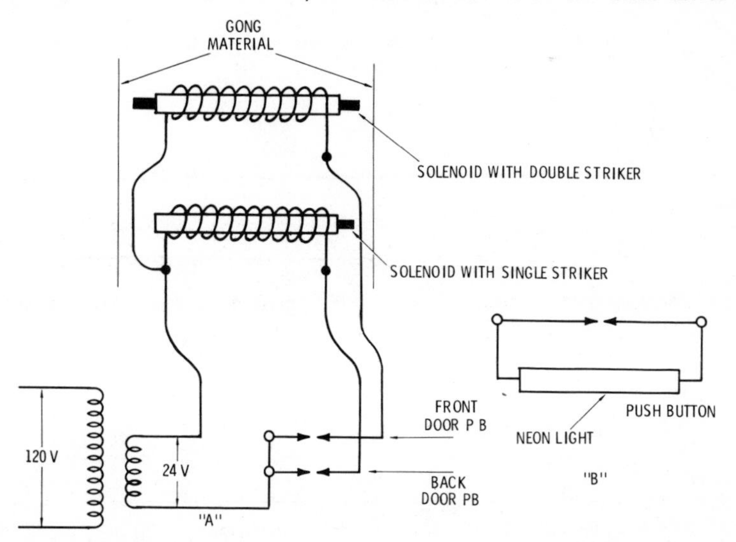

contacts, so when the button is not making contact the bulb is in series with the solenoid and transformer and lights up, this is illustrated in "B".

More expensive chimes have a timer which is energized by pushing the door bell button. The timer energizes different solenoids which strike different brass tubes and plays a tune, these tubes are tuned and must be the exact length or cut according to the manufacturer's specifications.

ELECTRIC DISCHARGE LIGHTING

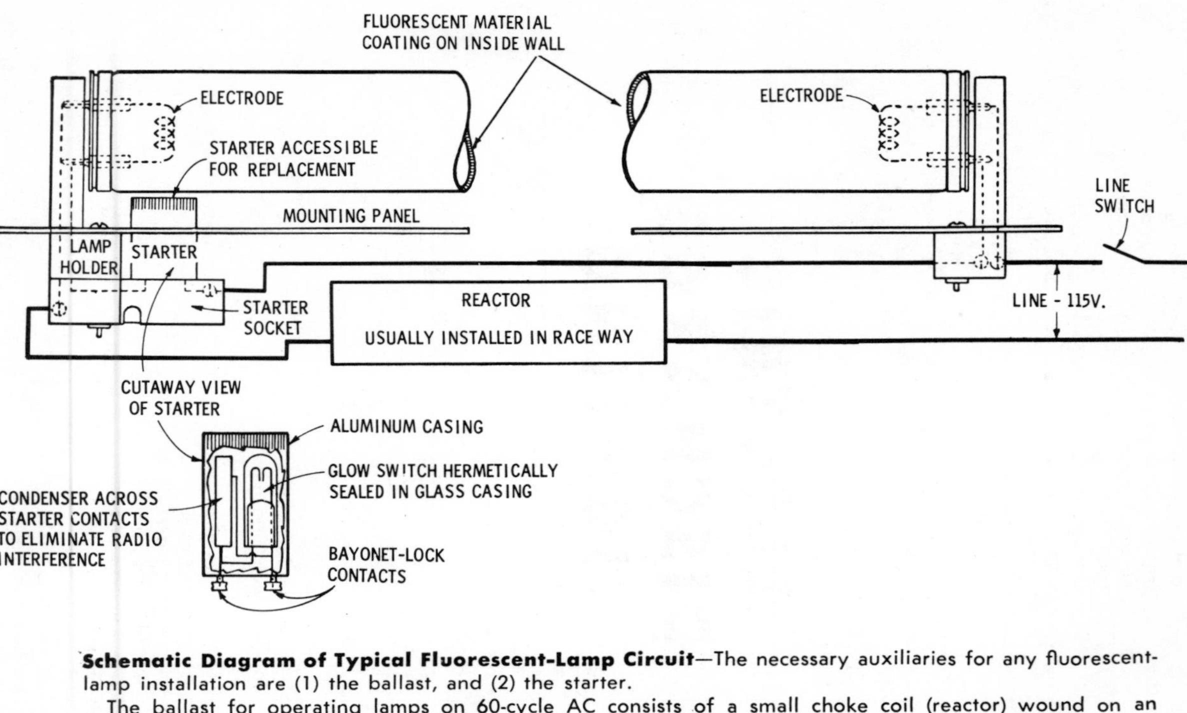

Schematic Diagram of Typical Fluorescent-Lamp Circuit—The necessary auxiliaries for any fluorescent-lamp installation are (1) the ballast, and (2) the starter.

The ballast for operating lamps on 60-cycle AC consists of a small choke coil (reactor) wound on an iron core.

The ballast serves three important functions, namely:

1. It preheats the electrode to make available a large supply of free electrons.
2. It provides a surge of relatively large potential to start the arc between the electrodes.
3. It prevents the arc current increasing beyond the limit set for each size of lamp.

FLUORESCENT-LAMP CIRCUITS

Ballasts—These may be designed for operation of a single lamp or, as is more common, for two lamps mounted in a single fixture. Certain practical advantages are obtained from the choice of an electrical circuit which combines under one cover the equipment for the control of two lamps.

Chief among the advantages are improved power factor, decreased stroboscopic effect and reduced auxiliary losses. Each lamp is operated through a separate choke coil. A capacitor is connected in series with one lamp and its choke coil to give a leading current. The leading and lagging currents will combine with a resulting line power factor of very nearly 100%.

When connecting lamps, ballast and starter into an electric circuit, it is of the utmost importance to observe the manufacturers' diagram usually labeled on the ballast. This diagram should be followed in each instance for proper operation of the lamp or lamps. Also it should be clearly understood that each lamp size must have a ballast designed for its particular wattage, voltage and frequency.

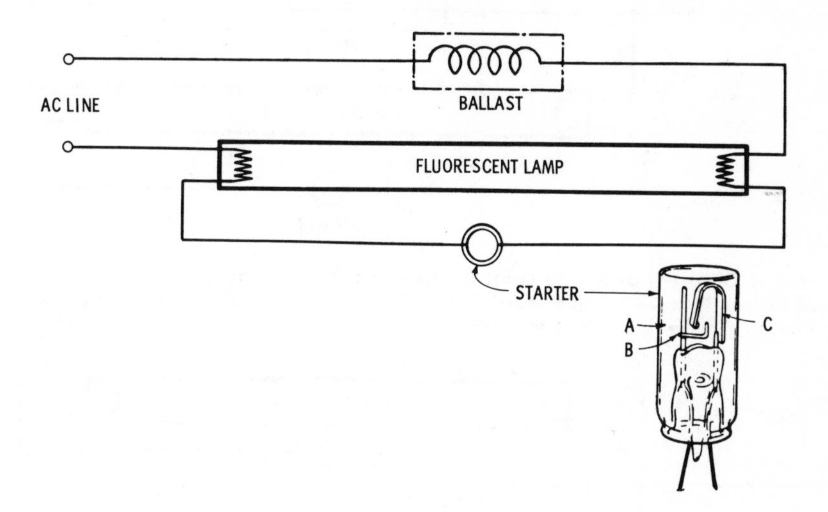

Wiring Diagram of Single Fluorescent Lamp—In the glow-type starter **A,** represents glass bulb filled with inert gas; **B,** fixed electrode; **C,** bimetal strip.

Starters—The starter is designed to act as a time-delay switch which will connect the two filament type electrodes in each end of the lamp in series with the ballast during the short preheating period when the lamp is first turned on and then open the circuit to establish the arc. This preheating causes the emission of electrons from the cathodes and thus makes it possible for the arc to strike without the use of excessively high voltage.

Operation—The switch is enclosed in a small glass bulb and consists of two electrodes, one of which is made from a bimetal strip, in an inert gas such as neon or argon. These electrodes are separated under normal conditions but when closed form part of a series circuit through the lamp electrodes and the choke coil (ballast).

When voltage is applied, a small current flows as a result of the glow discharge between two electrodes of the switch. Heating of the electrodes results which, by the expansion of the bimetallic element, causes the electrodes to touch. This short circuiting of the switch stops the glow discharge but allows a substantial flow of current to preheat the lamp electrodes. There is enough residual heat in the switch to keep it closed for a short period of time for the electrode preheating. The glow being quenched, the bimetal cools, the switch opens and the resultant high voltage surge starts normal lamp operation. If the lamp arc fails to strike, the cycle is repeated.

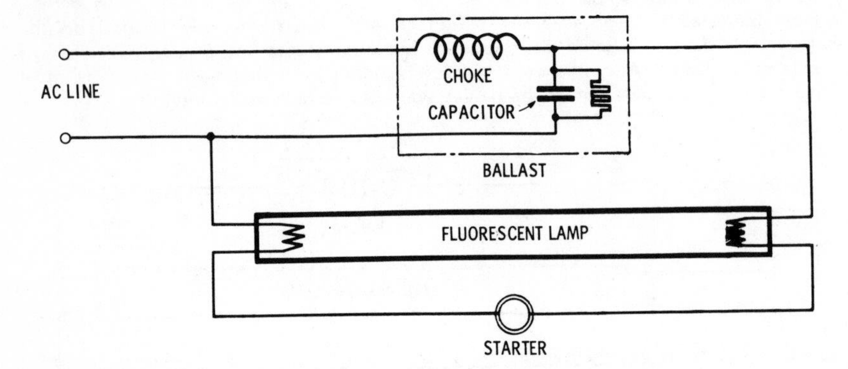

Wiring Diagram of Single Fluorescent Lamp with Capacitor for Improvement of Power Factor—For operation of the 13-, 30-, 40- and 100-watt lamps on 110- to 125-volt circuits, the ballast must include a transformer for stepping up the voltage.

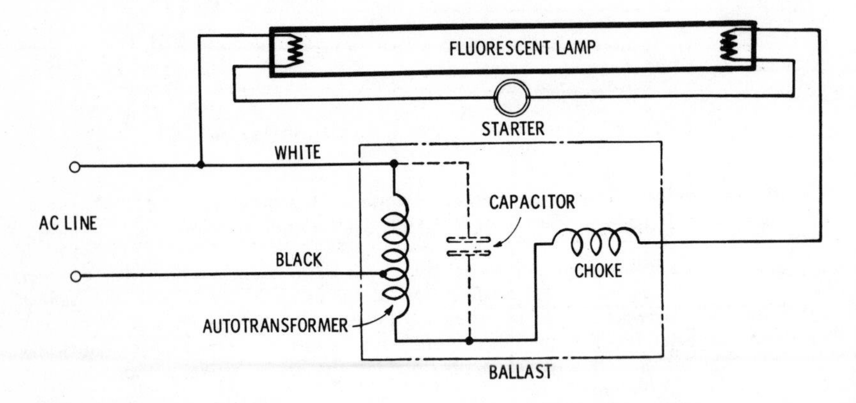

Wiring Diagram of Single Fluorescent Lamp with Power Factor-Corrected Ballast and Autotransformer for Stepping Up the Voltage.

FLUORESCENT-LAMP CIRCUITS

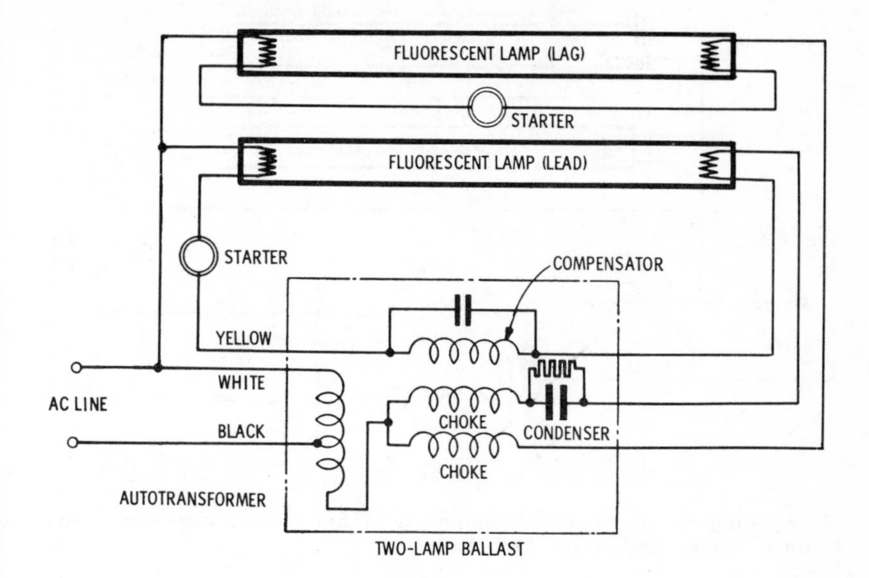

Two-Lamp Ballast with Built-In Starting Compensator and Autotransformer.

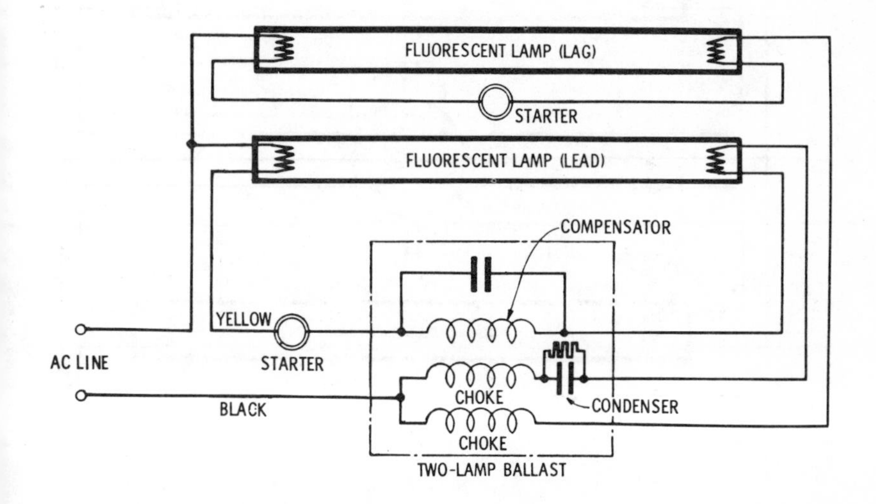

Wiring Diagram of Two-Lamp Ballast with Built-In Starting Compensator.

FLUORESCENT-LAMP CIRCUITS

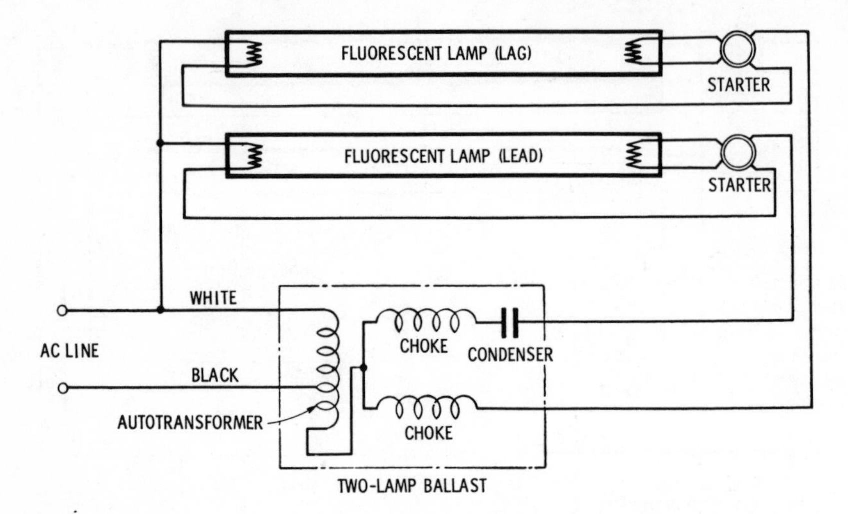

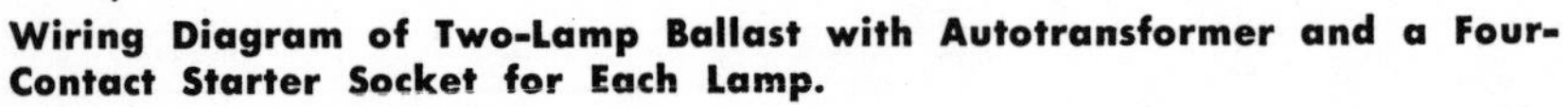

Wiring Diagram of Two-Lamp Ballast with Autotransformer and a Four-Contact Starter Socket for Each Lamp.

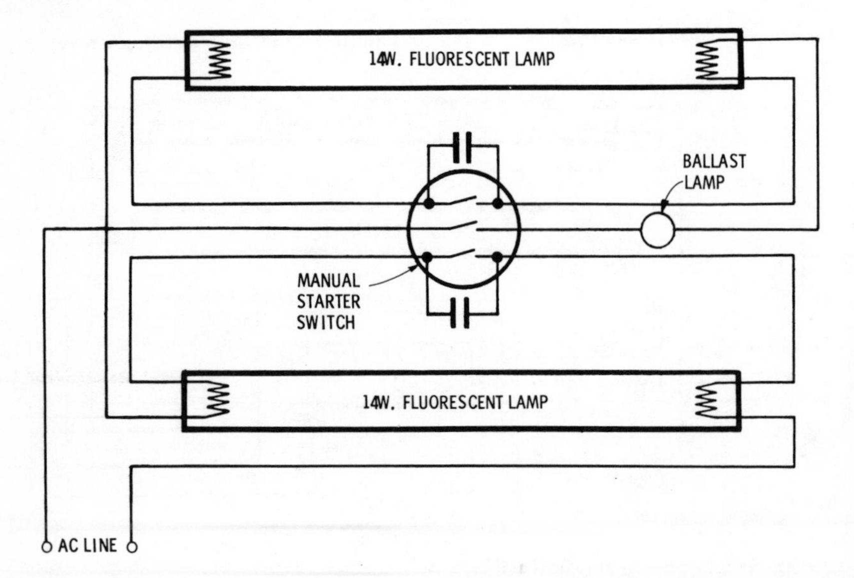

Wiring Diagram for Operating Two 14-Watt Fluorescent Lamps in Serie with a Special Incandescent Ballast Lamp.

FLUORESCENT-LAMP CIRCUITS

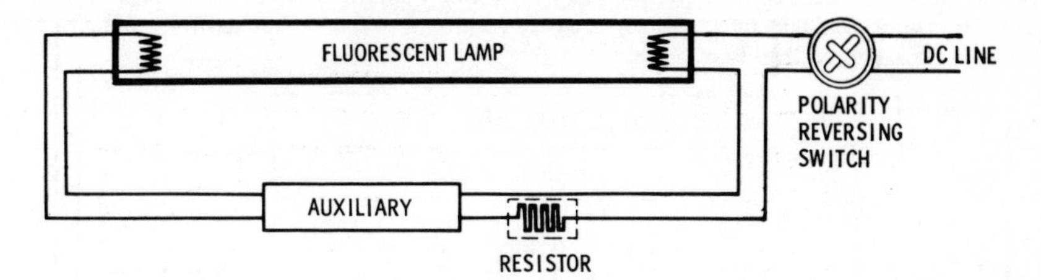

Wiring Diagram of Fluorescent Lamp for Operation on Direct Current—While the fluorescent lamp is basically an alternating-current lamp, it is also used on direct current where alternating current is not available. Due to the lack of voltage peaks when direct current is used, lamp starting is generally more difficult than on alternating current and special starting devices must be used. The thermal and manual switches in addition to a starting inductance are generally employed.

With fluorescent lamps, one end of the tube may become dim after operating a few hours on direct current. This is due to the bombardment of electrons in one direction only. By reversing the direction of current flow at certain intervals (once a day or more frequently if desired) by means of a special reversing switch, this dimming may be eliminated.

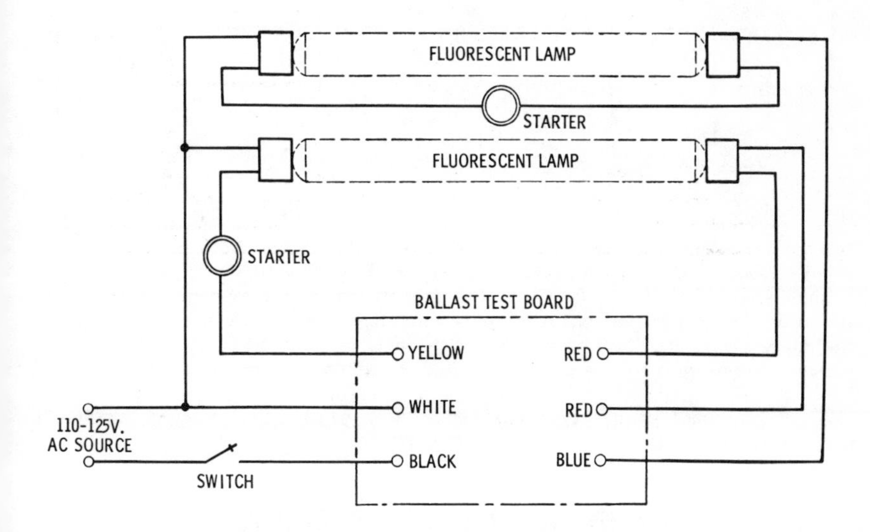

Wiring Arrangement for Ballast Test Board—By means of a circuit of this type, 40-watt, two-lamp ballasts for use on standard voltage may easily be tested. This is simply a two-lamp circuit with binding posts left at the point where the ballast must be connected in order that quick connections may be made. By providing socket spacing necessary to receive lamps of other sizes, such a test board can be used to check ballasts of any size, provided care is taken to make the connections through the proper binding posts.

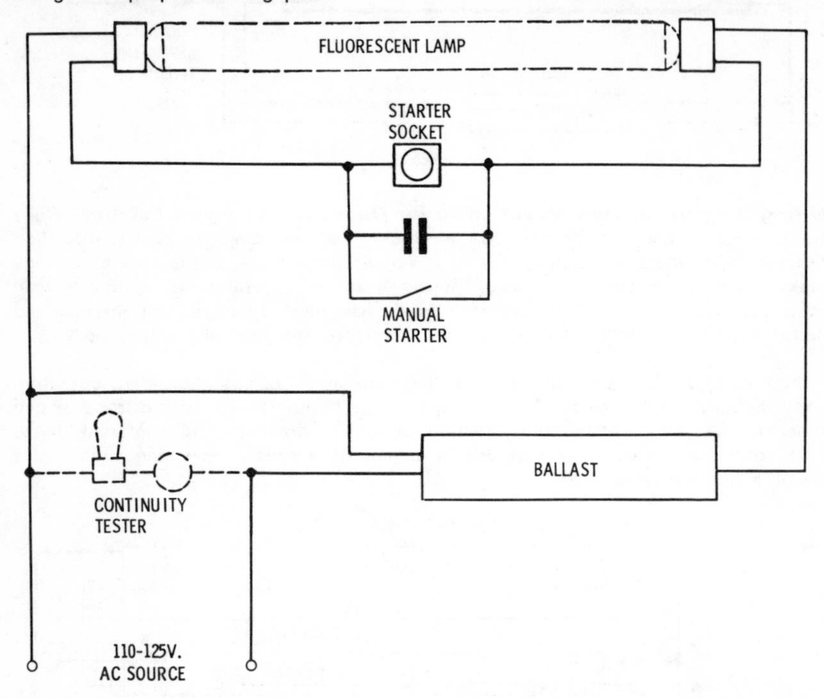

Wiring Diagram Illustrating a Simple Testing Board for Fluorescent Lamps— A test board of this type may be used for checking 40-watt lamps with the lamp holders spaced to receive lamps of the proper size and provided with a starter socket and manual starter switch properly connected to a suitable ballast. A filament-continuity checker can also be included if desired. This consists of a fluorescent-lamp socket in series with an incandescent lamp of 25-watt size or smaller.

A testing board of this type has been found helpful for checking fluorescent lamps and starters to see that they operate satisfactorily.

Lamp boards of this general type may also be made to check lamps of several different sizes by providing the necessary ballasts and the lampholders properly spaced to receive the lamps or one lamp socket mounted stationary and the other provided with pins so that it can be plugged into jacks located at the proper distance for taking lamps of various lengths.

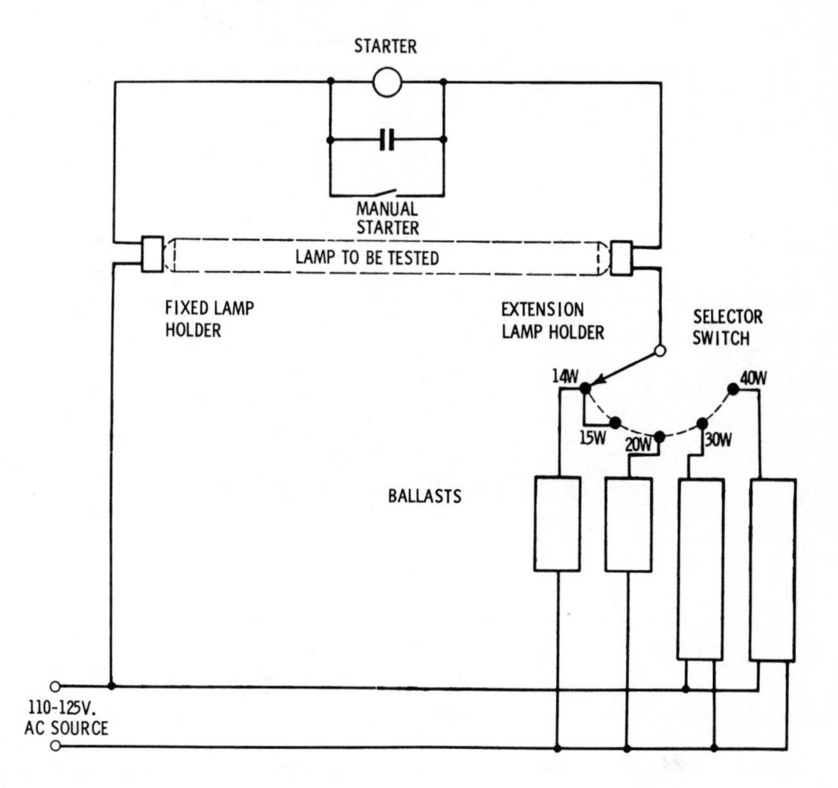

Wiring Diagram of Portable Test Kit for Checking Fluorescent Lamps and Starters—By means of a circuit arrangement of this type various size lamps and starters may be tested directly on the job. All that is required is that the kit be large enough to hold the required ballasts, as connection to one end of the lamp is made by means of a lamp holder on the end of an extension cord. A selector switch must be included for making connections to the proper ballast.

Fluorescent-Lamp Dimming—To dim fluorescent lamps, it is insufficient to merely lower the voltage because the lamp will be extinguished at the zero point of each cycle and the voltage will be insufficient to restrike the arc. The brightness of the lamp will be determined by the current flowing after the arc has been struck. It is necessary to provide an autotransformer to develop enough voltage to strike the arc and a dimming control circuit to control the current after the arc has been established.

FLUORESCENT-LAMP CIRCUITS

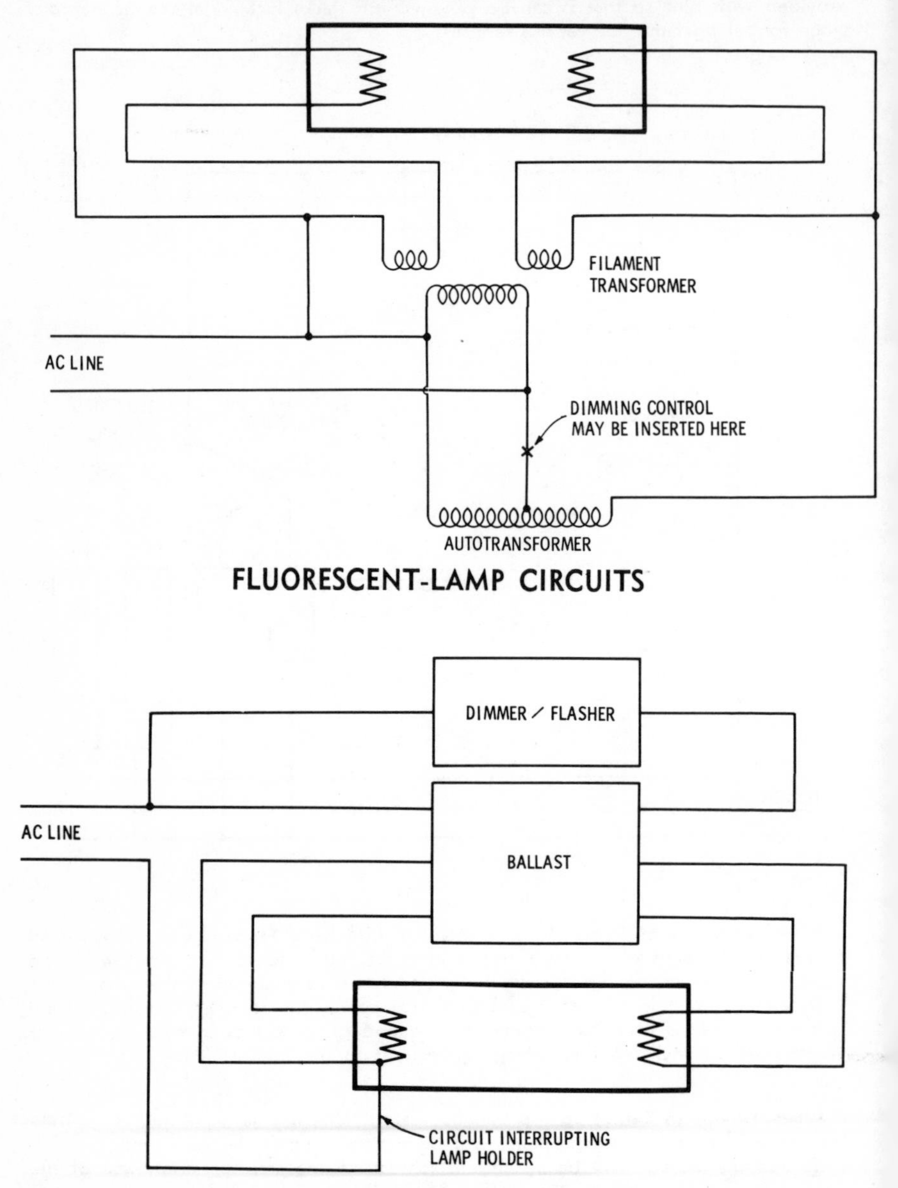

Wiring Diagram for Use of Fluorescent Lamp for Dimming or Flashing—Dimming ballasts and the proper control device make it possible to adjust the light level up or down quickly and easily. Fluorescent lamps used with dimmers offer several advantages over incandescent lamps—higher efficiency, lower operating costs, and better control of color over the dimming range. The flasher device permits the use of fluorescent lamps in display signs and the like.

FLUORESCENT LAMP SYSTEMS—INSTANT START CIRCUITS

On instant start lamps, sufficient voltage must be applied between the cathodes to break down the resistance of the lamp and strike the arc. The arc quickly heats

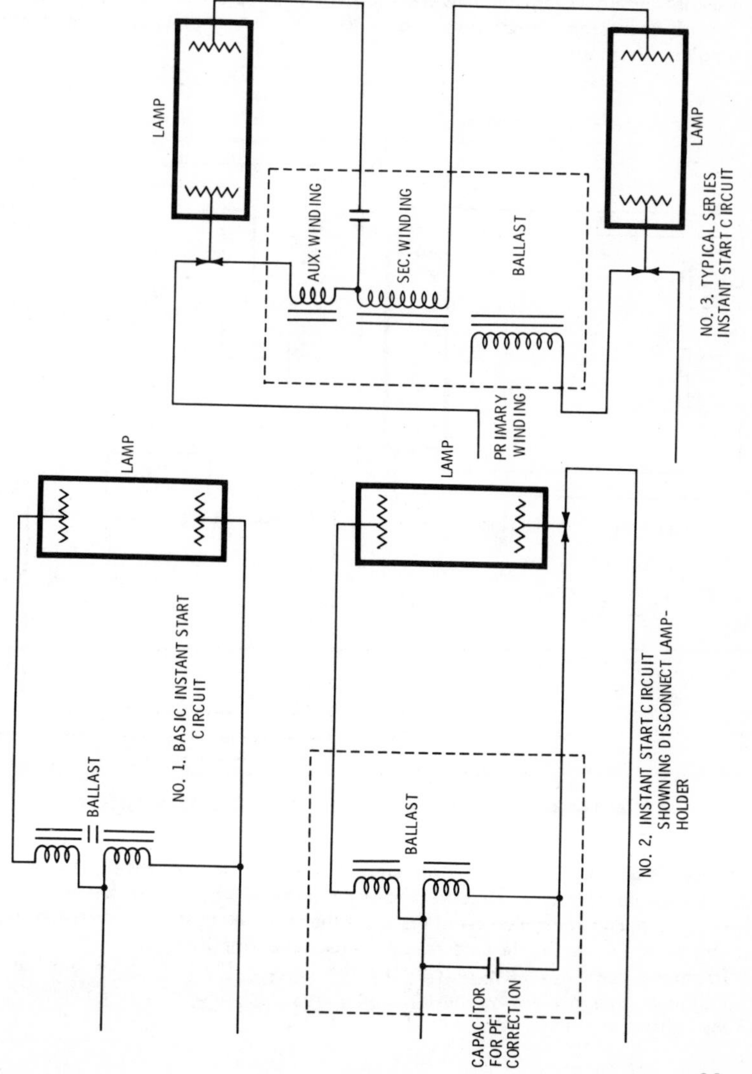

up the fine-wire cathodes, which then supply electrons to sustain the arcs. Slimline lamps have single-pin bases. Other lamps with bi-pin bases, have the leads from the pins connected together within the lamp, and are marked **instant start** so as not to confuse them with **rapid start lamps.**

FLUORESCENT LAMP SYSTEMS—RAPID START CIRCUITS

The rapid start principle utilizes low resistance cathodes which can be heated continuously with very low losses. This is probably the most popular and important fluorescent lighting systems on new installations. The rapid start principle also extends the use of fluorescent lamps into applications that were previously not possible—dimming and flashing.

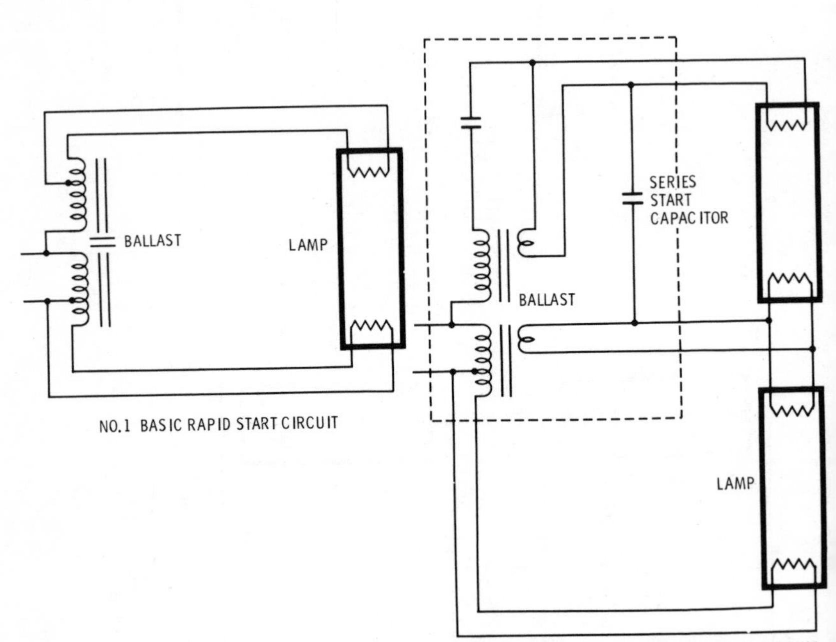

NO. 1 BASIC RAPID START CIRCUIT

NO. 2 TWO-LAMPS SERIES LEAD RAPID START CIRCUIT

FLUORESCENT LAMP SYSTEMS—DIMMING

With the development of rapid start fluorescent lamps, they may be dimm[ed] with a number of special circuits. These circuits use one of the following metho[ds:] thyratrons, silicon-controlled rectifiers and other solid-state devices, variable inducto[rs,] autotransformers, saturable core reactors, magnetic amplifiers, etc.

There are such a variety of controls that only a block diagram will be show[n.] The dimmer systems always have the circuitry included in the equipment and/ on the ballasts and dimmer.

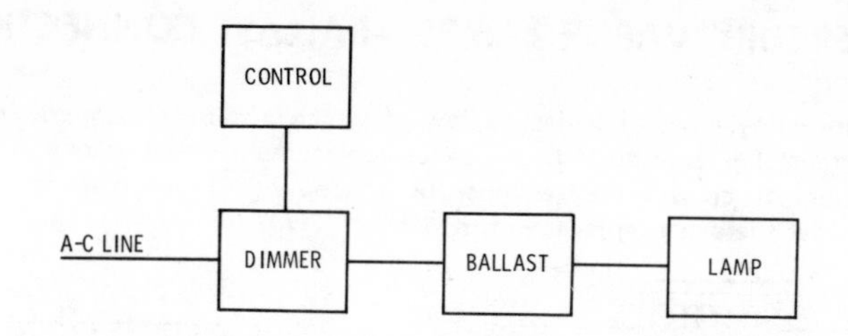

FLUORESCENT LAMP SYSTEMS—FLASHING

The life of fluorescent lamps is seriously affected by frequently turning them on and off. With rapid start lamps and flashers, this does not affect the life materially as the cathode temperatures are maintained, only the arc in the tube is cutoff.

As with dimming equipment, the diagrams for the particular flasher control will include the diagram for the circuits involved. A block diagram is shown.

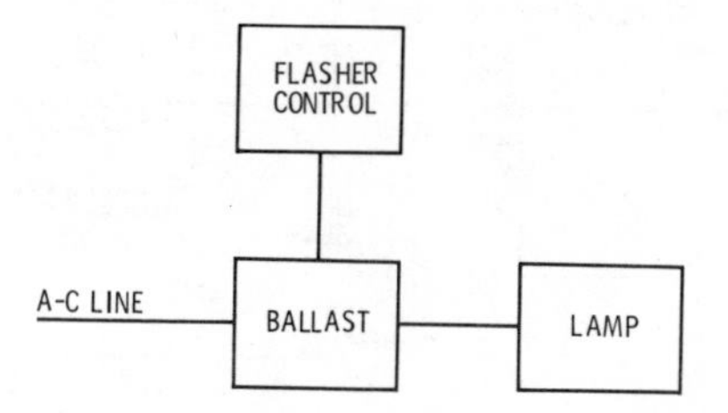

MERCURY-VAPOR LAMPS

The sketch shown here is the working parts of a mercury-vapor lamp. This is all enclosed inside an outer glass bulb. When the circuit is turned on, it takes all of 5 minutes for the lamp to reach full brilliancy and if shut off, the lamp has to cool before an arc will restrike and then the lamp comes up slowly again to full brilliancy.

Where these are used in a room, there should also be some rapid start fluorescent lamps in case the power goes off—there will be some light from the fluorescent bulbs when the power comes on until the mercury-vapor lamps strike again.

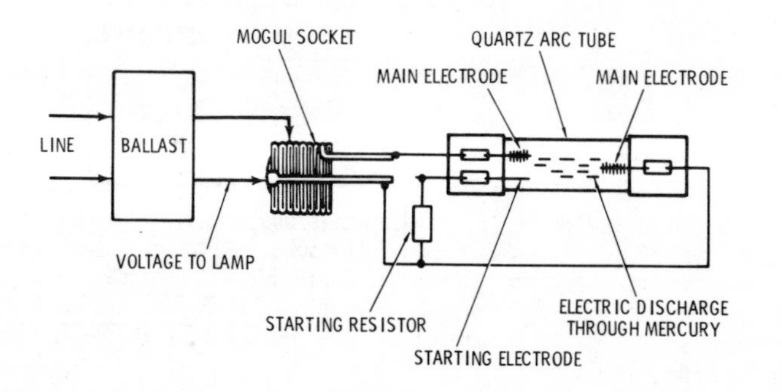

MERCURY VAPOR LAMPS—BALLAST CONNECTIONS

There are many types of ballast, a few of which will be illustrated. In replacing a ballast, it must be replaced with an exact replacement—that is one from the same manufacturer, or cross-reference must be made with other manufacturers' ballast numbers to get the exact replacement model.

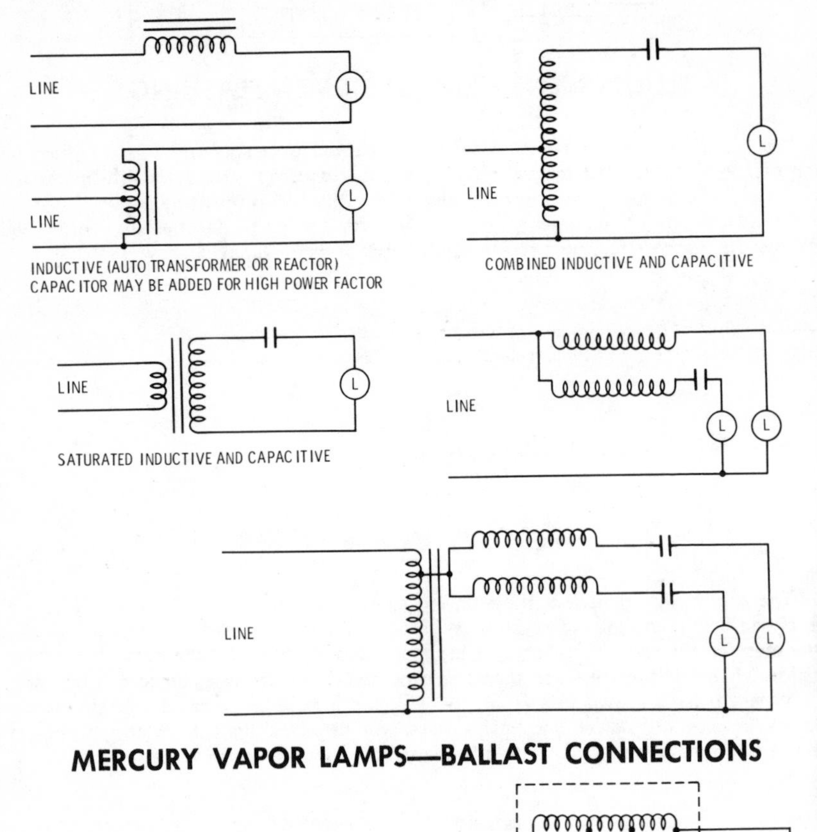

INDUCTIVE (AUTO TRANSFORMER OR REACTOR)
CAPACITOR MAY BE ADDED FOR HIGH POWER FACTOR

COMBINED INDUCTIVE AND CAPACITIVE

SATURATED INDUCTIVE AND CAPACITIVE

MERCURY VAPOR LAMPS—BALLAST CONNECTIONS

FOR LINE VOLTS	CONNECT LINE	ALSO CONNECT
220	1-2	2 TO 6
240	1-2	2 TO 5
265	1-2	2 TO 4
277	1-2	2 TO 3

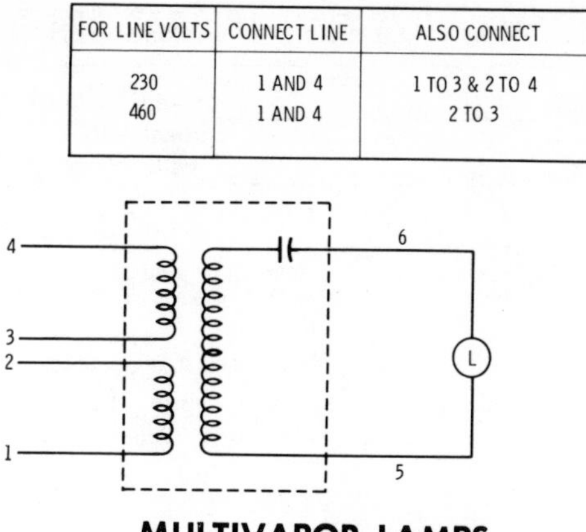

FOR LINE VOLTS	CONNECT LINE	ALSO CONNECT
230	1 AND 4	1 TO 3 & 2 TO 4
460	1 AND 4	2 TO 3

MULTIVAPOR LAMPS

Multivapor lamps are very similar to mercury-vapor lamps. There are metallic iodide additives to assist in cleaning the inside of the lamp. There is also a bimetal switch for starting. The ballast are very similar to those for mercury-vapor lamps.

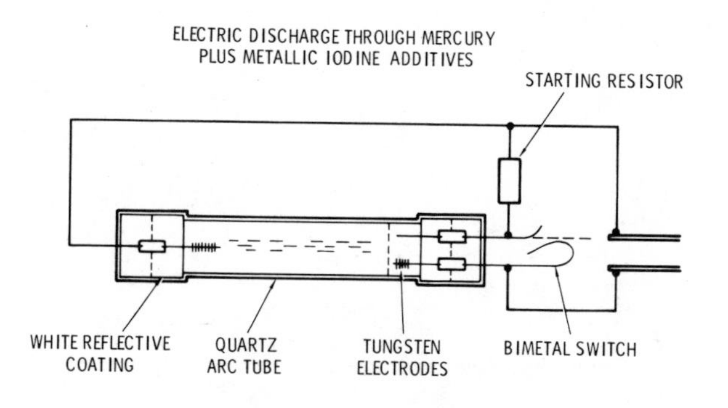

LUCALOX LAMPS

The lucalox lamp is a very high output lamp and is being used to a great extent on increasing light output over mercury-vapor lamps, without the increase in power consumption. Due to the sodium-vapor, the light is of a yellow color. The ceramic-arc tube will withstand 1300°C as compared to 800°C for quartz and 550°C for borosilicate glass. It will also transmit 92% of the visible wavelength. The ceramic is not affected by the corrosive effect of the sodium vapor.

There are presently two types of ballasts used, both of which must have a starter. One ballast is called the "Magnetic Design" and the other the "Phase-Control Design."

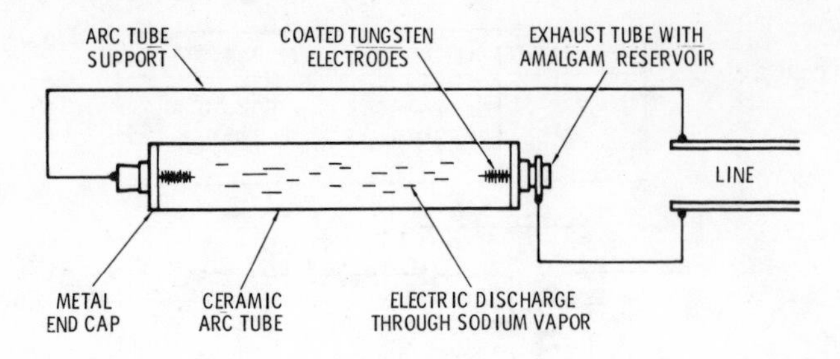

Both are 60 Hz design which provide a nominal RMS open-circuit voltage of abou 225-volts. Both contain an auxiliary starting circuit that supplies a high-voltag pulse of 2500-volts peak for the 400 watt lamp. Ballast losses usually range fror 15 to 20% of lamp voltage.

TYPICAL SERVICE ENTRANCE CONNECTIONS AND PROPER GROUNDING

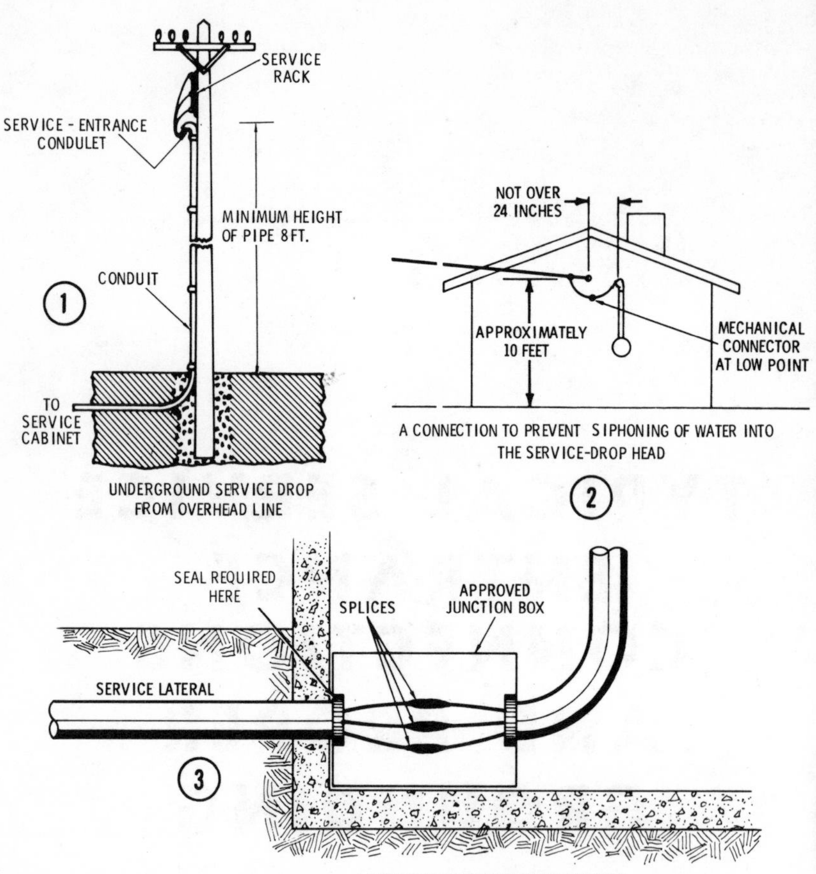

UNDERGROUND SERVICE DROP FROM OVERHEAD LINE

A CONNECTION TO PREVENT SIPHONING OF WATER INTO THE SERVICE-DROP HEAD

A SPLICE IN AN UNDERGROUND SERVICE

Fig. 1, illustrates an underground service supplied from a pole (Service Lateral) Fig. 2, illustrates a service lateral where it enters the building. Sometimes it comes up the outside of the building to a meter housing and then thru the wall to the service-equipment. Fig. 3 illustrates dimensions for service-drop and service-entrance If the mechanical connector is not used to prevent siphoning of water, the service entrance conduit shall be 24 inches above where the service-drop attaches to the building.

Figs. 4 and 5 show proper bonding and grounding connections at the meter housing and in the service-entrance equipment.

Figs. 6 and 7, illustrate service-entrance masts and the proper installation to get clearances above the ground.

Typical Wattmeter Connection in a Single-Phase Circuit—When connected as shown, the instrument is measuring power load plus losses in its own current coil circuit. If the instrument reads backwards, reverse the current leads. This registers

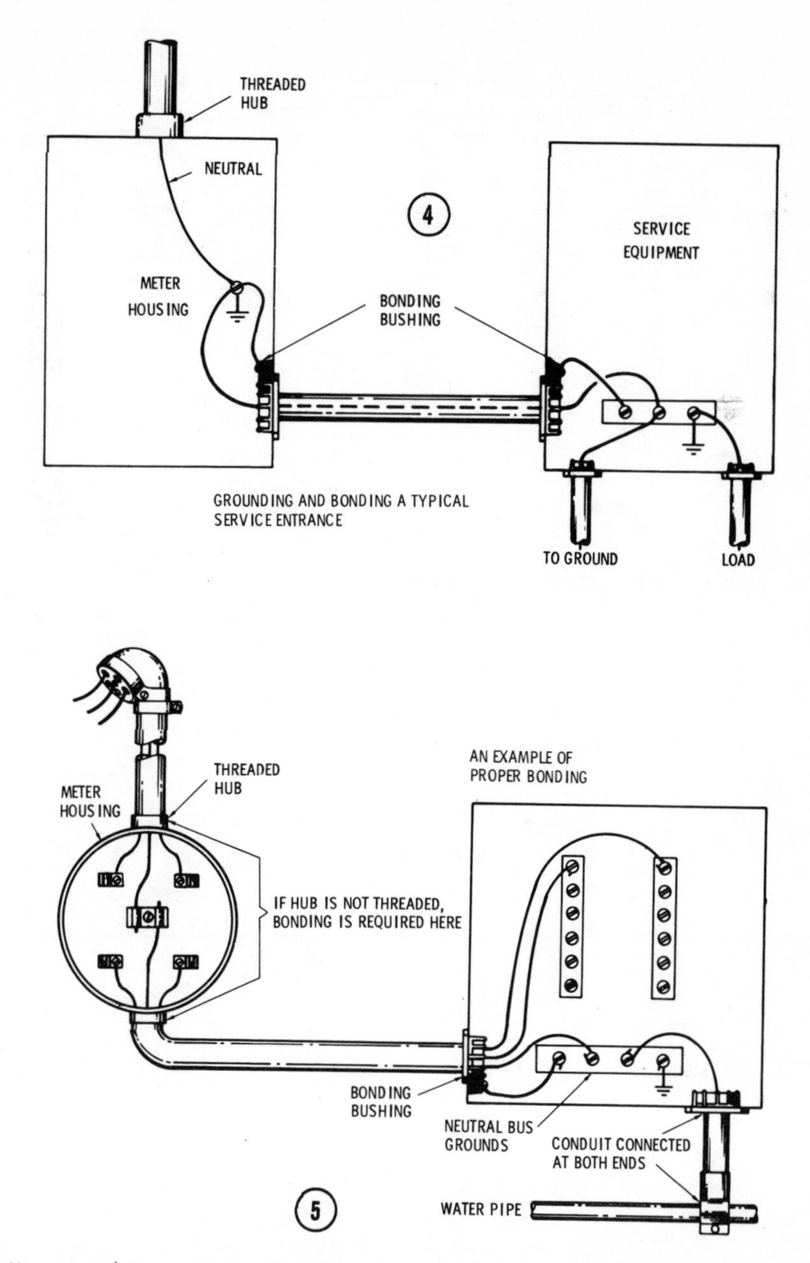

watts, not volt-amperes and neglects the consideration of the power factor of the circuit.

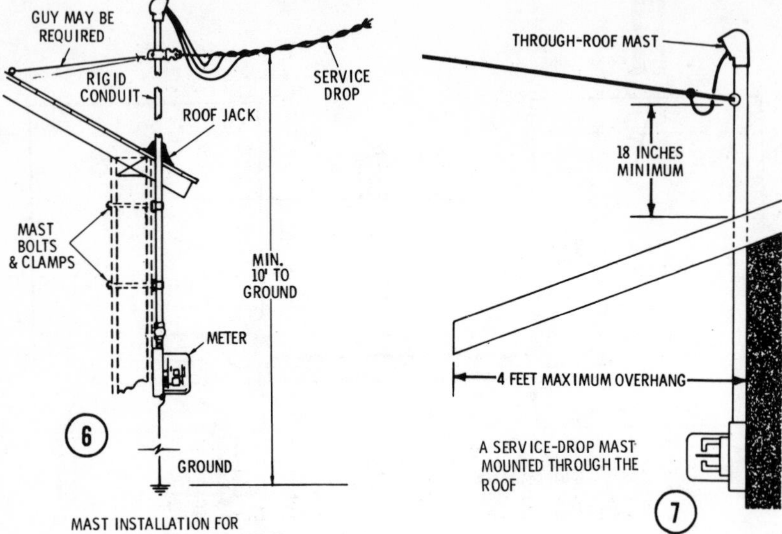

MAST INSTALLATION FOR PROPER SERVICE-DROP HEIGHT

A SERVICE-DROP MAST MOUNTED THROUGH THE ROOF

METERS AND CONNECTIONS

AMMETER CONNECTIONS

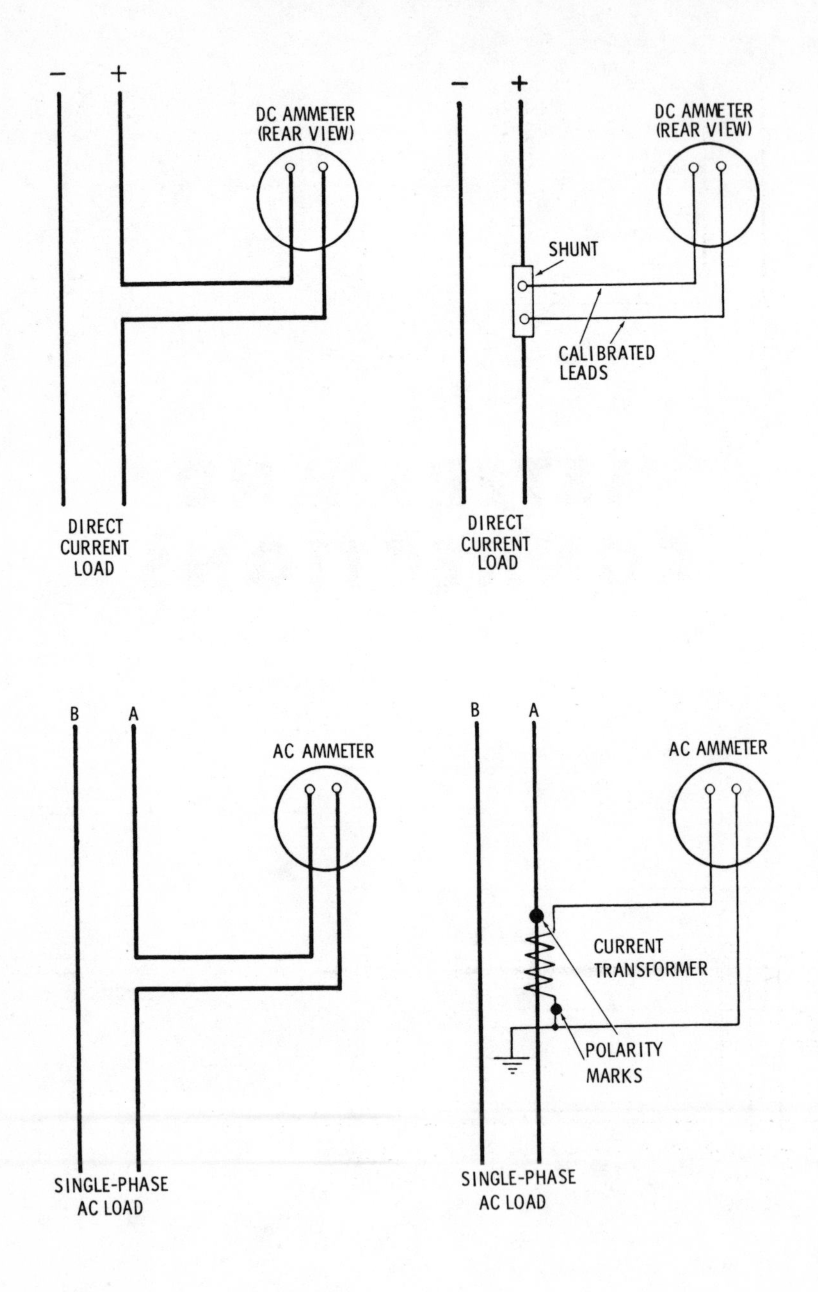

VOLTMETER AND AMMETER CONNECTIONS

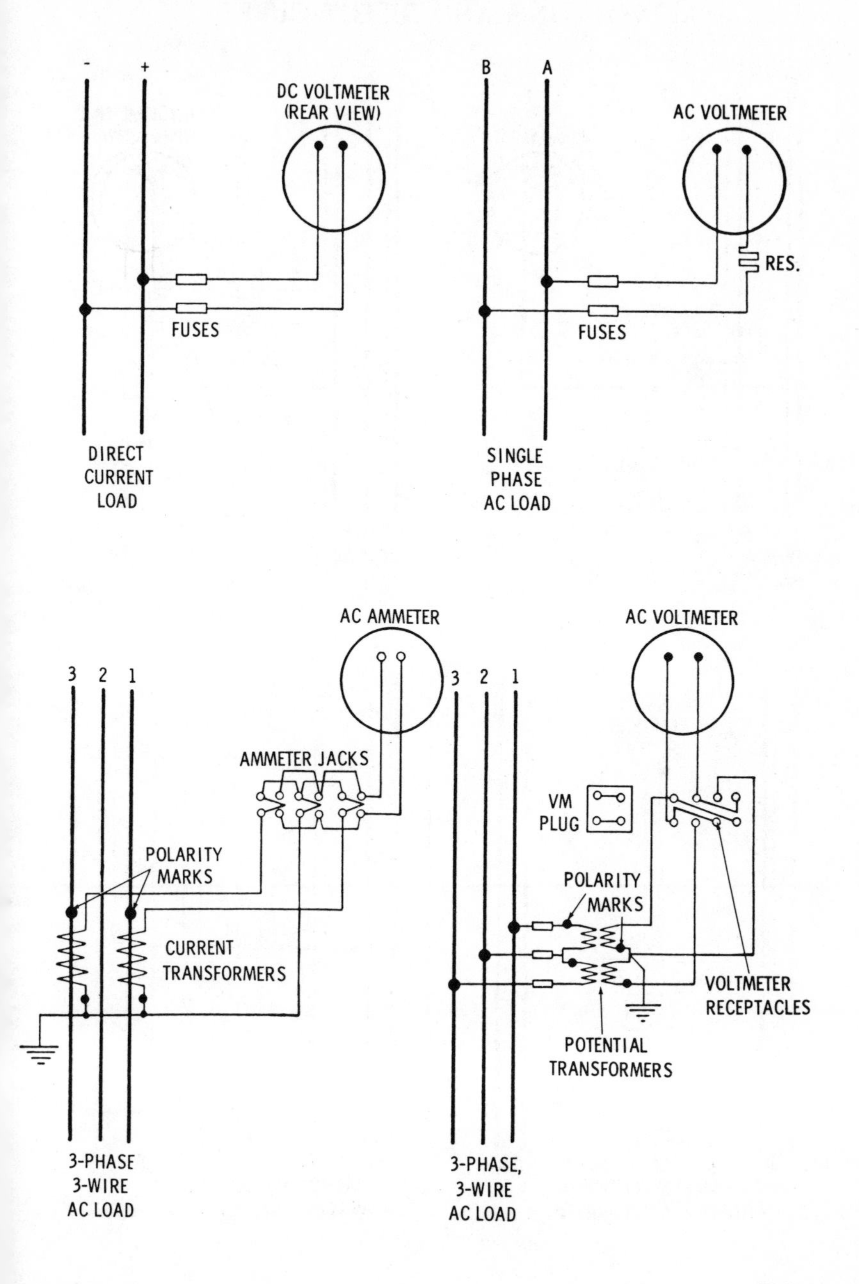

VOLTMETER CONNECTIONS FOR ALTERNATING AND DIRECT CURRENT

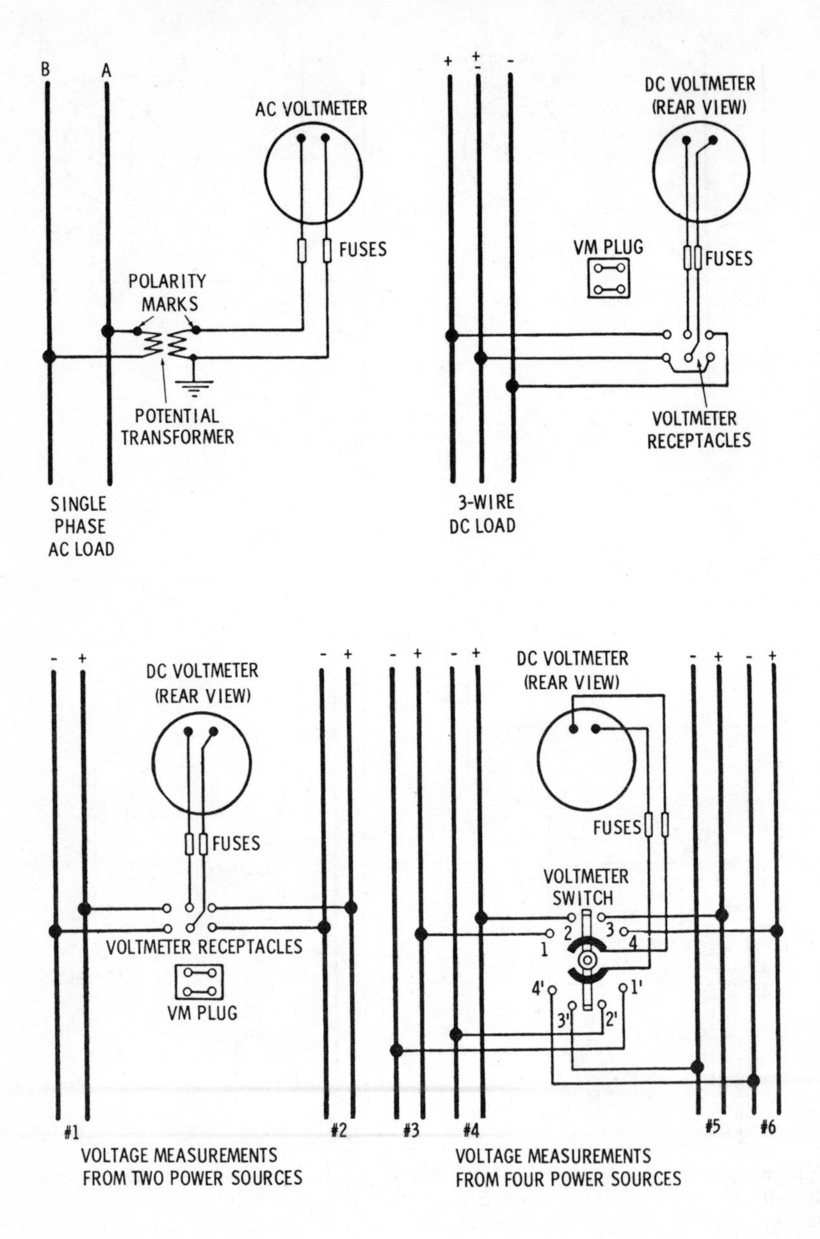

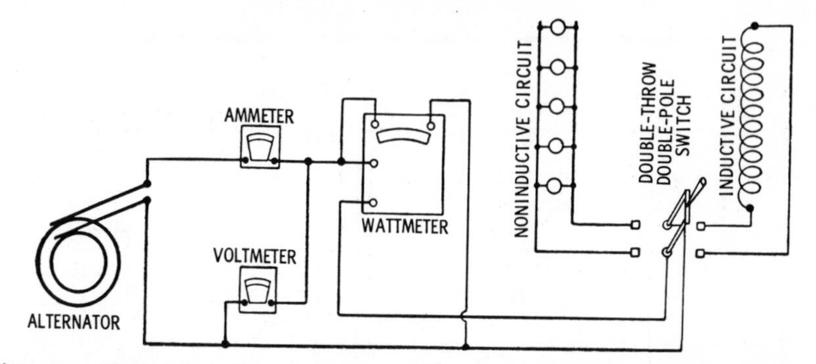

Diagram Illustrating Power-Factor Test on Noninductive and Inductive Circuits—The instruments are connected as shown and by means of the double-throw switch can be put on either the noninductive or inductive circuit. First turn switch to left so that current passes through the lamps; for illustration, the following readings are assumed: ammeter 10, voltmeter 110, and wattmeter 1,100. The power factor then is wattmeter reading ÷ volts × amperes = 1,100 actual watts ÷ 1,100 apparent watts = 1, that is, on noninductive circuits, the power factor is unity. Now the switch is thrown to the right connecting instruments with the inductive circuit, then for illustration, the following readings may be assumed: ammeter 8, voltmeter 110, and wattmeter 684. Now, as before, power factor = wattmeter reading ÷ volts × amperes = 684 ÷ (8 × 110) = 684 ÷ 880 = .78.

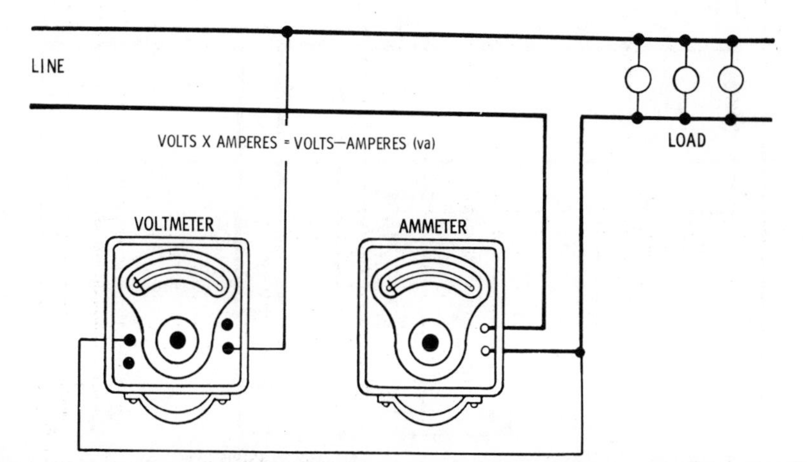

Typical Connection Diagram of Voltmeter and Ammeter in a Single-Phase Circuit—When the instruments are connected as shown, the voltmeter measures load voltage, not line voltage. The ammeter measures load current plus voltmeter current.

METER CONNECTIONS

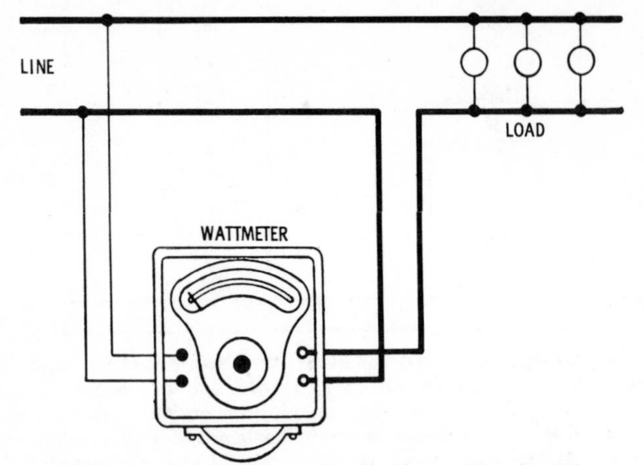

Typical Wattmeter Connection in a Single-Phase Circuit—When connected as shown, the instrument is measuring power load plus losses in its own current coil circuit. If the instrument reads backwards, reverse the current leads.

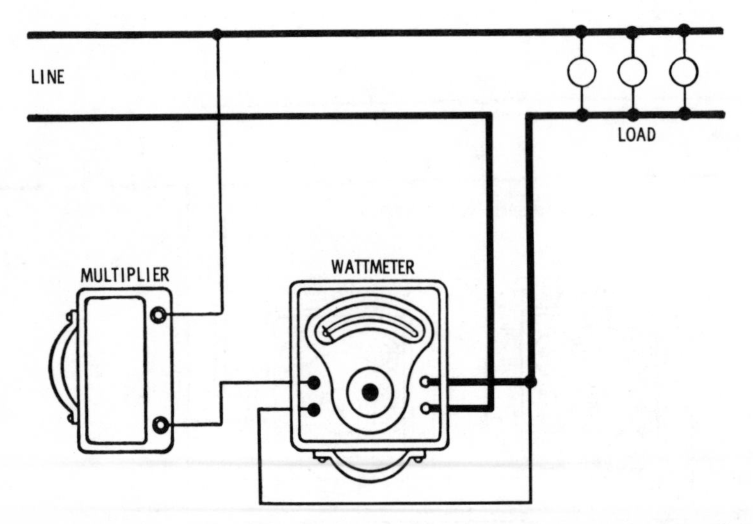

Typical Wattmeter Connection in a Single-Phase Circuit, When Used with Potential Multiplier—When connected as shown the instrument is measuring power load plus losses in its own potential and multiplier circuit.

METER CONNECTIONS

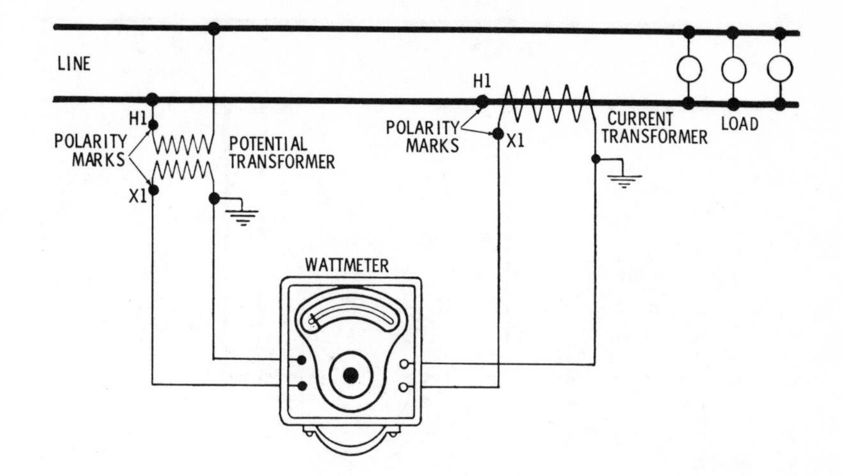

Typical Wattmeter Connections in a Single-Phase Circuit, When Used with Potential and Current Transformer.

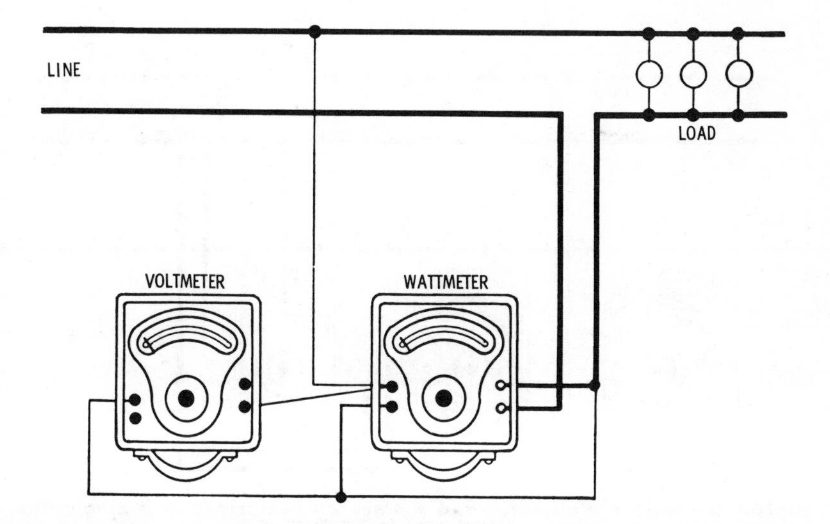

Typical Wattmeter and Voltmeter Connections in a Single-phase Circuit— When connected as shown the wattmeter measures power load plus losses in the voltmeter and wattmeter potential circuits.

METER CONNECTIONS

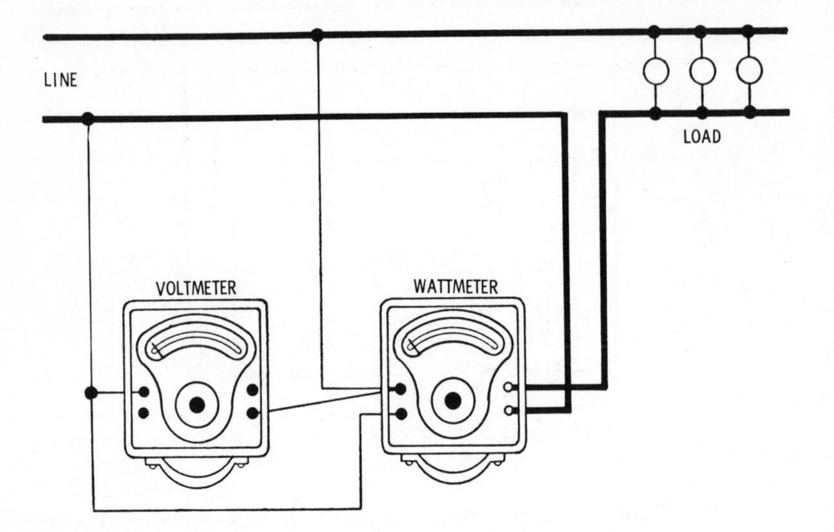

Typical Voltmeter and Wattmeter Connections in a Single-Phase Circuit—When connected as shown, the wattmeter measures power load plus losses in its own current-coil circuit.

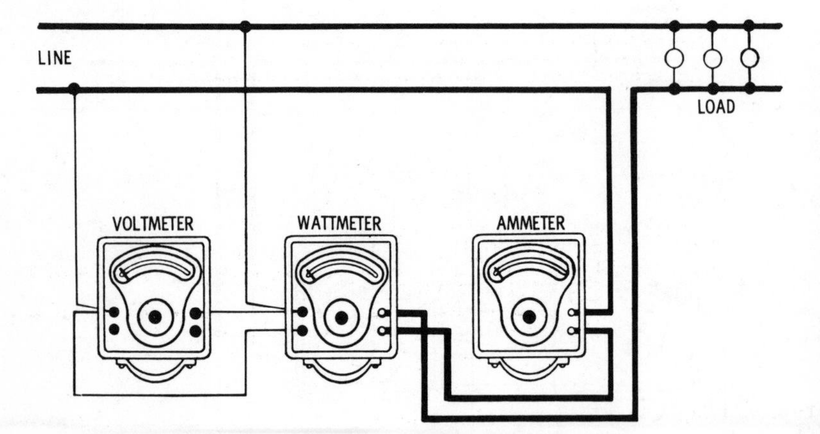

Typical Wattmeter, Voltmeter and Ammeter Connections in a Single-Phase Circuit—In a meter combination of this type the power factor of the circuit may easily be determined by dividing the wattmeter reading by the product of the voltmeter and ammeter readings. When connected as shown, the wattmeter measures power load plus losses in the ammeter and wattmeter current-coil circuit.

METER CONNECTIONS

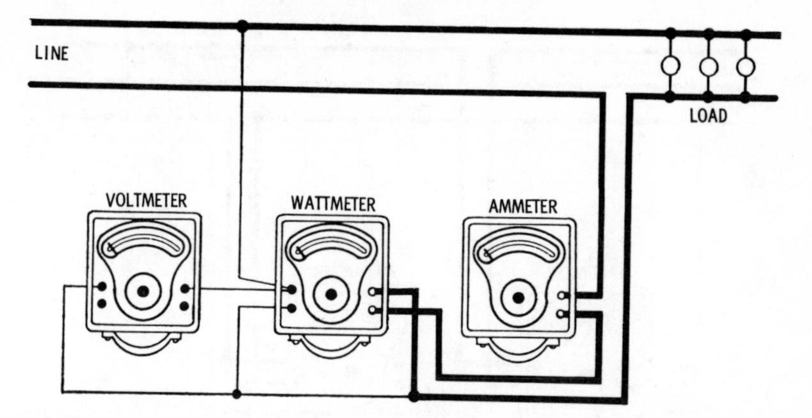

Typical Wattmeter, Voltmeter and Ammeter Connections in a Single-Phase Circuit—In a meter combination of this type, the power factor of the circuit may easily be determined by dividing the wattmeter reading by the product of the voltmeter and ammeter readings. When connected as shown the wattmeter measures the sum of the power losses of the load, the potential circuit of the wattmeter and the voltmeter. The power factor may be calculated. P.F. = Watts/VA

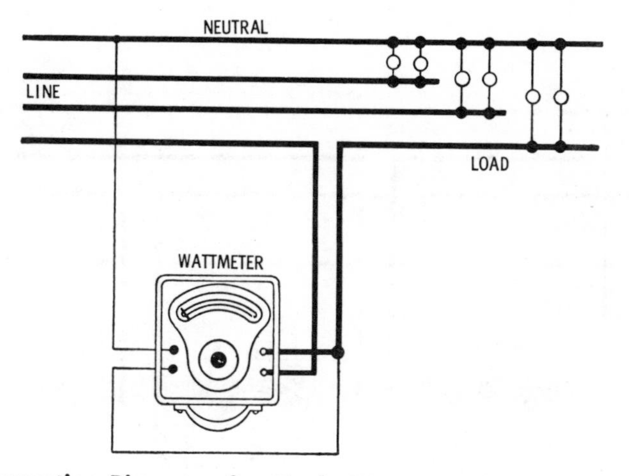

Typical Connection Diagram of a Single Wattmeter in a Balanced Three-Phase, Four-Wire Circuit—While only the wattmeter connections are shown, the voltmeter and ammeter may be connected as illustrated previously. When connected as shown, the power of the system is three times the indication of the single wattmeter. The wattmeter indicates its own potential losses plus the power in one phase of the load.

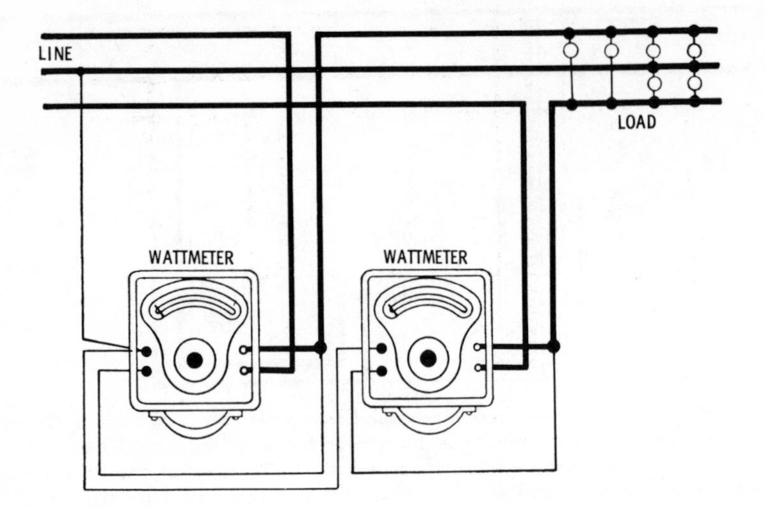

Typical Connection Diagram of Voltmeter and Ammeter in a Single-Phase Circuit—When the instruments are connected as shown, the voltmeter measures line voltage, not load voltage. The ammeter measures load current only.

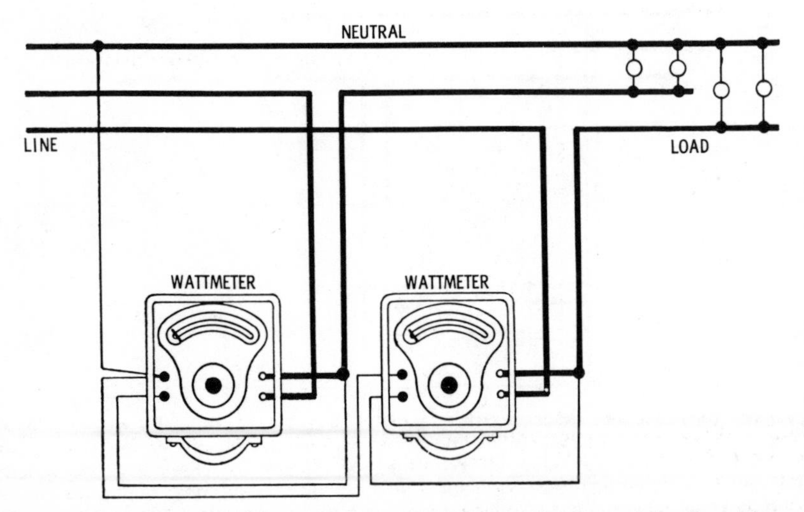

Typical Wiring Diagram Showing Two Wattmeters Connected for Two-Phase, Three-Wire Balanced or Unbalanced Load.

METER CONNECTIONS

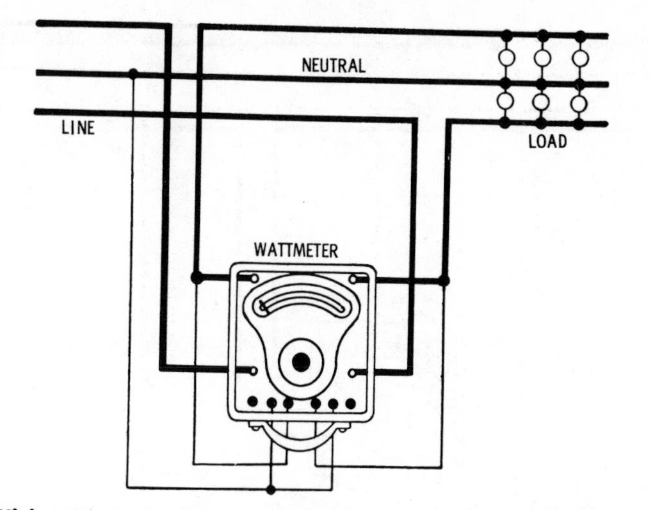

Typical Wiring Diagram Showing Connection of a Polyphase Wattmeter in a Two-Phase, Three-Wire Circuit, Balanced or Unbalanced Voltage or Load.

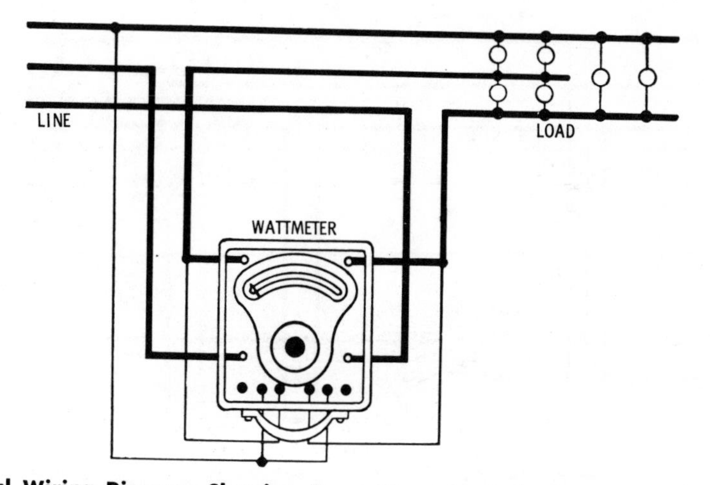

Typical Wiring Diagram Showing Connection of a Polyphase Wattmeter in a Three-Phase, Three-Wire Circuit—It should be observed that the accuracy of tests made with single-phase wattmeters will be somewhat higher than those made with polyphase wattmeters. Voltage ranges can be extended by the use of multipliers or transformers. To obtain high accuracy the instruments should be used at 40% of rated current or above.

METER CONNECTIONS

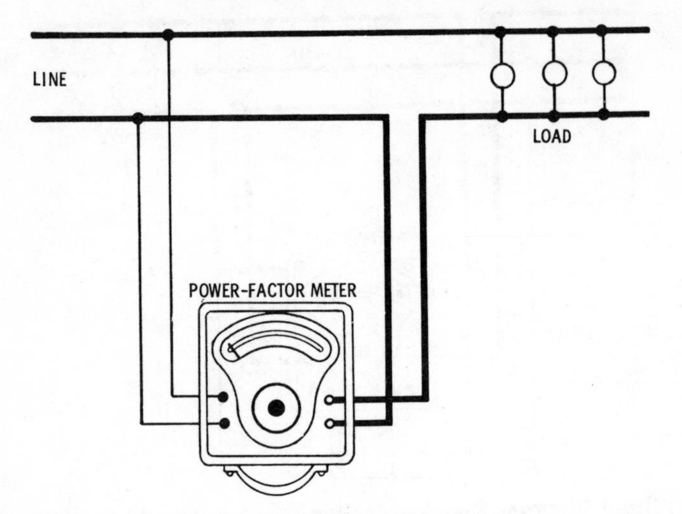

Typical Power-Factor-Meter Connection When Used on a Single-Phase Circuit—Single-phase, power-factor meters should be used only at the calibrated frequency.

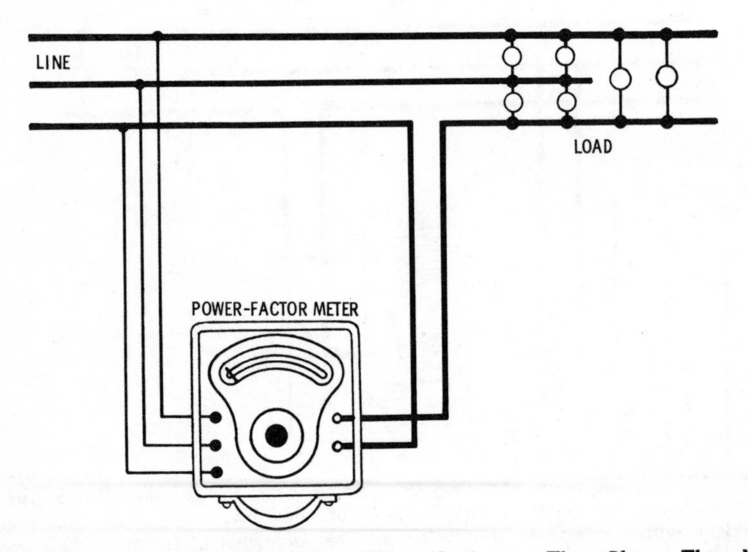

Typical Power-Factor-Meter Connection When Used on a Three-Phase, Three-Wire Circuit.—This will not give the power factor of all three-phases, as the wattmeter is only registering amperes in one-phase.

WATTMETER CONNECTIONS FOR SINGLE-PHASE AC CIRCUITS

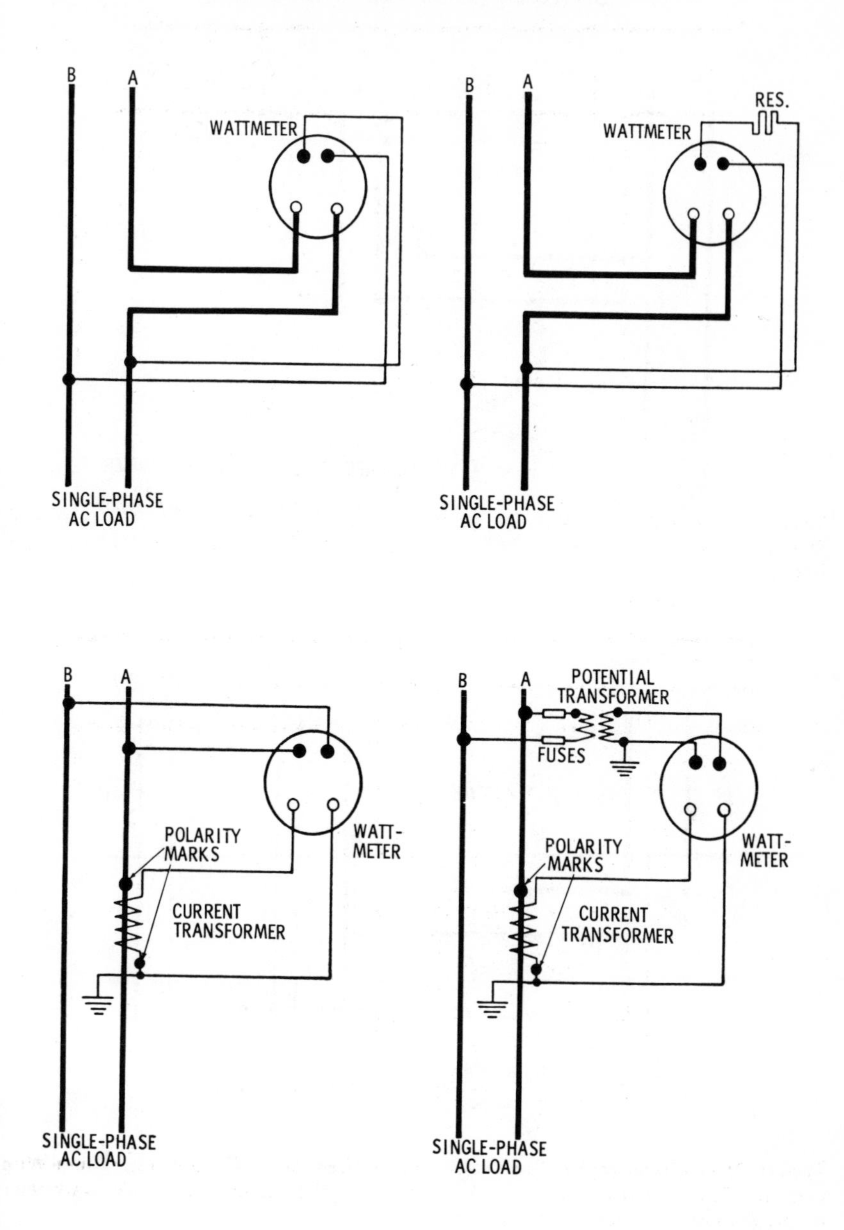

WATTMETER AND VOLTMETER CONNECTIONS IN SINGLE-PHASE AC CIRCUITS

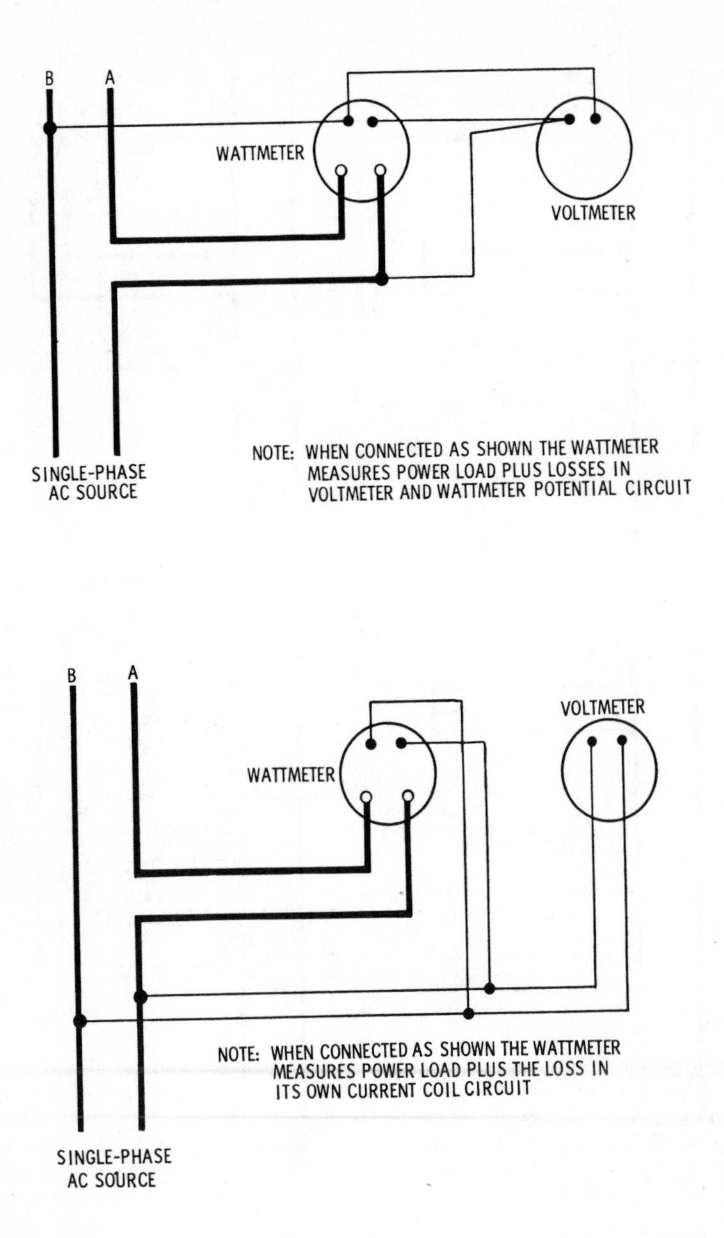

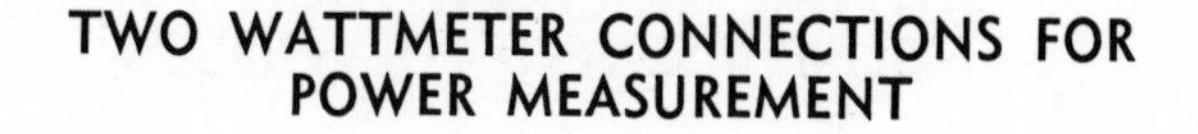

TWO WATTMETER CONNECTIONS FOR POWER MEASUREMENT

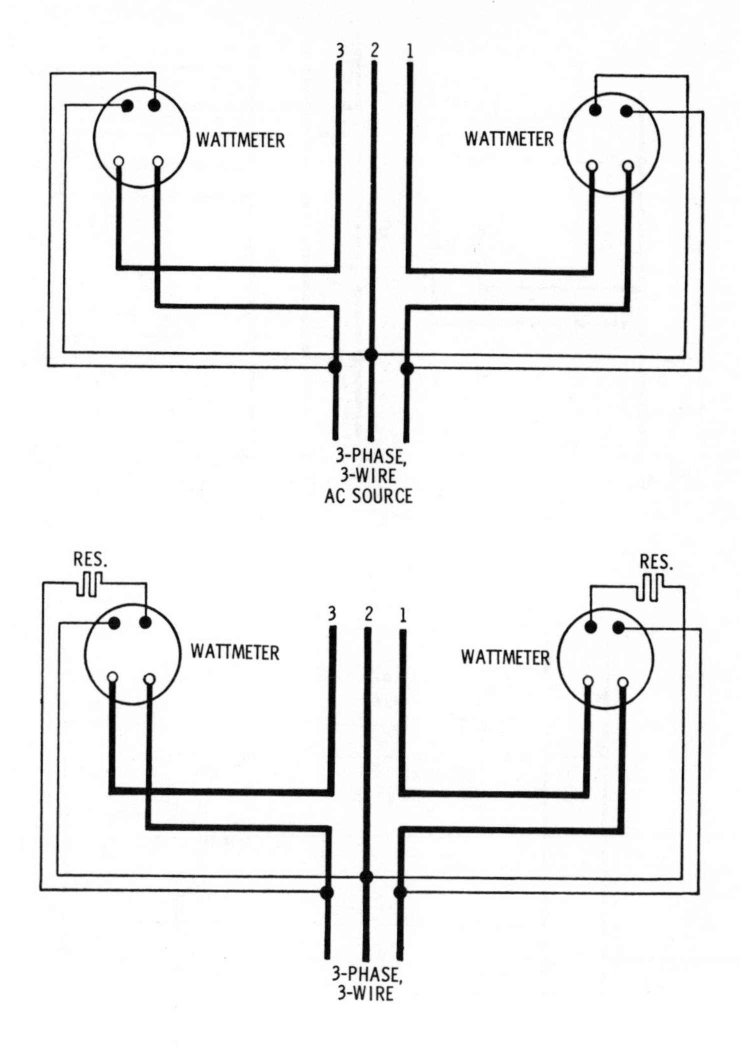

Connections of Two Wattmeters for Measurement of Power—When connected as illustrated, the two wattmeters will not indicate alike even if the load is balanced. Above 50% power factor, the three-phase power is the sum of the two readings. Below 50% power factor, it is necessary to reverse the reading of one wattmeter (by reversing its current leads) and then take the difference between the readings of the two meters.

WATTMETER CONNECTIONS, SINGLE-PHASE AND POLYPHASE

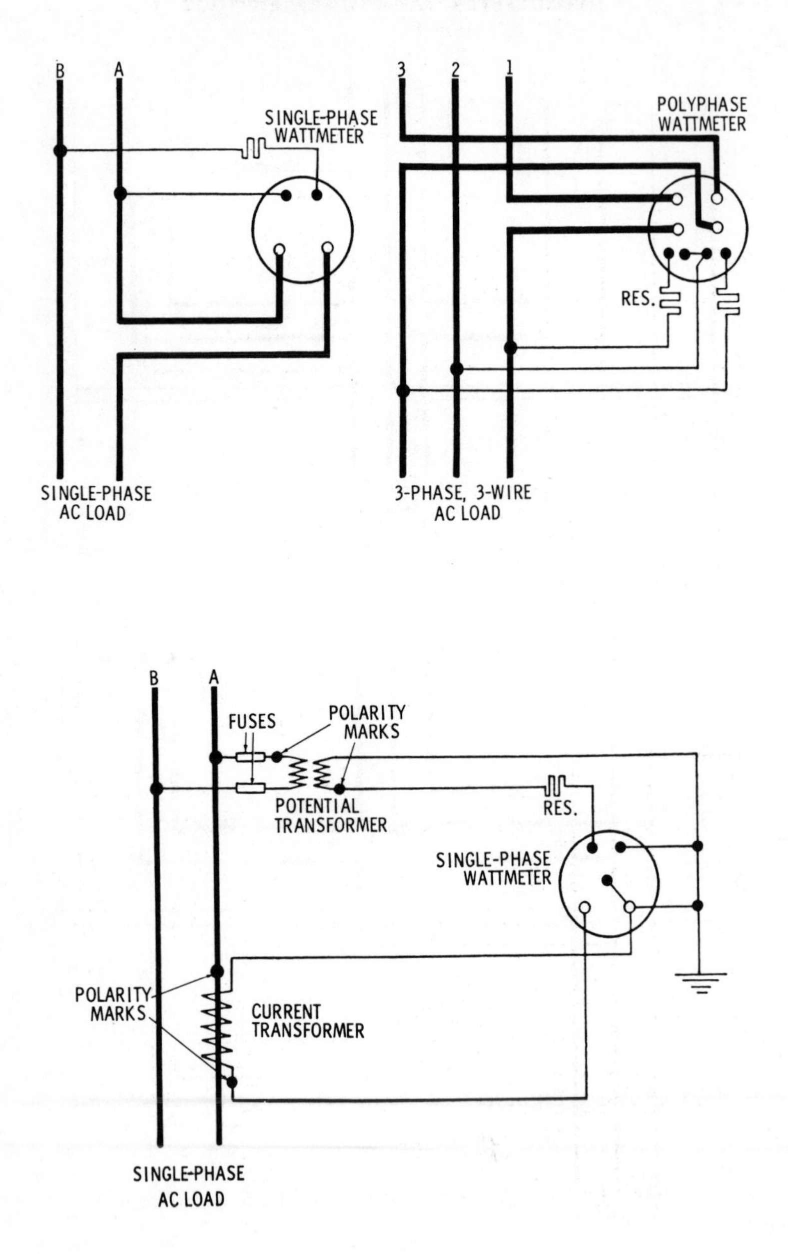

WATTMETER CONNECTIONS, SINGLE-PHASE AND POLYPHASE

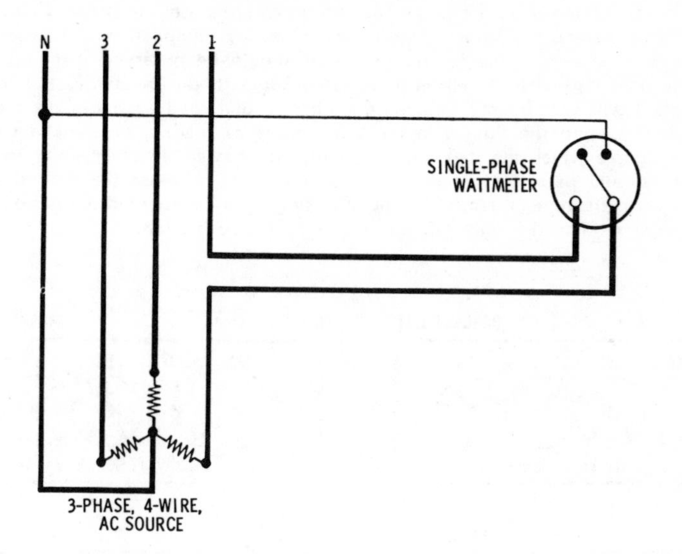

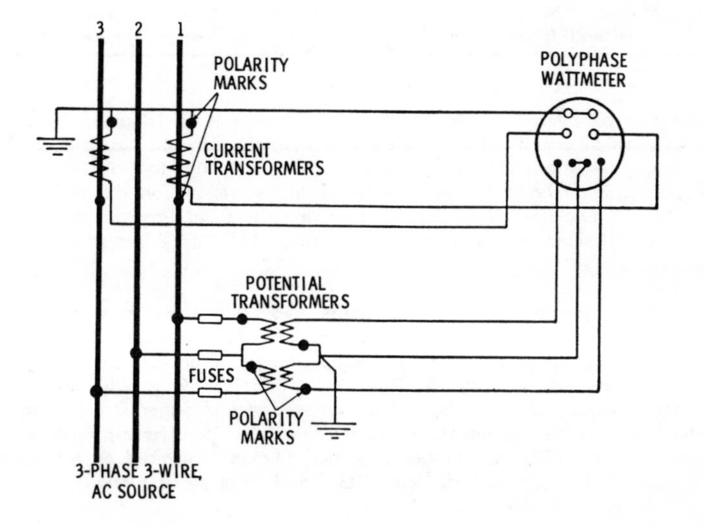

ELECTRIC-METER READING

How to Read an Electric Meter—At certain intervals of time—usually once a month—a consumer of electricity is billed for the amount of electrical energy in kilowatt-hours registered by the meter located on the premises. In order to facilitate the reading of meters, the front of the meter is usually equipped with four equally divided dials as shown in the figures below. It should be observed that each division on the **first right-hand** dial represents one kilowatt-hour or unit. (One kilowatt-hour equals 1,000 watt-hours.) Beginning with this dial read each dial to the left in succession, placing the figures in the same order as read; always make sure to take those figures which the dial finger actually has passed. If uncertain if the dial finger has actually passed a certain figure or not, note whether the next dial has passed its zero (0), remembering that no dial finger has completed a division until the dial finger next to the right has made a complete revolution.

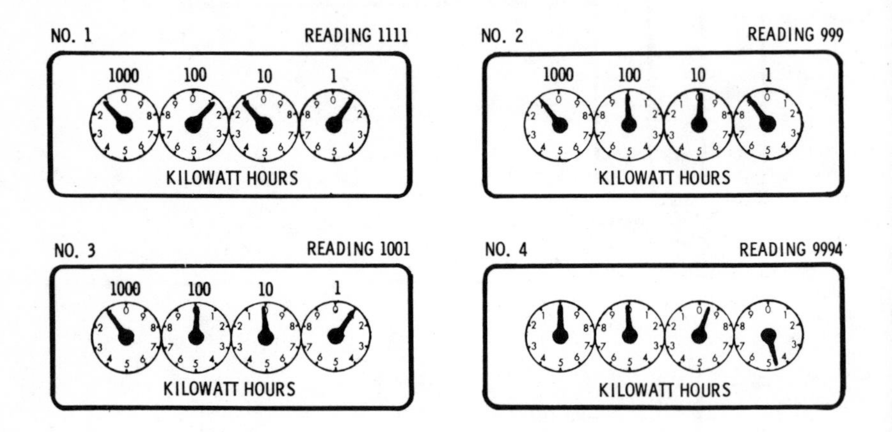

The relation between the speeds of all dial fingers is ten to one, i.e. one complete revolution of one dial hand indicates one division on the next dial to the left. If the above precautions are observed, it is a simple matter to read any meter. For example, the meter shown in example No. 1, indicates one on the dial at the extreme right, the two next following indicate one each and finally the last dial also indicates one, making the total register reading 1111 or a registration of 1111 kilowatt hours.

The reading example represented by meter No. 2 in a similar manner indicates 9 on the dial at the extreme right; the second dial finger rests on 0, but since the first rests only on 9 and has not as yet completed its revolution, it follows that the second dial finger also indicates 9. This 9 placed before 9 already obtained gives 99. The same is true about the third dial. The second dial finger at 9 has not as yet completed its revolution so the third has not completed its division; hence another 9 is obtained making 999. The number of kilowatt-hours registered on meters Nos. 3 and 4 will similarly be obtained, being 1001 and 9994 respectively.

Some registers of kilowatthour meters are, "The Cyclometer Type KWH Registers." This is read the same as the mileage indicator on your automobile. These are used extensively on Rural Electrification Association meters, where the customer reads the meter monthly and sends the reading in to the company. The company sends out a meter reader periodically to check the accuracy of the readings that are sent in. You read, from the right-hand side, units, tens, hundreds, thousands, and tens of thousands.

On industrial services, where high-voltage and high-amperages are involved, potential and current-transformers are used. For instance, assume a 13,800 volt service, with up to a 875 ampere supply. The meters are usually 115 volt meters, so 115/13,800 would give the potential transformers a 120 to 1 ratio. The current rating of the watthour meter is usually 5 amperes full load and the possible load is 875 amperes, so the current transformers would be 5/875 or a 1 to 175 ratio. The register reading has to be multiplied by the product of the two ratios, or 120 × 175 = 21,000. Thus, if the difference between this meter reading and the last meter reading was 135, this 135 would have to be multiplied by 21,000 (or K = 21,000) to obtain the kilowatthours used in that period. Thus 135 × 21,000 = 2,835,000 kilowatt-hours used.

WATTHOUR METER CONNECTIONS

Watthourmeters may be self contained, that is large enough to carry the full current and voltage of the service, or if the service capacity is too large, current transformers (CT's) and/or potential transformers (PT's) may be required.

The configurations for meters and meter housings are standardized—one manufacturer's meter will fit anothers meter housing, etc.

The following information and diagrams are taken from: "Instruction Manual-Alternating Current Watthour Meters" by **Sangamo Electric Co.** Permission to use this material was given by "Mr. H.M. Coulson, District Manager, **Sangamo Electric Co.,** Denver, Colo." This is not proprietory information, but is standard information. Where meter types appear, these happen to be the **Sangamo** meter types.

INSTRUMENT TRANSFORMERS

In these service connection diagrams, instrument transformers are denoted by the following symbols:

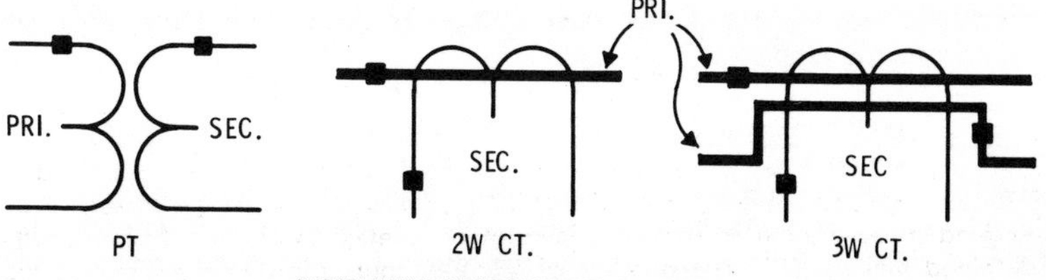

INSTRUMENT TRANSFORMER SYMBOLS

The polarity marks of a potential transformer indicate that when a voltage is applied from the polarity terminal to the non-polarity terminal of the primary, a voltage in phase with the primary voltage and proportional in magnitude to it will exist from the polarity to the non-polarity terminal of the secondary.

The polarity marks of a current transformer indicate that when a primary current enters at the polarity terminal of the primary, a current in phase with the primary current and proportional to it in magnitude will leave the polarity terminal of the secondary.

Polarity marks merely define a relation between primary and secondary windings and, in general, denote no particular location or relation to the circuit in which the transformer is used. Accordingly, it is usually possible to reverse the primary connections of a transformer from those shown in a diagram, without affecting the metering, **provided the secondary is similarly reversed.** Exceptions to this occur where transformers include tapped windings or where primary terminals are not equally insulated.

In each of these service diagrams all potential transformers are assumed to have the same ratio and all current transformers are assumed to have the same ratio unless otherwise noted. Where both 2-wire and 3-wire CT's are involved, 400:5 for a 2-wire, for example, is considered the same ratio as 400 and 400:5 for a 3-wire CT.

The above current transformer symbols are applicable to either window type or to bar or wound-type transformers. The manner in which it is used determines whether a window type transformer is considered 2-wire or 3-wire.

Window (or "through") type current transformers are rated on the basis of a single primary conductor through the window. Thus a transformer rated 400:5 amperes produces 5 amperes secondary current when a single conductor through the window carries 400 amperes (400 ampere-turns through the core).

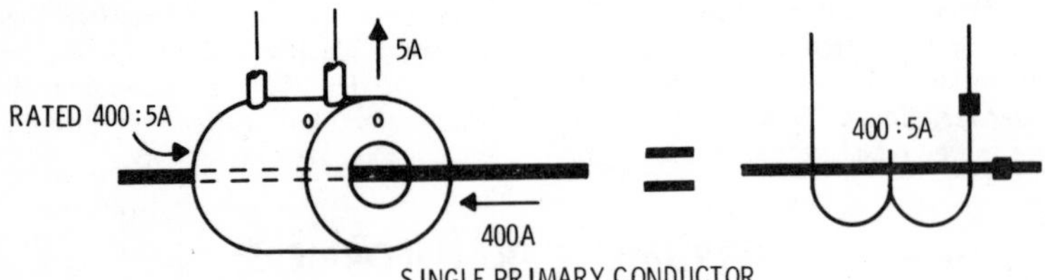

SINGLE PRIMARY CONDUCTOR

The actual ratio of a window type current transformer may be reduced by looping the primary conductor through the window more than once. Thus if the primary conductor passes through the window twice in the same direction it requires only 200 amperes (but still 400 ampere-turns through the core) to produce 5 amperes secondary current in a transformer rated 400:5 amperes; and the transformer will then have an effective ratio of 200:5 amperes.

If more than one primary conductor is passed through the window, the transformer will produce a secondary current proportional to the vector sum of the currents in the primary conductors. Thus if two lines of a single phase three-wire supply are passed through a window type transformer (in opposite directions, since the line currents are of opposite polarity) rated 400:5 amperes, the transformer becomes the equivalent of a three-wire transformer rated 200 and 200:5 amperes; and with 200 amperes in each primary conductor, the secondary current will be 5 amperes.

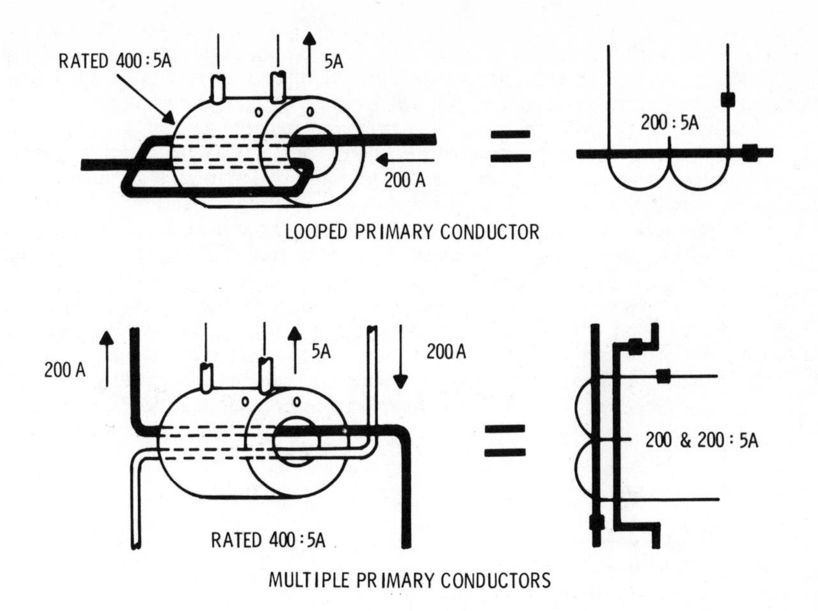

SYMBOLS

Symbols and notation used throughout these diagrams conform to EEI publication number 59-71, "Symbols for Metering Diagrams," and should require no explanation.

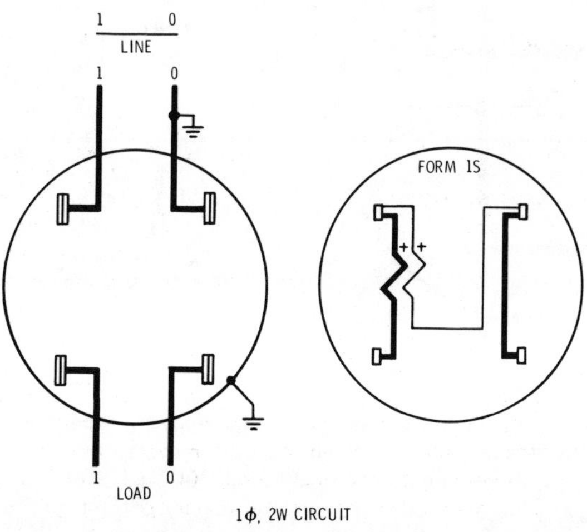

1φ, 2W CIRCUIT

1 STATOR, 1φ, 2 W TYPE JS, J2S OR J3S METER, SELF-CONTAINED

The internal wiring diagrams defining each form number conform to MSJ-10 but, in addition, carry "+" polarity marks indicating the relative direction of the windings. When the voltage from the polarity to the non-polarity end of the potential coil is substantially in phase with the current entering the polarity end of a current coil, the meter tends to run in the forward direction. Thus, for example, for a resistive line-to-line load, the current in line 1 is in the same direction as the voltage 1 - 2 across the potential coil and so produces forward rotation. The return current, in line 2, is from load to line but since it enters its current coil at a polarity mark, it also combines with the voltage to produce positive, or forward, torque on the meter, adding to that of the current in line 1.

Also, in conformance with MSJ-10 and to aid in identification, potential disconnect links are shown open although in actual service they must be closed. (An exception to this occurs in the case of 1 ϕ, 3 wire meters used to meter 2 wire service where, the link at the right hand side **must** be left open or, preferably, removed entirely).

Terminals marked K, Y, and Z are reserved for bringing out leads from pulse initiating devices.

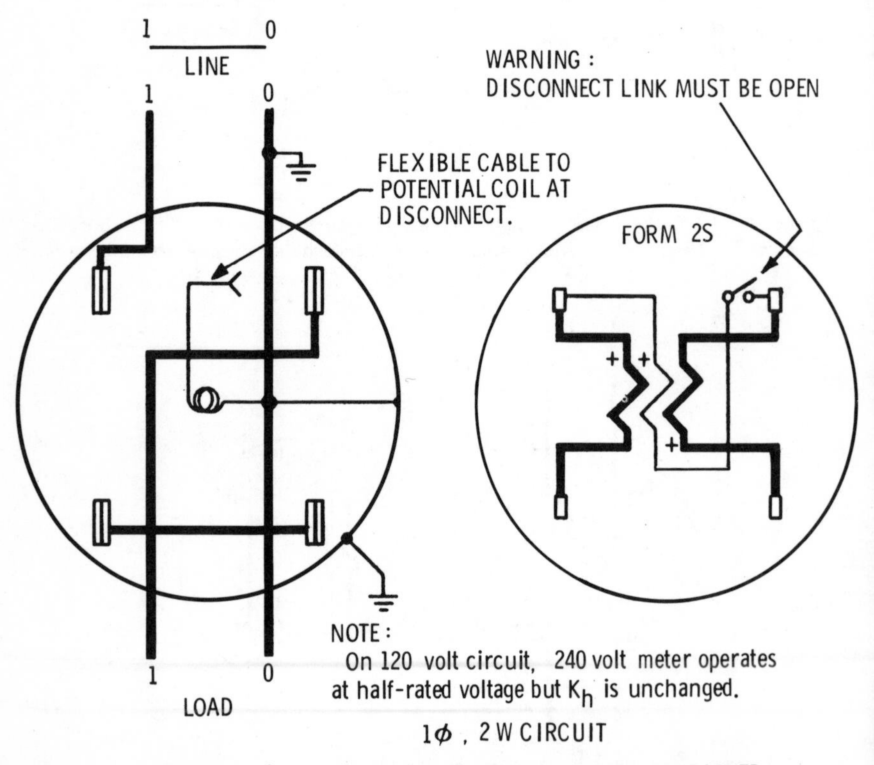

1 STATOR, 1ϕ, 3 W TYPE J2S OR J3S METER, SELF-CONTAINED

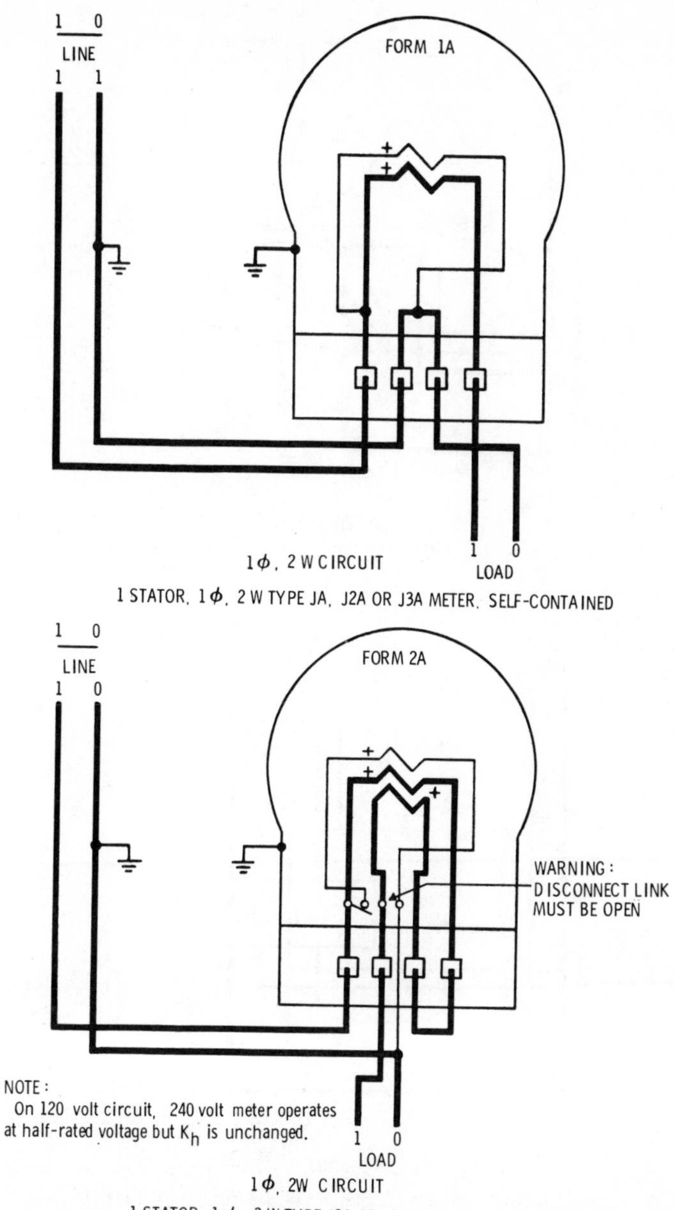

1 0
LINE
1 1
FORM 1A
1 0
LOAD
1φ, 2 W CIRCUIT
1 STATOR, 1φ, 2 W TYPE JA, J2A OR J3A METER, SELF-CONTAINED
1 0
LINE
1 0
FORM 2A
WARNING:
DISCONNECT LINK
MUST BE OPEN
NOTE:
On 120 volt circuit, 240 volt meter operates
at half-rated voltage but Kh is unchanged.
1 0
LOAD
1φ, 2W CIRCUIT
1 STATOR, 1φ, 3 W TYPE J2A OR J3A METER, SELF-CONTAINED

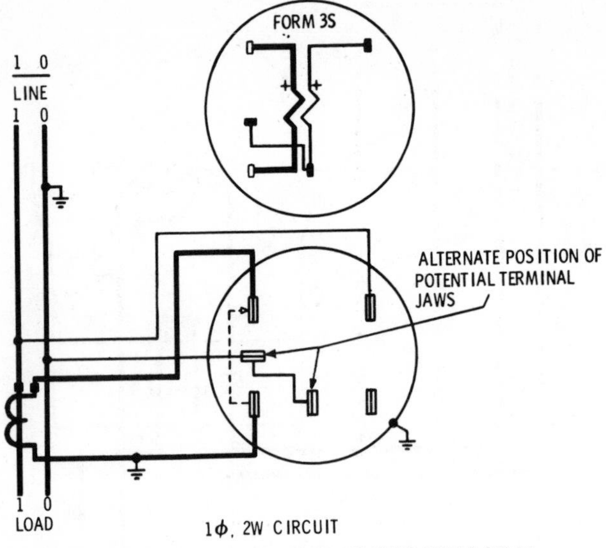

1ϕ, 2W CIRCUIT

1 STATOR, 1ϕ, 2 W TYPE JS, J2S, J3S OR J4S METER WITH 3-2 W CT

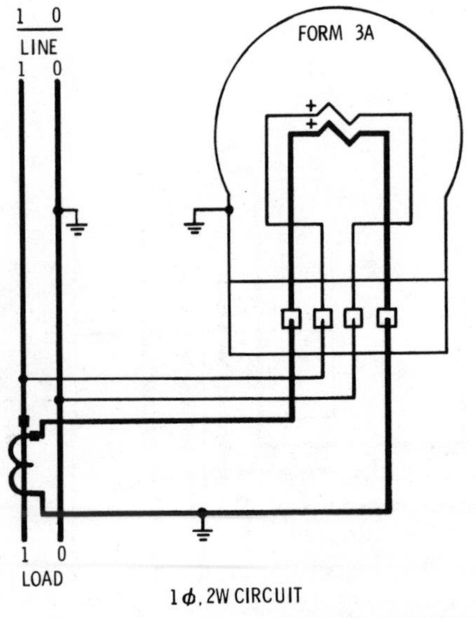

1ϕ, 2W CIRCUIT

1 STATOR, 1ϕ, 2 W TYPE JA, J2A OR J3A METER WITH 1-2 W CT

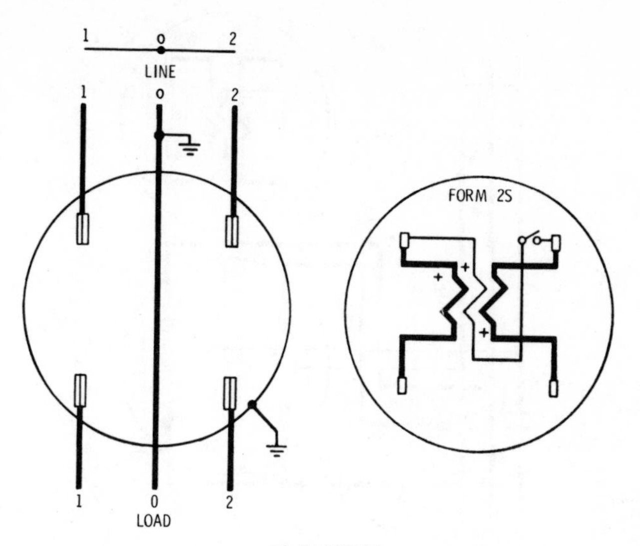

1ϕ, 3W CIRCUIT

1 STATOR, 1ϕ, 3 W TYPE JS, J2S, J3S OR J4S METER, SELF-CONTAINED

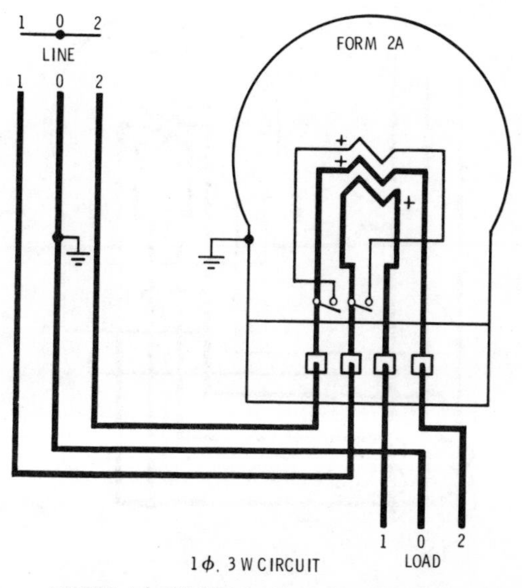

1ϕ, 3 W CIRCUIT

1 STATOR, 1ϕ, 3 W TYPE JA, J2A OR J3A METER, SELF-CONTAINED

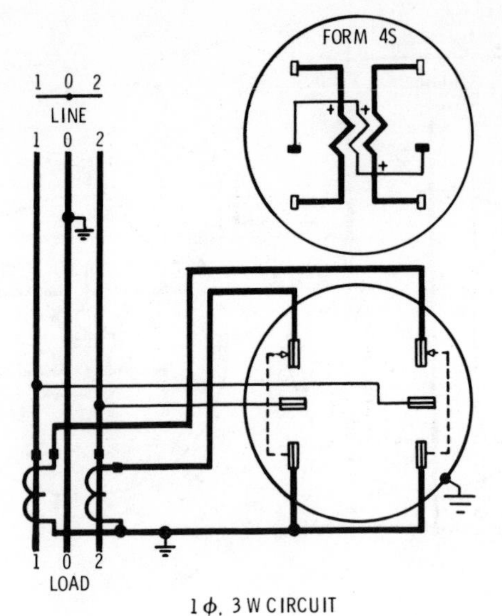

1 ϕ, 3 W CIRCUIT

1 STATOR, 1 ϕ, 3 W TYPE JS, J2S, J3S OR J4S METER WITH 2-2 W CT'S

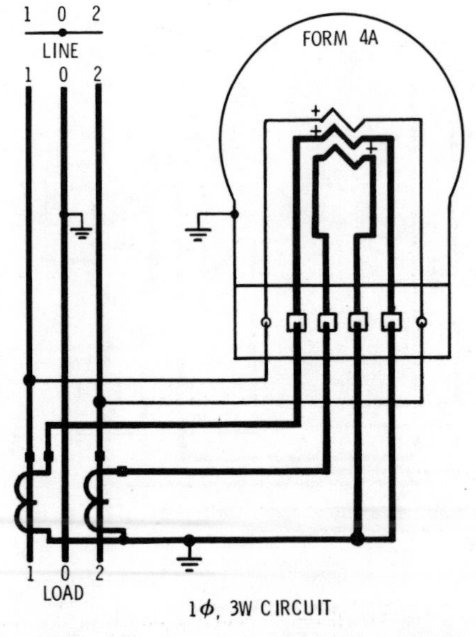

1 ϕ, 3W CIRCUIT

1 STATOR, 1 ϕ, 3 W TYPE JA, J2A OR J3A METER WITH 2-2 W CT'S

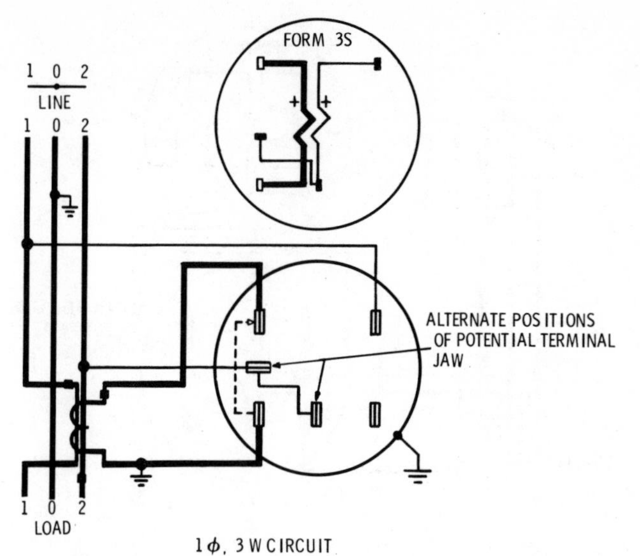

1ϕ, 3 W CIRCUIT

1 STATOR, 1 , 2 W TYPE JS, J2S, J3S OR J4S METER WITH 1-3 W CT

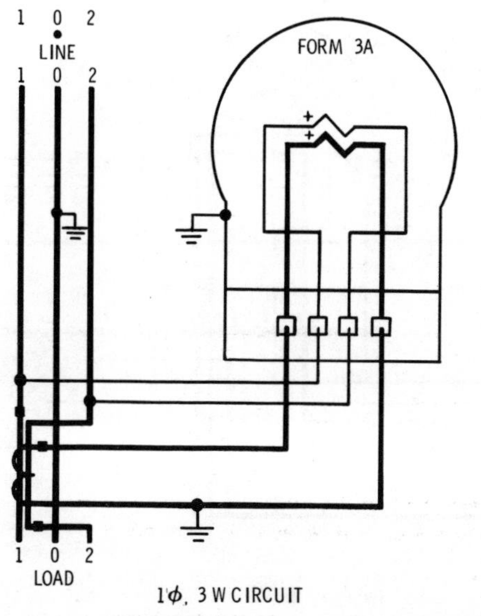

1ϕ, 3 W CIRCUIT

1 STATOR, 1ϕ, 2 W TYPE JA, J2A OR J3A METER WITH 1-3 W CT

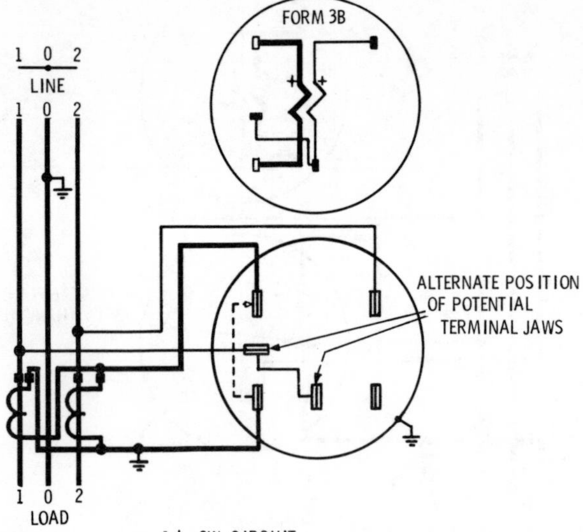

1ϕ, 3W CIRCUIT

1 STATOR, 1 ϕ, 2 W TYPE JS, J2S, J3S OR J4S METER WITH 2-2 W CT'S IN PARALLEL

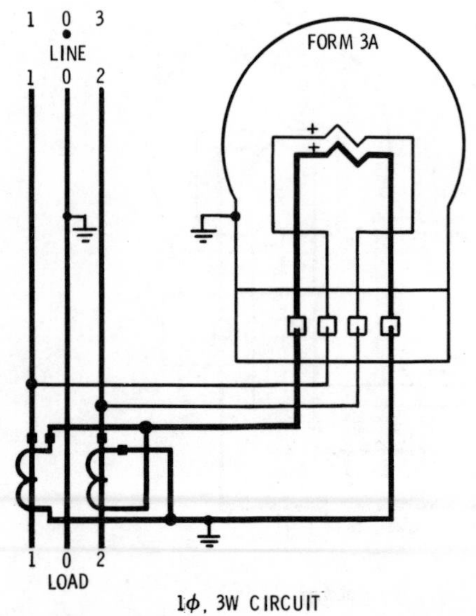

1ϕ, 3W CIRCUIT

1 STATOR, 1 ϕ, 2 W TYPE JA, J2A OR J3A METER WITH 2-2

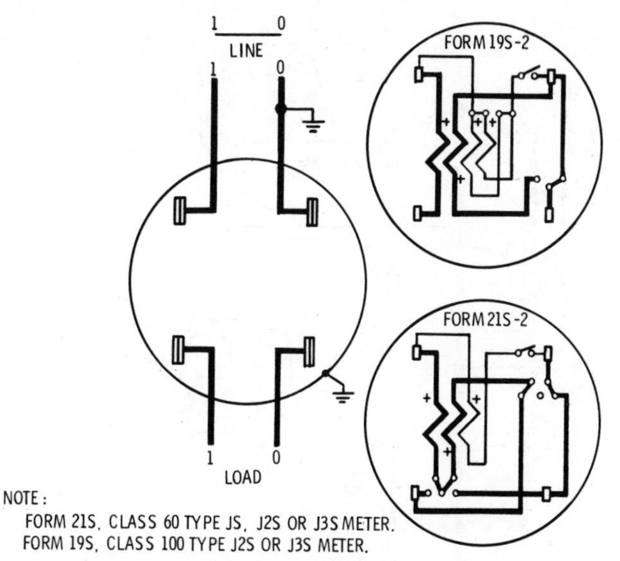

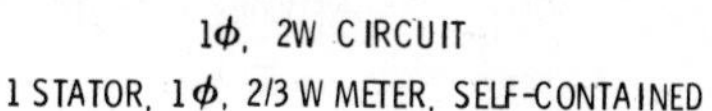
1φ, 2W CIRCUIT

1 STATOR, 1φ, 2/3 W METER, SELF-CONTAINED

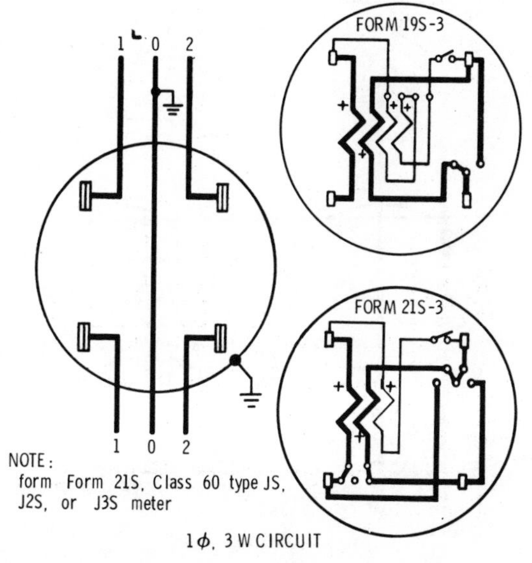

1φ, 3 W CIRCUIT

1 STATOR, 1φ, 2/3 W METER, SELF-CONTAINED

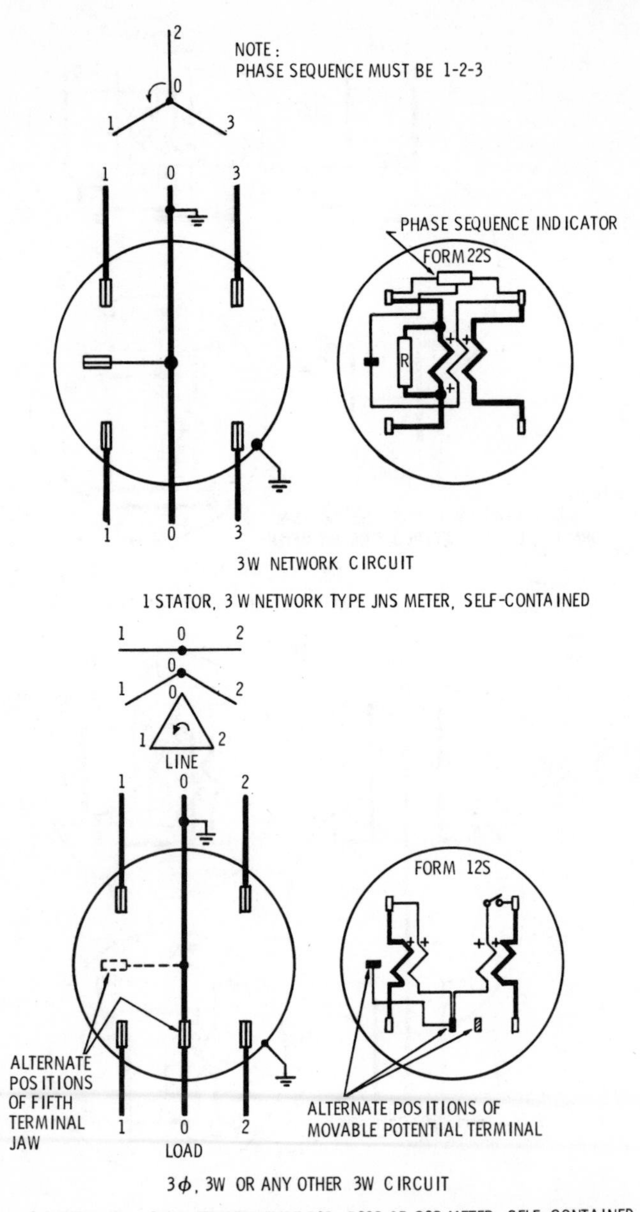

3W NETWORK CIRCUIT

1 STATOR, 3 W NETWORK TYPE JNS METER, SELF-CONTAINED

3ϕ, 3W OR ANY OTHER 3W CIRCUIT

2 STATOR, 3ϕ, 3 W (NETWORK) TYPE P2S, P20S OR S2S METER, SELF-CONTAINED

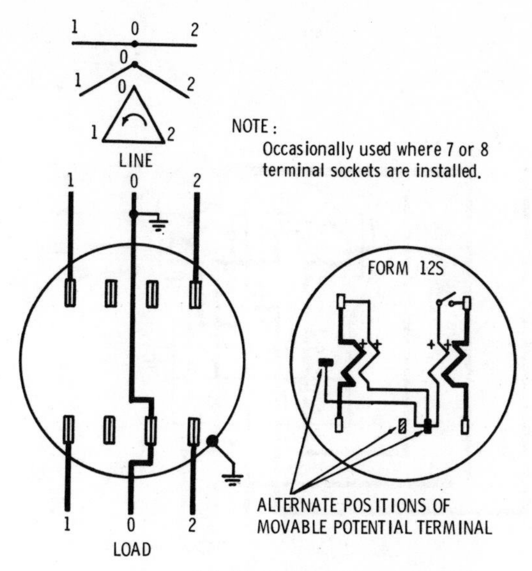

3ϕ, 3W OR ANY OTHER 3W CIRCUIT

2 STATOR, 3ϕ, 3 W (NETWORK) TYPE P2S, P20S OR S2S METER, SELF-CONTAINED

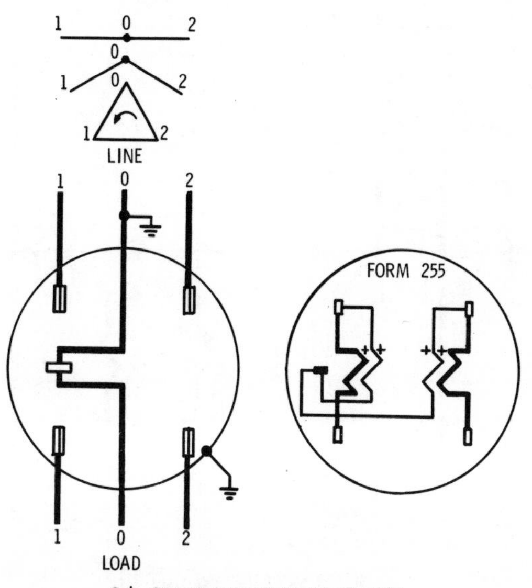

3ϕ, 3 W OR ANY OTHER 3 W CIRCUIT

2 STATOR, 3 W URBAN NETWORK TYPE P20SU OR S12S METER, SELF-CONTAINED

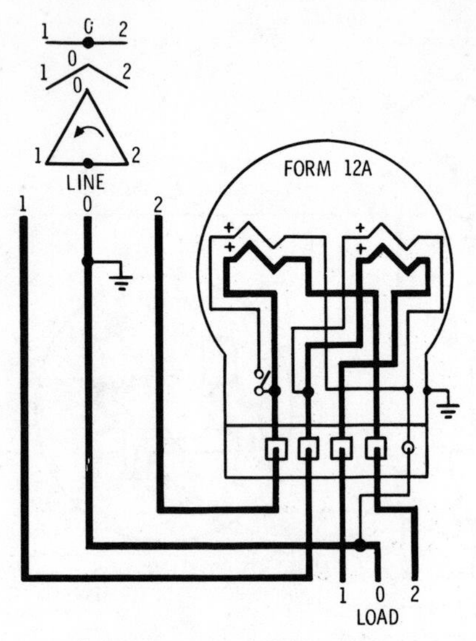

3ϕ, 3 W OR ANY OTHER 3 W CIRCUIT

2 STATOR, 3ϕ, 3 W (NETWORK) TYPE P2A, P20A OR S2A METER, SELF-CONTAINED

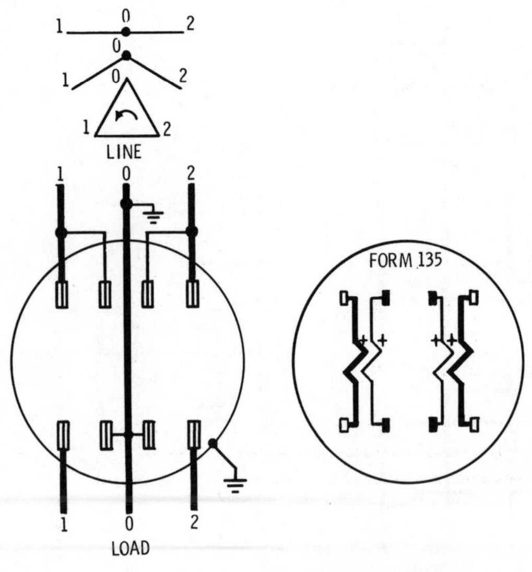

3ϕ, 3W OR ANY OTHER 3W CIRCUIT

2 STATOR, 3ϕ, 3 W TYPE P2SP, P20SP OR S3S METER, SELF-CONTAINED

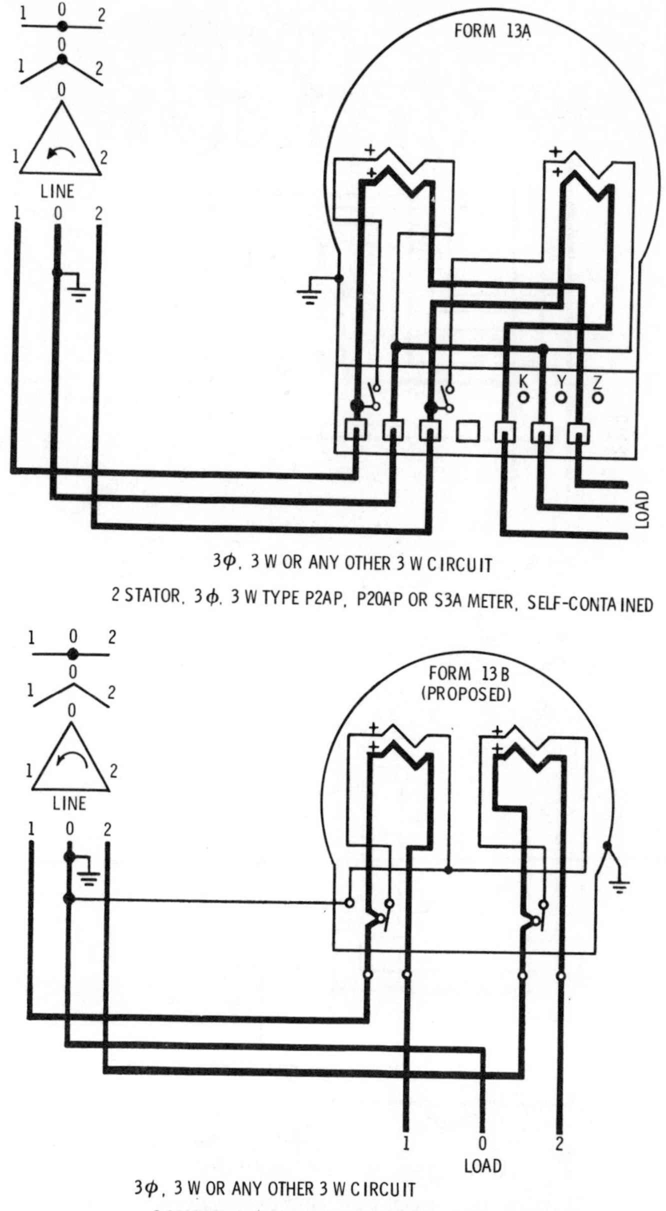

3ϕ, 3 W OR ANY OTHER 3 W CIRCUIT

2 STATOR, 3ϕ, 3 W TYPE P2AP, P20AP OR S3A METER, SELF-CONTAINED

3ϕ, 3 W OR ANY OTHER 3 W CIRCUIT

2 STATOR, 3 ϕ, 3 W TYPE S3B METER, SELF-CONTAINED

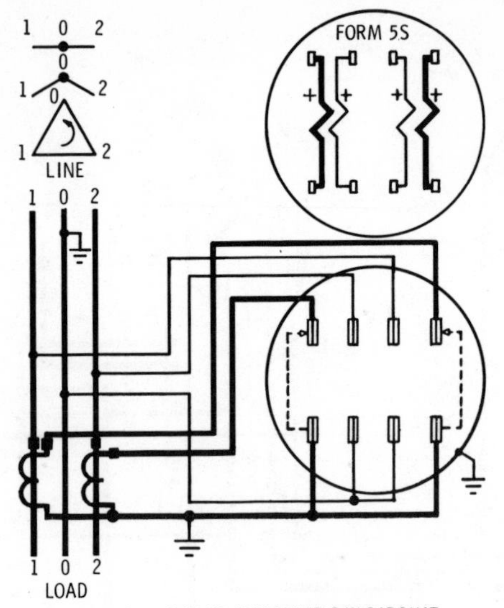

3 ϕ, 3 W OR ANY OTHER 3 W CIRCUIT

2 STATOR, 3 ϕ, 3 W TYPE P2SP, P20SP OR S3S METER WITH 2-2 W CT'S

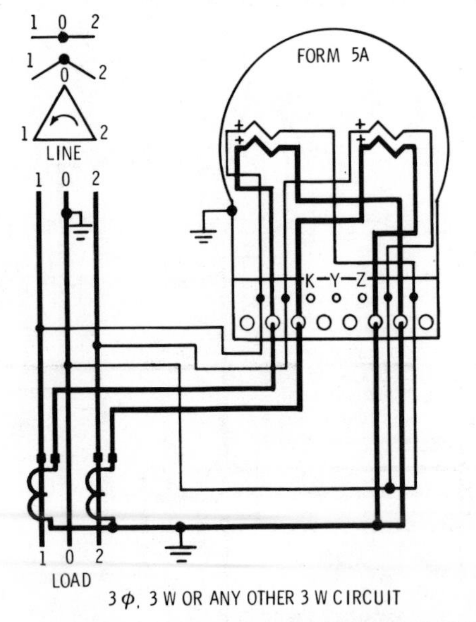

3 ϕ, 3 W OR ANY OTHER 3 W CIRCUIT

2 STATOR, 3 ϕ, 3 W TYPE P2AP, P20AP OR S3A METER WITH 2-2 W CT'S

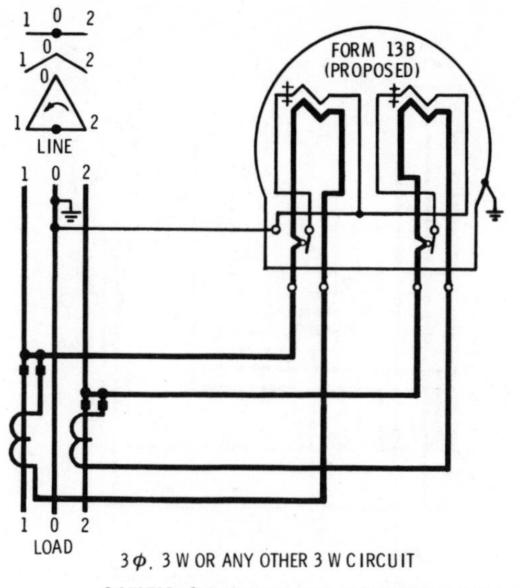

3ϕ, 3 W OR ANY OTHER 3 W CIRCUIT

2 STATOR, 3ϕ, 3 W TYPE S3B METER WITH 2-2 W CT'S

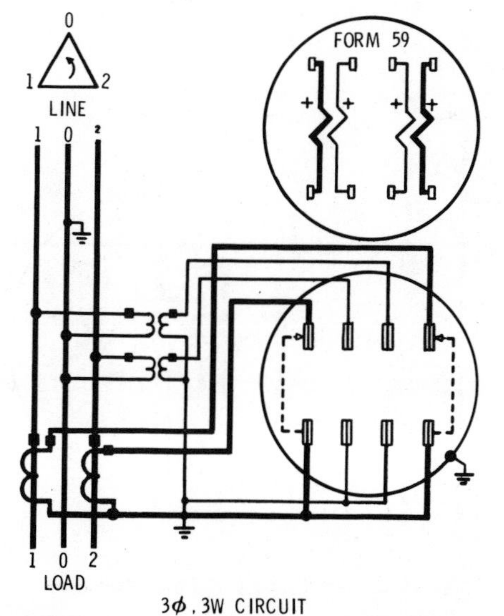

3ϕ, 3W CIRCUIT

2 STATOR, 3ϕ, 3 W TYPE P2SP, P20SP OR S3S METER WITH 2-2 W CT'S and 2 PT'S

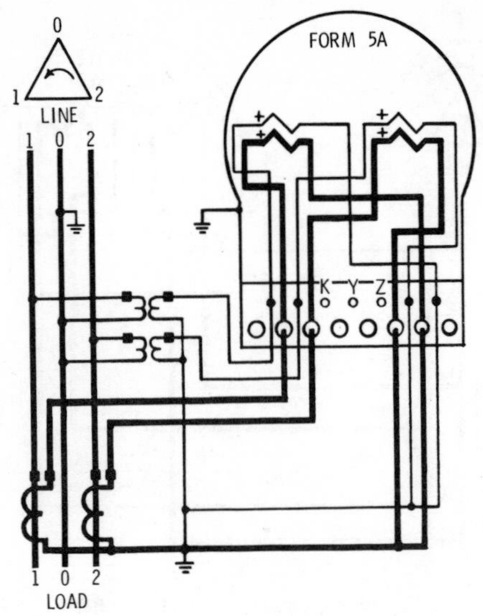

3ϕ, 3W CIRCUIT

2 STATOR, 3ϕ, 3 W TYPE P2AP, P20AP OR S3A METER WITH 2-2 W CT'S AND 2 PT'S

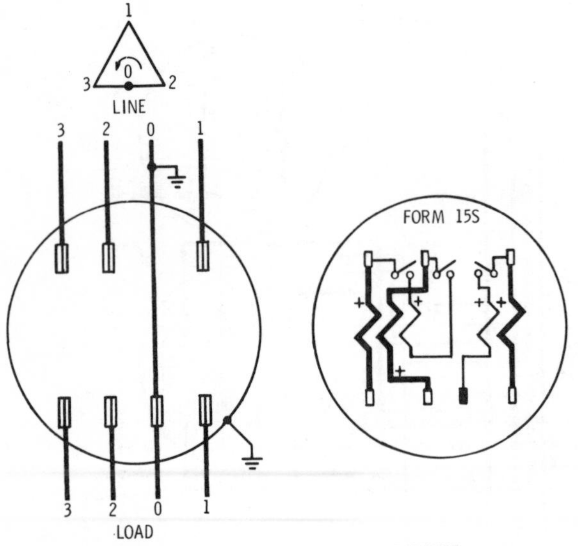

3ϕ, 4 W, Δ CIRCUIT

2 STATOR, 3ϕ, 4 W, Δ, TYPE P2SΔ, P20SΔ, OR S6S METER, SELF-CONTAINED

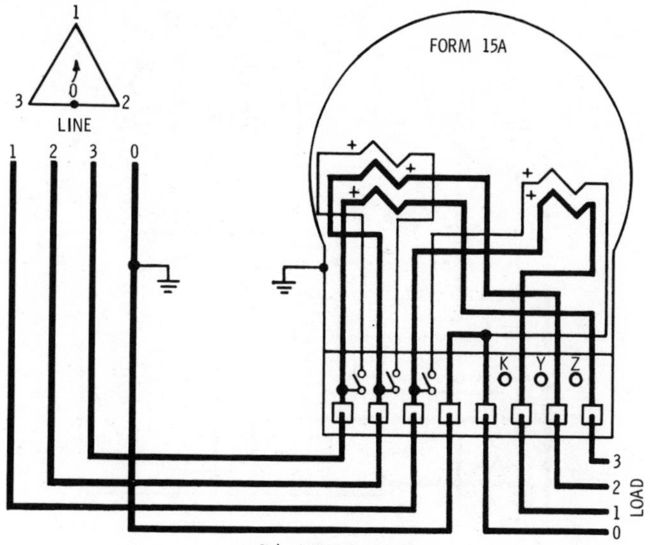

3 φ, 4 W, Δ CIRCUIT

2 STATOR, 3 φ, 4 W, Δ, TYPE P2A Δ, P20A Δ, OR S6A METER, SELF-CONTAINED

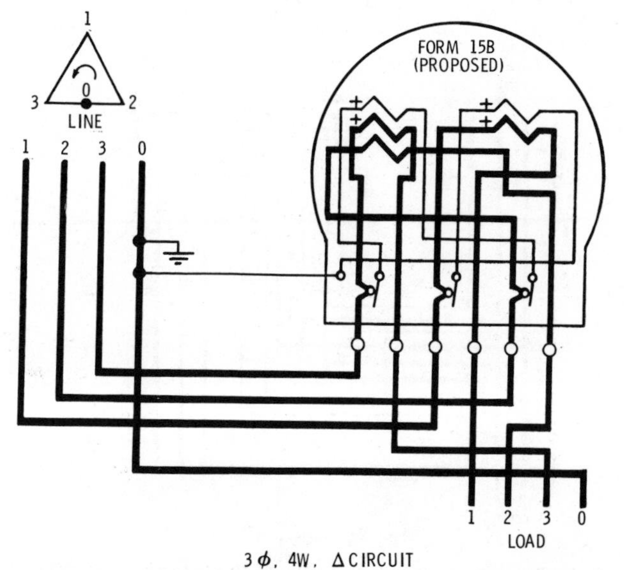

3 φ, 4W, Δ CIRCUIT

2 STATOR, 3 φ, 4 W, Δ, TYPE S6B METER, SELF-CONTAINED

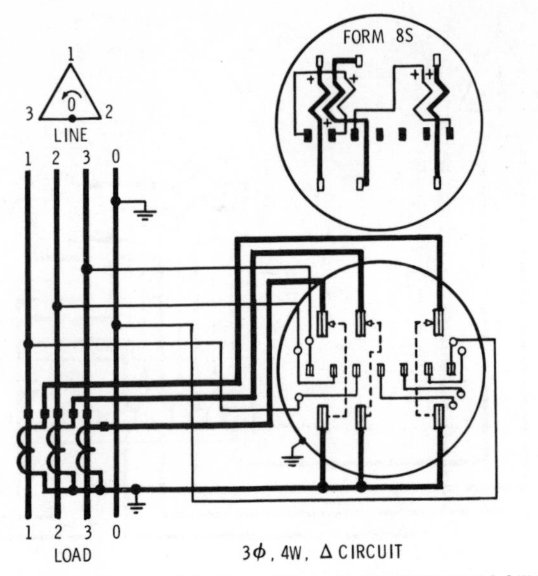

2 STATOR, 3ϕ, 4 W, Δ, TYPE P2S Δ, P20S Δ, OR S6S METER WITH 3-2 W CT'S

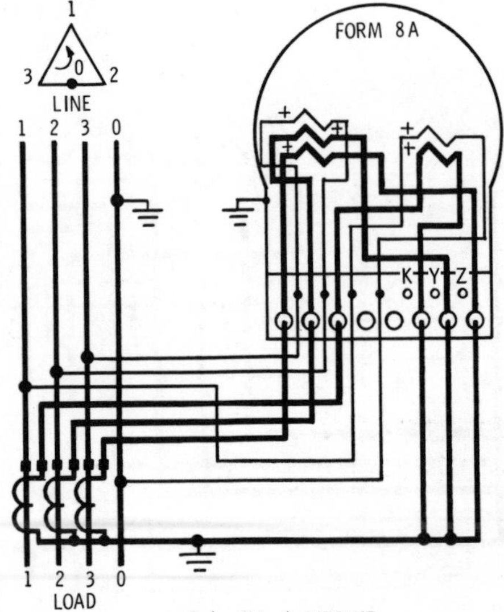

2 STATOR, 3ϕ, 4 W, Δ, TYPE P2A Δ, P20A Δ, OR S6A METER WITH 3-2 W CT'S

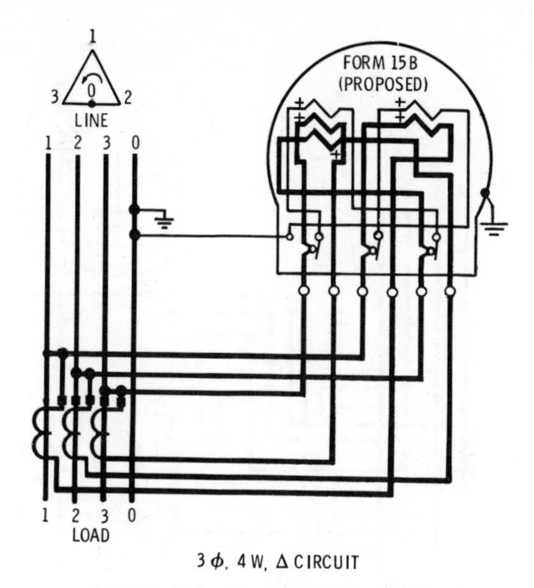

3 ϕ, 4 W, Δ CIRCUIT

2 STATOR, 3 ϕ, 4 W, Δ, TYPE S6B METER WITH 3-2 W CT'S

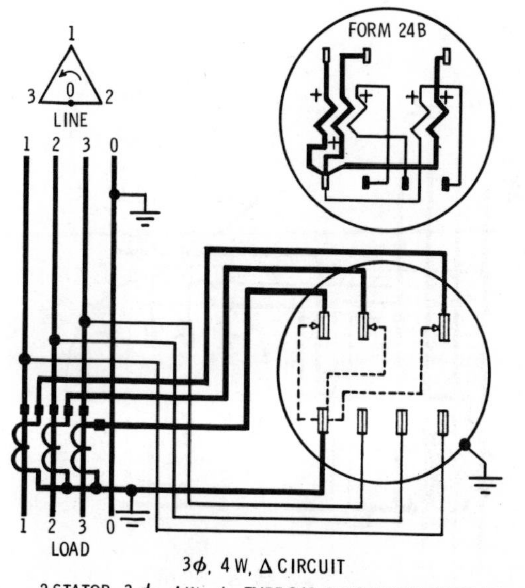

3ϕ, 4 W, Δ CIRCUIT

2 STATOR, 3 ϕ, 4 W, Δ, TYPE S6S (7 TERMINAL) METER WITH 3-2 W CT'S

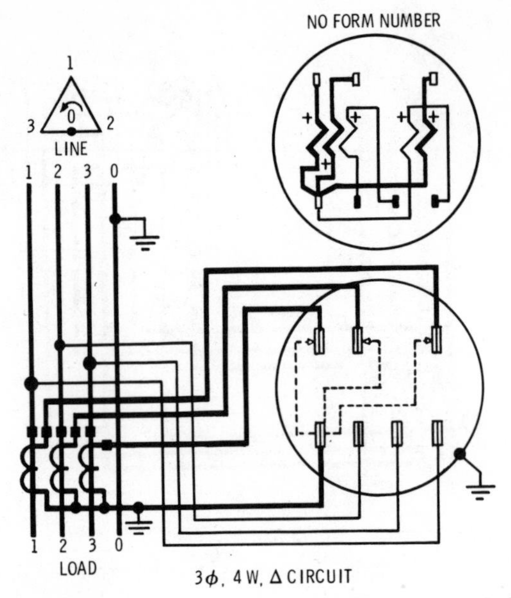

2 STATOR, 3 φ, 4 W, Δ, TYPE P2SΔ OR P20SΔ (7 TERMINAL) METER WITH 3-2 W CT'S

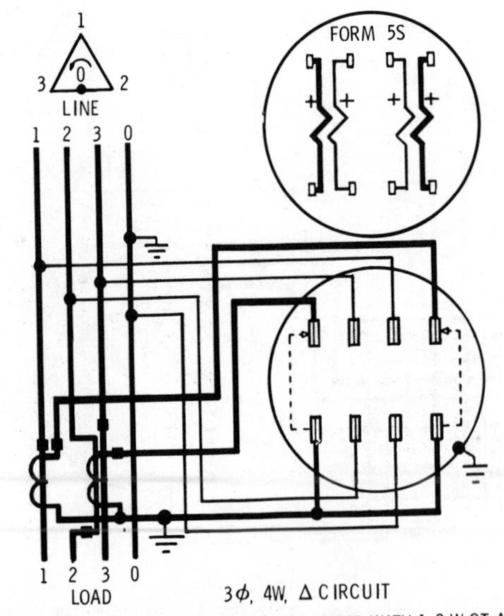

2 STATOR, 3 φ, 3 W, TYPE P2SP, P20SP OR S3S METER WITH 1-2 W CT AND 1-3 W CT

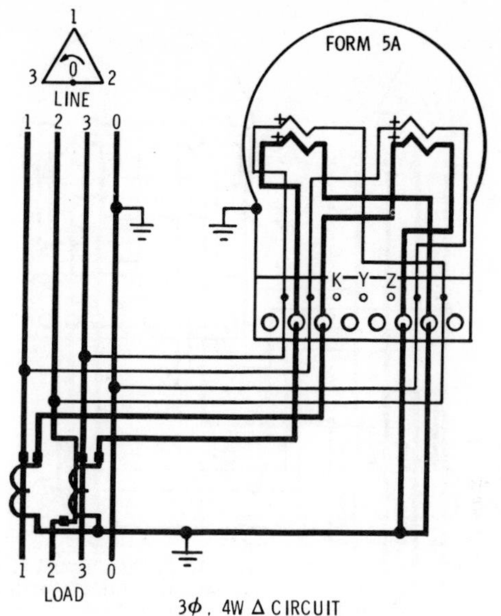

3ϕ, 4W Δ CIRCUIT

2 STATOR, 3ϕ, 3 W, TYPE P2AP, P20AP OR S3A METER WITH 1-3 W CT AND 1-2 W CT

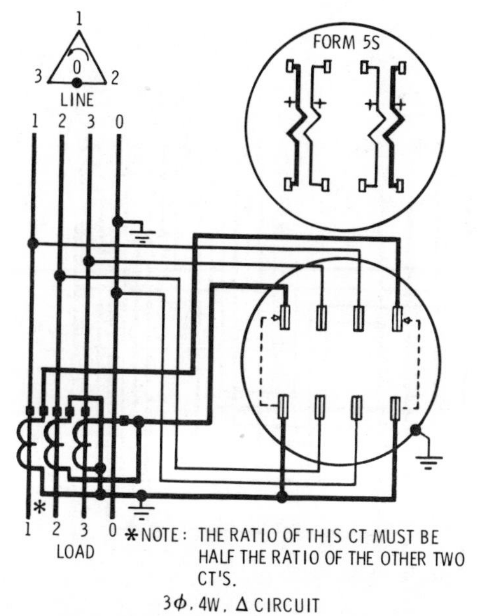

*NOTE: THE RATIO OF THIS CT MUST BE HALF THE RATIO OF THE OTHER TWO CT'S.

3ϕ, 4W, Δ CIRCUIT

2 STATOR, 3ϕ, 3 W, TYPE P2SP, P20SP OR S3S METER WITH 3-2 W CT'S (2 CT'S IN PARALLEL)

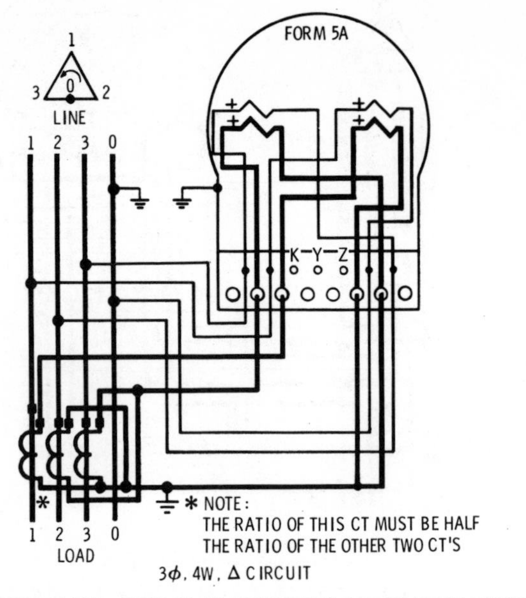

2 STATOR, 3ϕ, 4 W TYPE P2AP, P20AP OR S3A METER WITH 3-2 W CT'S (2 CT'S IN PARALLEL)

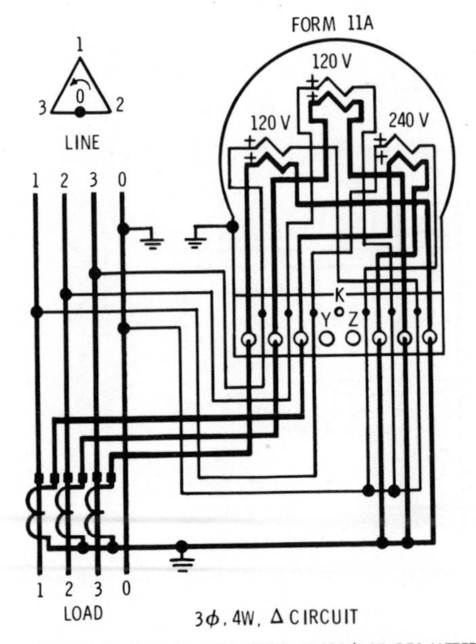

3 STATOR, 3ϕ, 4 W, Δ, TYPE P3AΔ, P30AΔ OR S7A METER WITH 3-2 W CT'S

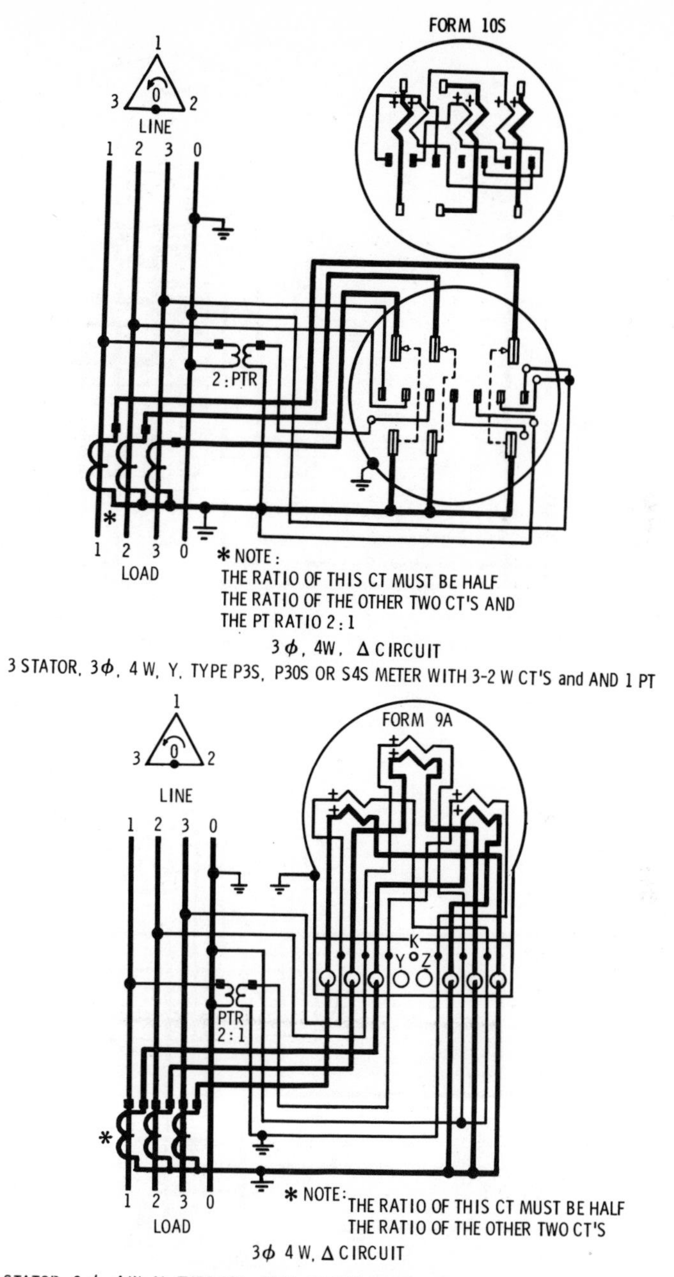
FORM 10S
1
3
0
2
LINE
1 2 3 0
2:PTR
*
1 2 3 0
LOAD
* NOTE:
THE RATIO OF THIS CT MUST BE HALF
THE RATIO OF THE OTHER TWO CT'S AND
THE PT RATIO 2:1
3 ϕ, 4W, Δ CIRCUIT
3 STATOR, 3ϕ, 4 W, Y, TYPE P3S, P30S OR S4S METER WITH 3-2 W CT'S and AND 1 PT
1
3
0
2
LINE
FORM 9A
1 2 3 0
K
Y
Z
PTR
2:1
*
1 2 3 0
LOAD
* NOTE: THE RATIO OF THIS CT MUST BE HALF
THE RATIO OF THE OTHER TWO CT'S
3ϕ 4 W, Δ CIRCUIT
3 STATOR, 3 ϕ, 4 W, Y, TYPE P3A, P30A, OR S4A METER WITH 3-2 W CT'S AND 1 PT

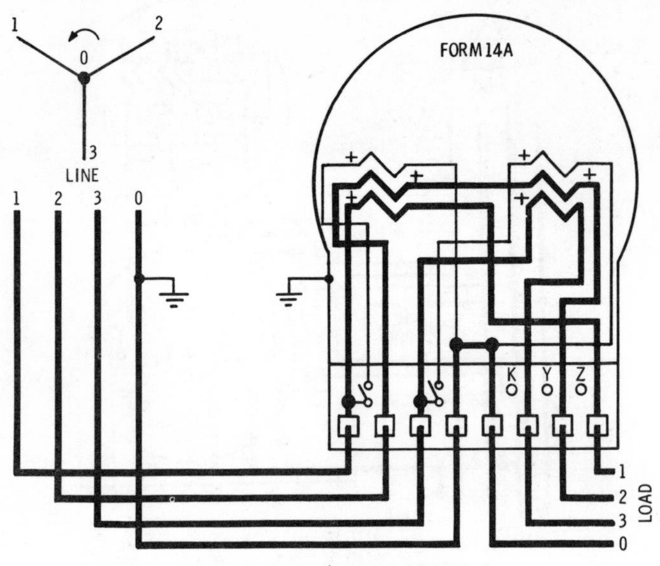

3 ϕ, 4 W, Y CIRCUIT

2 STATOR, 3ϕ, 4 W, Y, TYPE P2AY, P20AY, OR S5A METER, SELF-CONTAINED

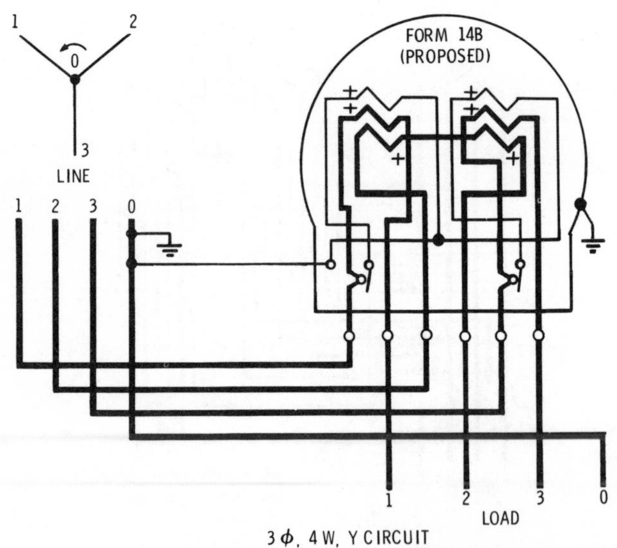

3 ϕ, 4 W, Y CIRCUIT

2 STATOR, 3 ϕ, 4 W, Y, TYPE S5B METER, SELF-CONTAINED

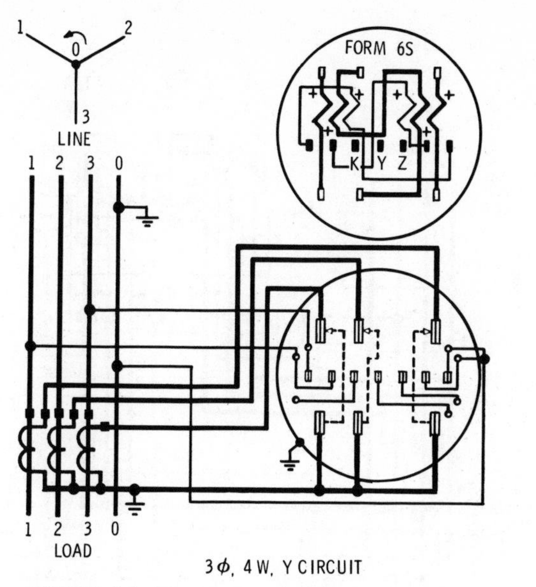

3ϕ, 4 W, Y CIRCUIT

2 STATOR, 3ϕ, 4 W, Y, TYPE P2SY, P20SY OR S5S METER WITH 3-2 W CT'S

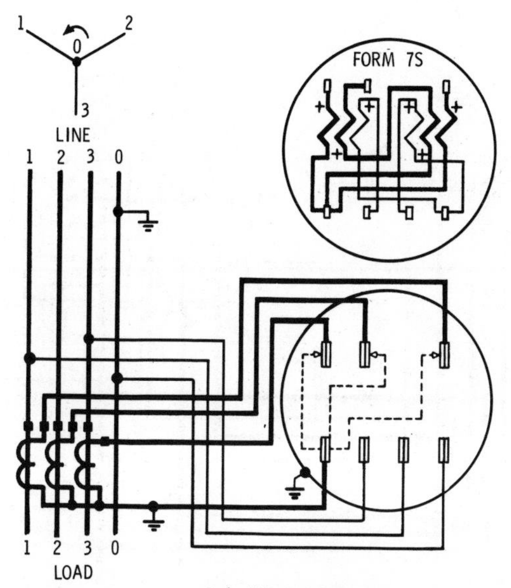

3ϕ, 4 W, Y CIRCUIT

2 STATOR, 3 , 4 W, Y TYPE P2SY, P20SY OR S5S (7 TERMINAL) METER WITH 3-2 W CT'S

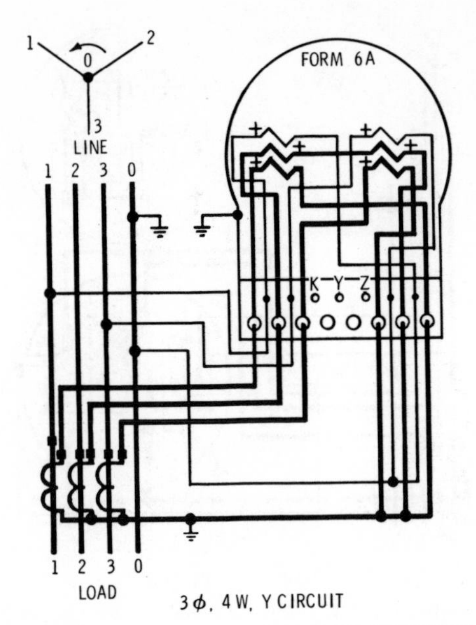

3 ϕ, 4 W, Y CIRCUIT

2 STATOR 3 ϕ, 4 W, Y, TYPE P2AY, P20AY OR S5A METER WITH 3-2 W CT'S

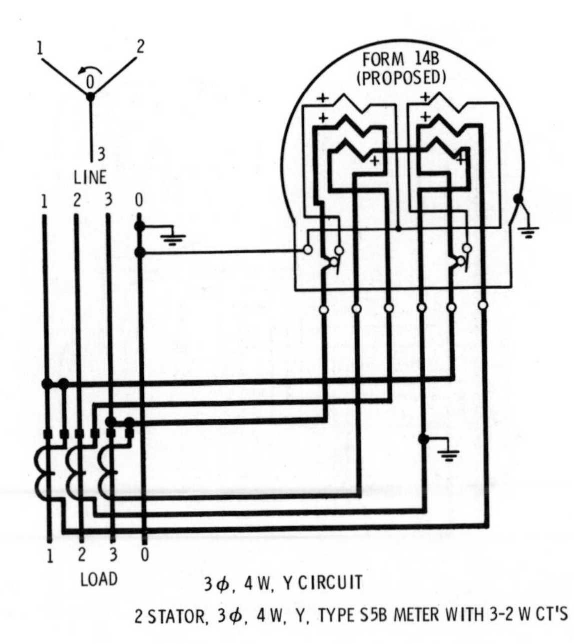

3 ϕ, 4 W, Y CIRCUIT

2 STATOR, 3 ϕ, 4 W, Y, TYPE S5B METER WITH 3-2 W CT'S

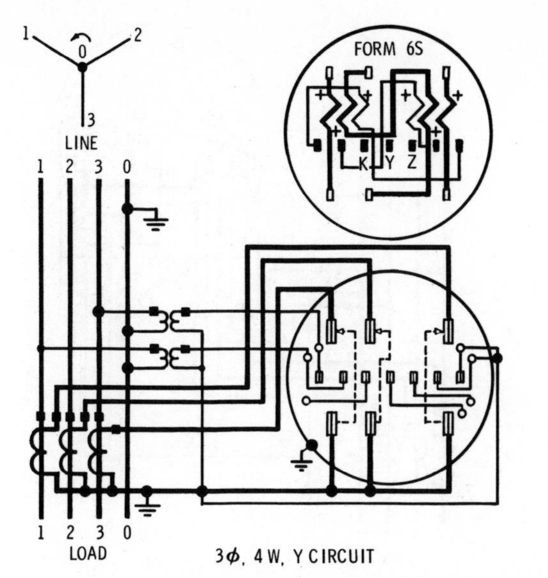

2 STATOR, 3ϕ, 4 W, Y, TYPE P2SY, P20SY OR S5S METER WITH 3-2 W CT'S AND 2 PT'S

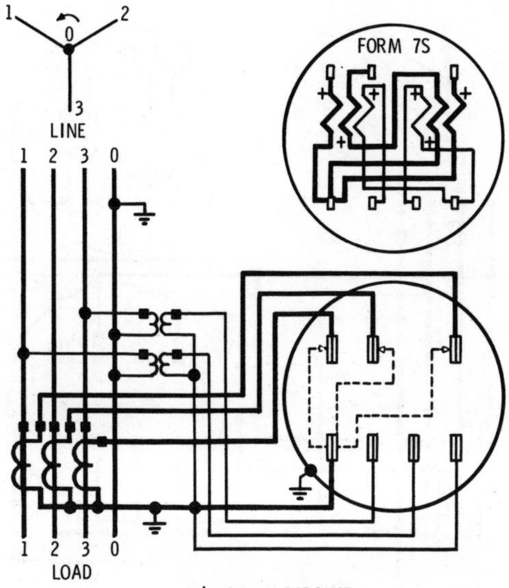

2 STATOR, 3ϕ, 4 W, Y TYPE P2SY, P20SY OR S5S (7 TERMINAL) METER WITH 3-2 W CT'S AND 2 PT'S

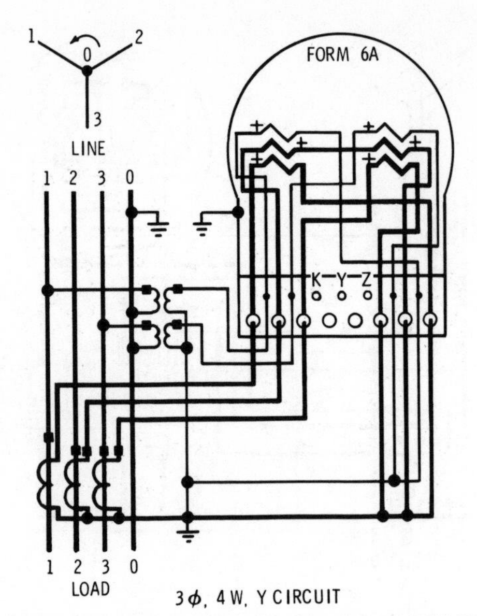

3ϕ, 4 W, Y CIRCUIT

2 STATOR, 3ϕ, 4 W, Y, TYPE P2AY, P20AY OR S5A METER WITH 3-2 W CT'S AND 2 PT'S

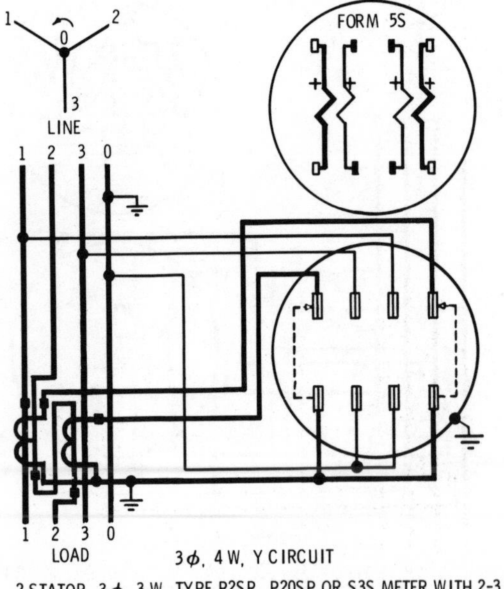

3ϕ, 4 W, Y CIRCUIT

2 STATOR, 3ϕ, 3 W, TYPE P2SP, P20SP OR S3S METER WITH 2-3 W CT'S

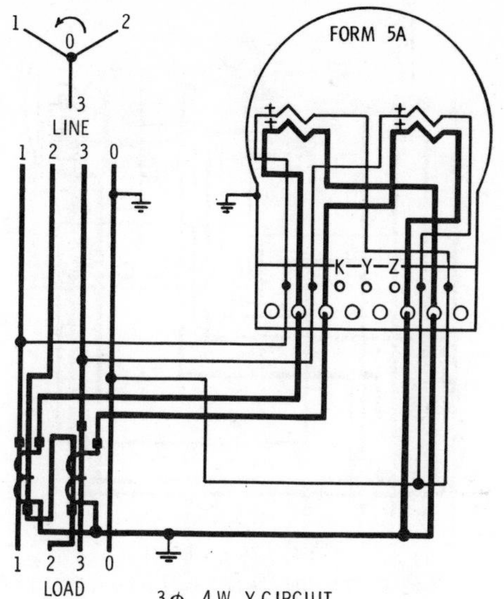

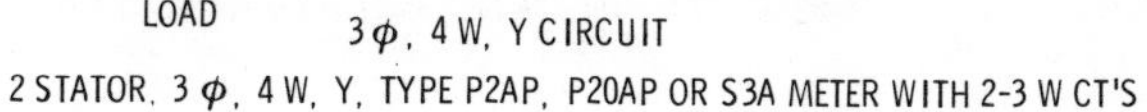
3φ, 4 W, Y CIRCUIT

2 STATOR, 3 φ, 4 W, Y, TYPE P2AP, P20AP OR S3A METER WITH 2-3 W CT'S

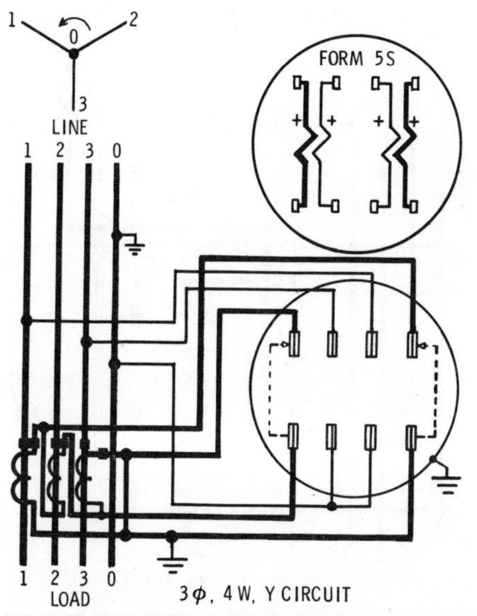

3φ, 4 W, Y CIRCUIT

2 STATOR, 3 φ, 3 W, TYPE P2SP, P20SP OR S3S METER WITH 3-2 W CT'S CONNECTED IN DELTA

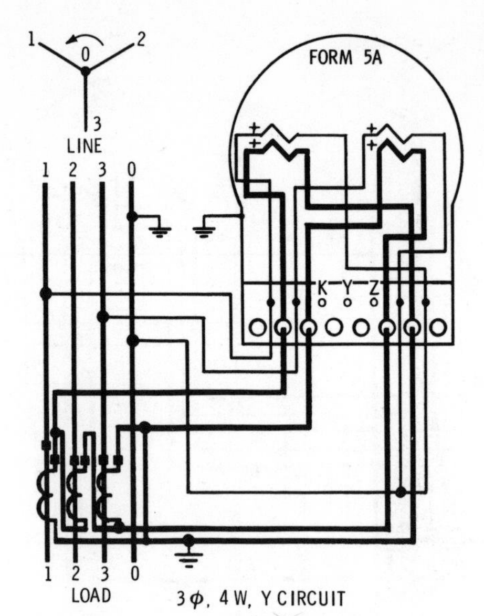

3ϕ, 4 W, Y CIRCUIT

2 STATOR, 3ϕ, 3 W, TYPE P2AP, P20AP OR S3A METER WITH 3-2 W CT'S CONNECTED IN DELTA

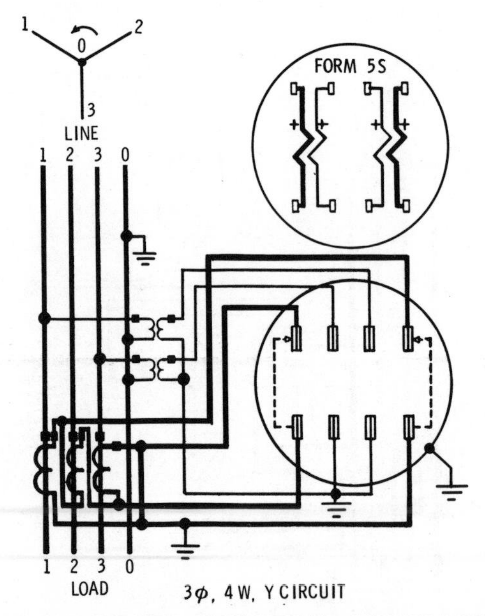

3ϕ, 4 W, Y CIRCUIT

2 STATOR, 3ϕ, 3 W, TYPE P2SP, P20SP, OR S3S METER WITH 3-2 W CT'S IN DELTA AND 2 PT'S

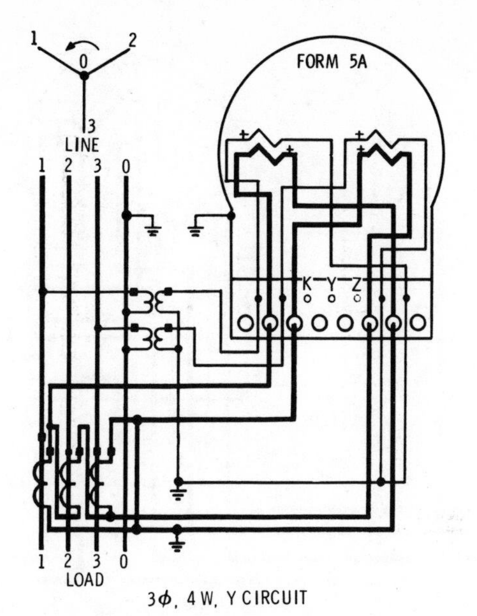

3 ϕ, 4 W, Y CIRCUIT

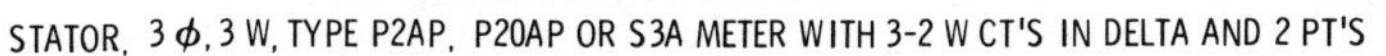
STATOR, 3 ϕ, 3 W, TYPE P2AP, P20AP OR S3A METER WITH 3-2 W CT'S IN DELTA AND 2 PT'S

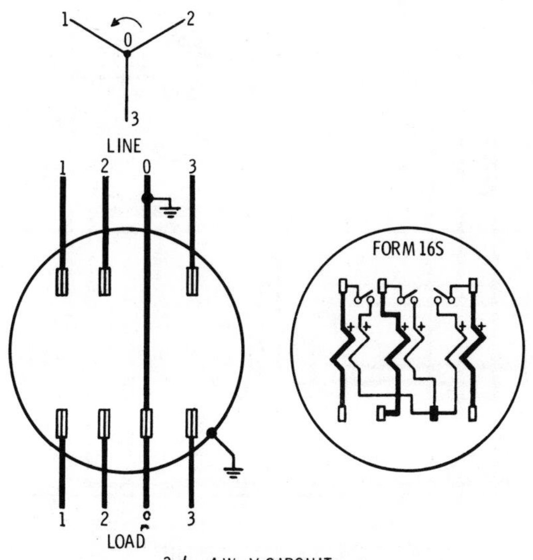

3 ϕ, 4 W, Y CIRCUIT

3 STATOR, 3 ϕ, 4 W, Y, TYPE P3S, P30S OR S4S METER, SELF-CONTAINED

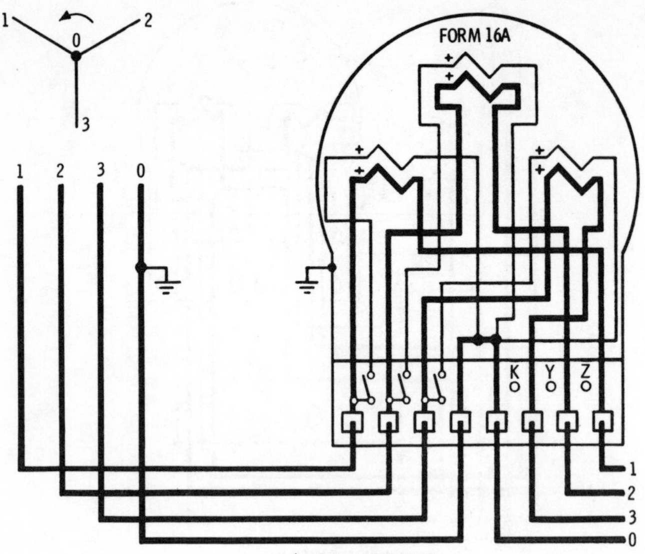

3ϕ, 4 W, Y CIRCUIT

3 STATOR, 3 , 4 W, Y, TYPE P3A, P30A OR S4A METER, SELF-CONTAINED

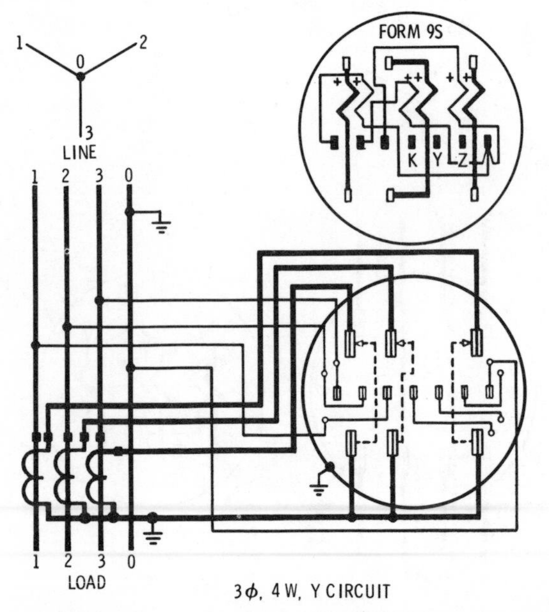

3ϕ, 4 W, Y CIRCUIT

3 STATOR, 3ϕ, 4 W, Y, TYPE P3S, P30S OR S4S METER WITH 3-2 W CT'S

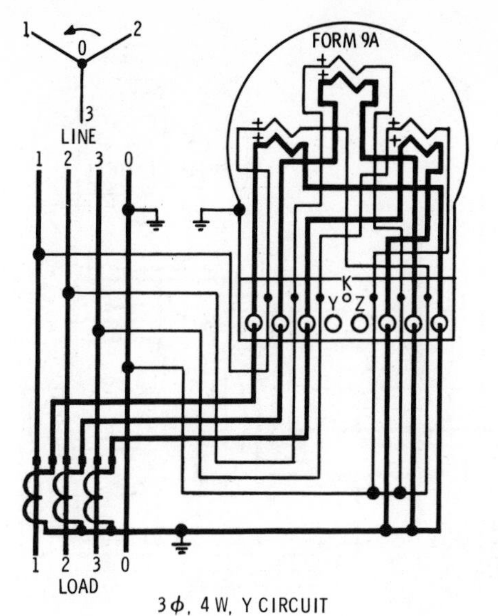

3ϕ, 4 W, Y CIRCUIT

3 STATOR, 3ϕ, 4 W, Y, TYPE P3A, P30A OR S4A METER WITH 3-2 W CT'S

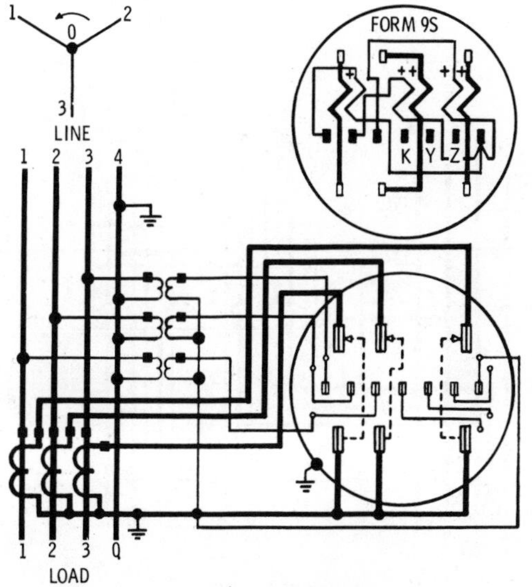

3ϕ, 4 W, Y CIRCUIT

3 STATOR, 3ϕ, 4 W, Y, TYPE P3S, P30S OR S4S METER WITH 3-2 W CT'S AND 3 PT'S

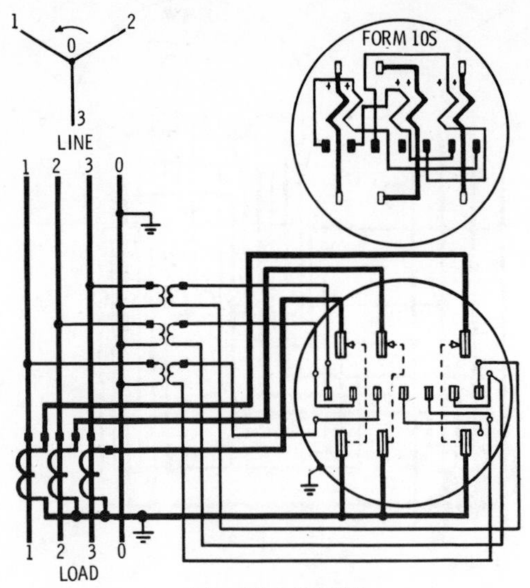

3ϕ, 4 W, Y CIRCUIT

3 STATOR, 3 ϕ, 4 W, Y, TYPE P3S, P30S OR S4S METER WITH 3-2 W CT'S AND 3 PT'S (SEPARATE PT'S)

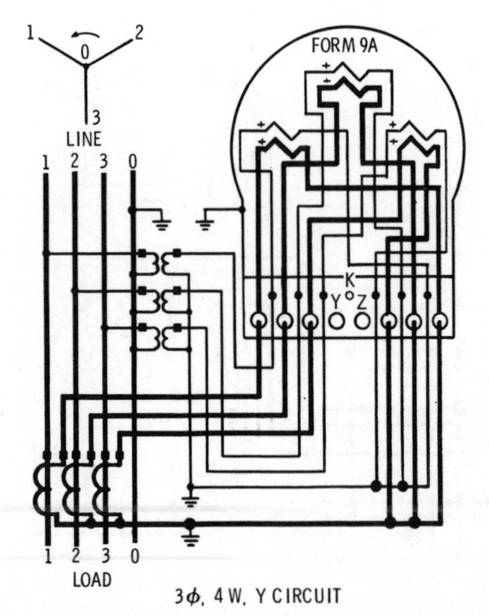

3ϕ, 4 W, Y CIRCUIT

3 STATOR, 3 ϕ, 4 W, Y, TYPE P3A P30A OR S4A METER WITH 3-2 W CT'S AND 3 PT'S

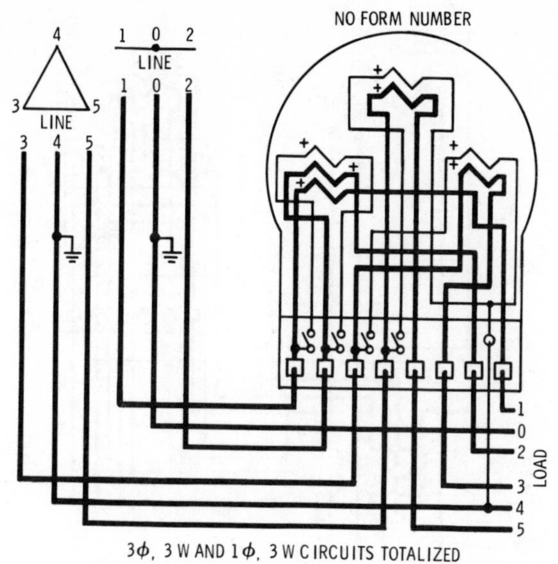

3φ, 3 W AND 1φ, 3 W CIRCUITS TOTALIZED

3 STATOR, MULTICIRCUIT TYPE P30AM METER, SELF-CONTAINED

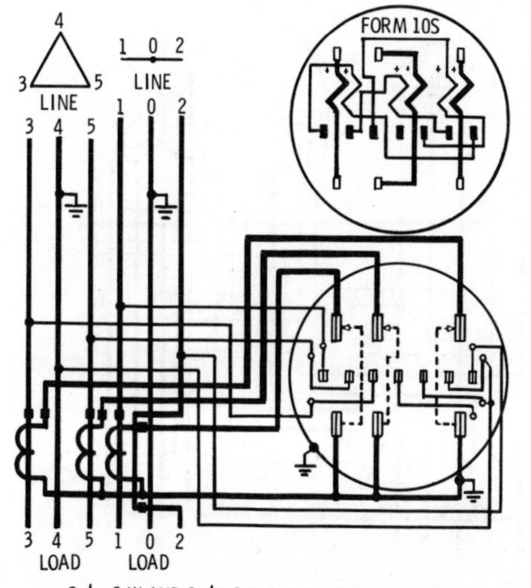

3φ, 3 W AND 1φ, 3 W CIRCUITS TOTALIZED

3 STATOR, 3φ, 4 W, Y, TYPE P30S OR S4S METER WITH 1-3 W CT AND 2-2 W CT'S

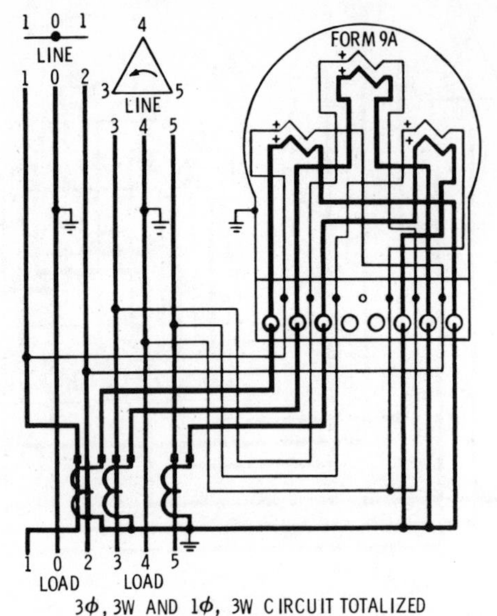

3ϕ, 3W AND 1ϕ, 3W CIRCUIT TOTALIZED

3 STATOR, 3ϕ, 4 W, Y. TYPE P3A, P30A OR S4A METER WITH 1-3 W CT AND 2-2 W CT'

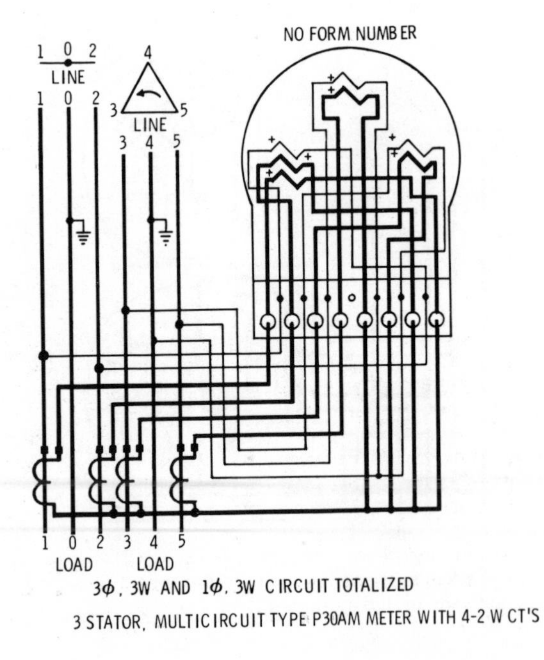

3ϕ, 3W AND 1ϕ, 3W CIRCUIT TOTALIZED

3 STATOR, MULTICIRCUIT TYPE P30AM METER WITH 4-2 W CT'S

WATTHOUR-METER TEST

A watthour meter is used for measurement of electric energy. Principally it consists of an electric motor with associated windings so arranged that the mechanical torque produced indicates the electrical power.

One winding of the meter is usually connected in series with the load and the other across the circuit. The torque of such a motor will be proportional to the power and the total revolutions of the motor will be a measure of energy consumed by the load.

In addition a watthour meter is equipped with a register which records the revolutions of the meter shaft and a magnetic brake whose function it is to retard the revolutions of the motor.

Due to the inability of most meters to record the energy consumption correctly over a period of time, periodic test schedules are usually followed where each meter in service is compared with a **portable standard** meter, that is, a meter in which the error has been reduced to a minimum.

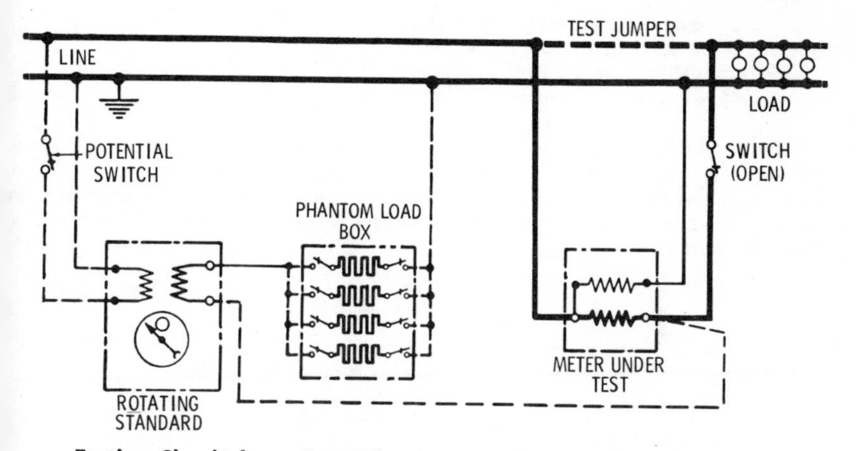

Testing Circuit for a Two-Wire Meter, Using a Resistance Load.

The size of error permitted may vary in different parts of the country but is usually around ± 3 per cent of its rated load.

When setting up a meter for test, the current coils of the two meters, that is, of the portable standard and the meter under test, are connected in series, whereas their potential coils are connected in parallel.

During the test period, the revolutions of the standard are compared with the meter under test for the same interval of time, allowance being made in the calculations for the disc constant of the two meters.

WATTHOUR-METER TEST

In cases where the load is unknown it may be determined by timing the standard with a stop watch and comparing the value of the watts from the expression,

$$\text{True watts} = \frac{3{,}600 \times \text{Revolutions} \times \text{Watthour Constant}}{\text{Time in seconds}} \tag{1}$$

With reference to the test circuits the rotating standard is operated by a potential switch which stops and starts the standard, a reading of the standard is taken at the beginning and at the end of the test, and the difference between these two readings gives the number of revolutions of the standard.

If no correction is to be applied to the rotating standard, the per cent accuracy of the watthour meter under test is obtained from equation,

$$\text{Per cent accuracy} = \frac{k_h \times r}{K_h \times R} \tag{2}$$

Where

r = revolutions of meter under test.
R = revolutions of rotating standard.
k_h = watthour constant of meter under test.
K_h = watthour constant of rotating standard.

The method shown may be facilitated by introducing an additional symbol, value for which may be given to the tester in tabular form. Thus, if R_o = the number o revolutions the rotating standard should make when the tested meter is correct, th number of revolutions of two watthour meters for a given load vary inversely a their disc constants, then

$$\frac{R_o}{r} = \frac{k_h}{K_h} \text{ or } R_o = \frac{k_h \times r}{K_h} \tag{3}$$

Substituting R_o in the equation for per cent accuracy, we obtain

$$\% \text{ accuracy} = \frac{R_o \times 100}{r}$$

Example—In a certain test the rotating standard has a constant $K_h = 0.05$ and th watthour meter under test has a constant $k_h = 0.5$. If r, the number of revolution of the meter under test, = 2, determine the number of revolutions of the rotatin standard.

Solution—Substituting values in formula (3) we obtain

$$R_o = \frac{k_h \times r}{K_h} = \frac{0.5 \times 2}{0.05} = 20.$$

That is, for 2 revolutions of the meter under test, the standard should make 2 revolutions.

WATTHOUR-METER TEST

Example—Assume the rotating standard in the previous example actually made 20.24 revolutions, what is the accuracy of the watthour meter under test?

Solution

$$\text{Per cent accuracy} = \frac{R_0 \times 100}{r} = \frac{20 \times 100}{20.24} = 98.8\%.$$

This actually means that the meter is 1.2% slow and should be speeded up slightly.

Example—In a test of a DC 15 ampere watthour meter, the corrected average volt and ampere readings are 220 and 14.75, respectively. During the test interval 38 revolutions are counted in 53.5 seconds and the meter constant is 1.25. What is the per cent accuracy of the meter at this load?

Solution—Average standard watts

$$W_1 = 14.75 \times 220 = 3{,}245$$

Inserting our values in equation (1) we obtain the average meter watts as

$$W = \frac{3{,}600 \times 38 \times 1.25}{53.5} = 3{,}196$$

$$\text{Meter accuracy} = \frac{W}{W_1} = \frac{3{,}196}{3{,}245} = 0.985 \text{ or } 98.5\%.$$

Other well known methods used in testing of watthour meters are: (1) **the indicating instrument method** and (2) **the stroboscopic method.** In the former, load is applied to the meter and watthours are measured by means of indicating instruments and timing devices such as stop watches or chronographs.

The ratio between the indicated or meter watthours and true watthours represents the accuracy of the meter under test, and is usually expressed in per cent.

The stroboscopic method involves the comparison of the speed of two similar discs, and utilizes a light source, a lens system, a photoelectric cell and amplifying equipment. This method of meter testing finds application in meter shops having a large number of meters to be tested. It is not limited to type of meter to be tested except that marking or slotting of the disc is necessary in order to obtain pulsating light.

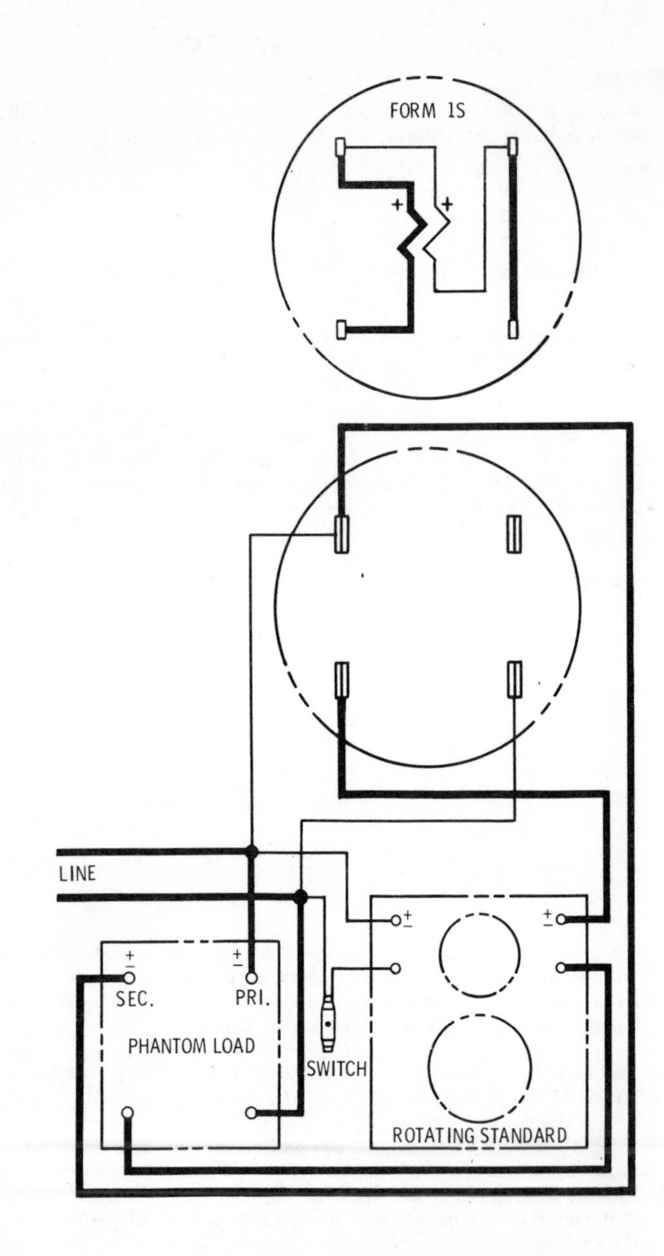

1 STATOR, 1 , 2 W, SELF-CONTAINED, FORM 1S

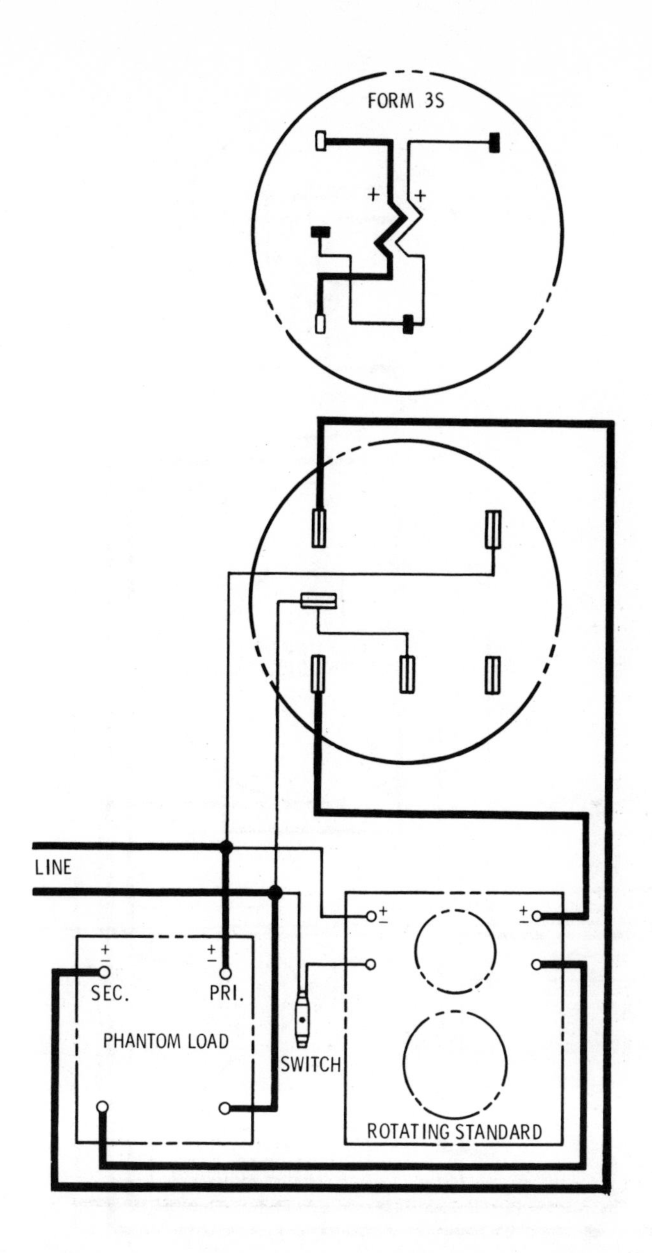

1 STATOR, 1 ϕ, 2 W, TRANSFORMER RATED, FORM 3S

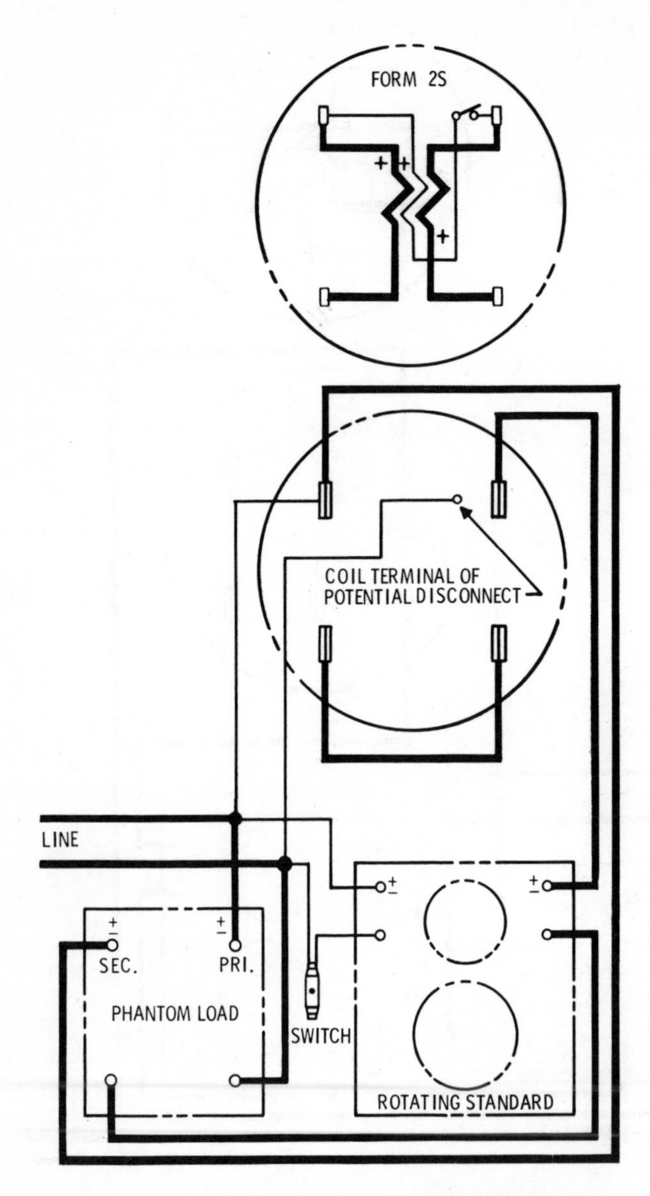

1 STATOR, 1 ϕ, 3 W, SELF-CONTAINED, FORM 2S

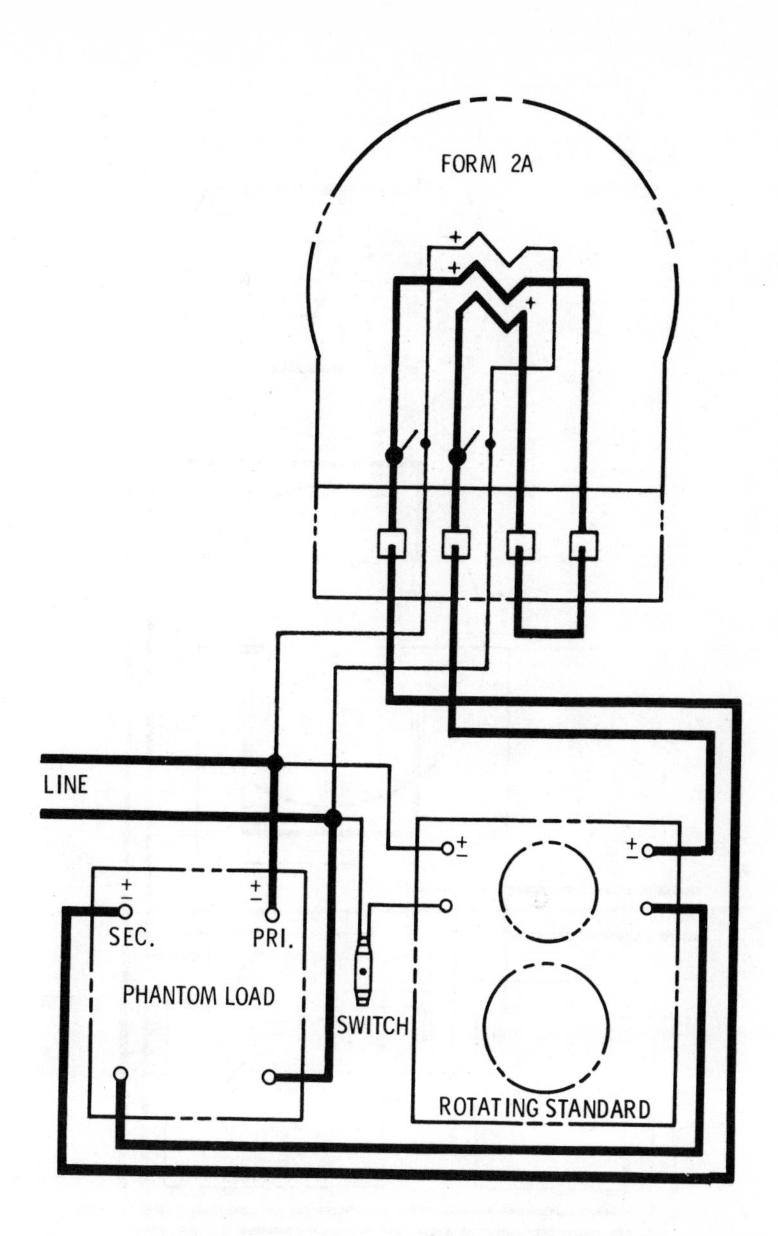

1 STATOR, 1ϕ, 3 W, SELF-CONTAINED, FORM 2A

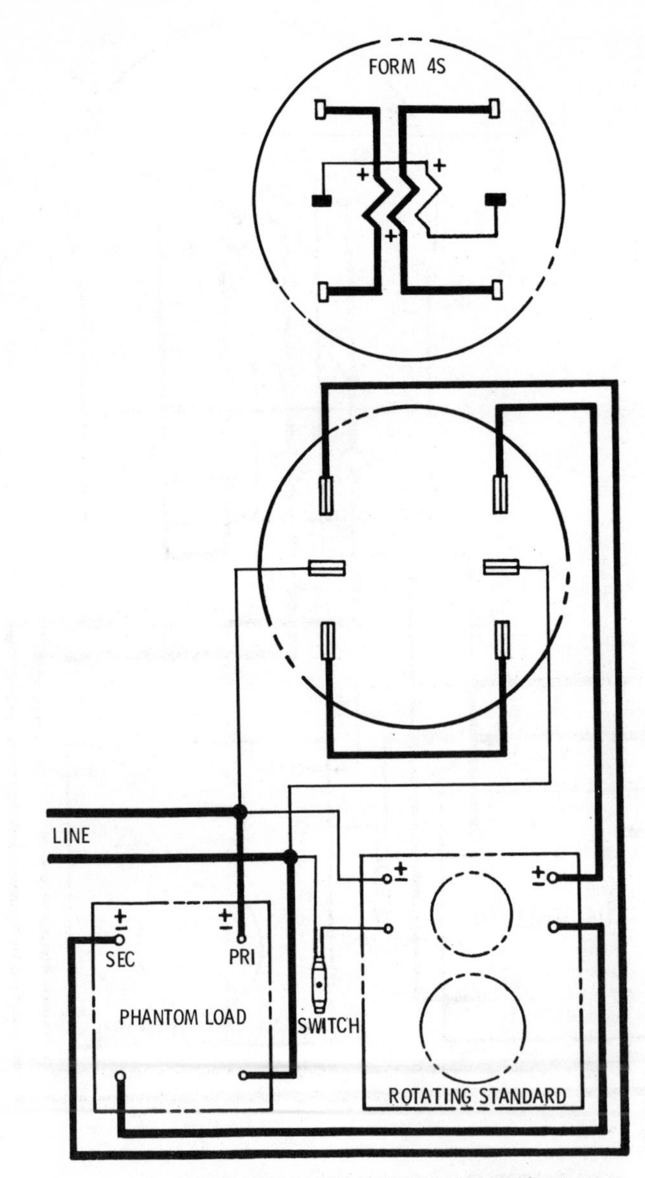

1 STATOR, 1ϕ, 3 W, TRANSFORMER RATED, FORM 4S

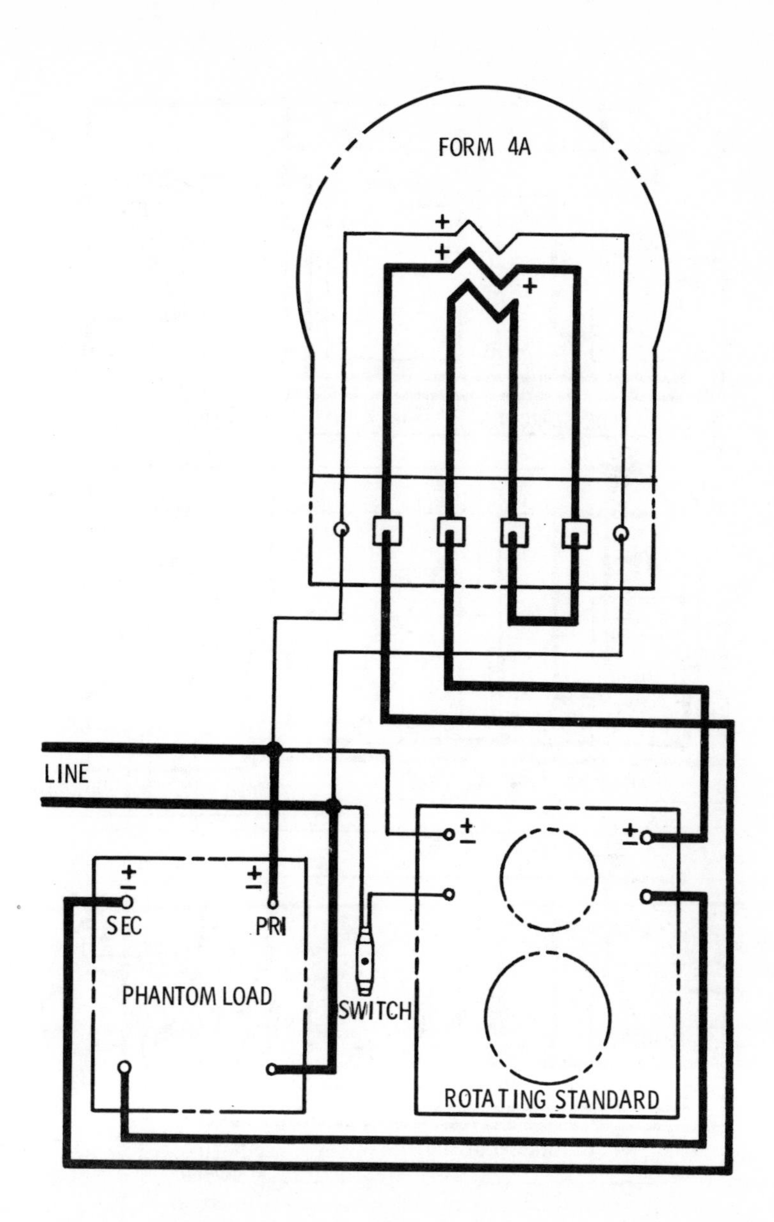

1 STATOR, 1ϕ, 3 W, TRANSFORMER RATED, FORM 4A

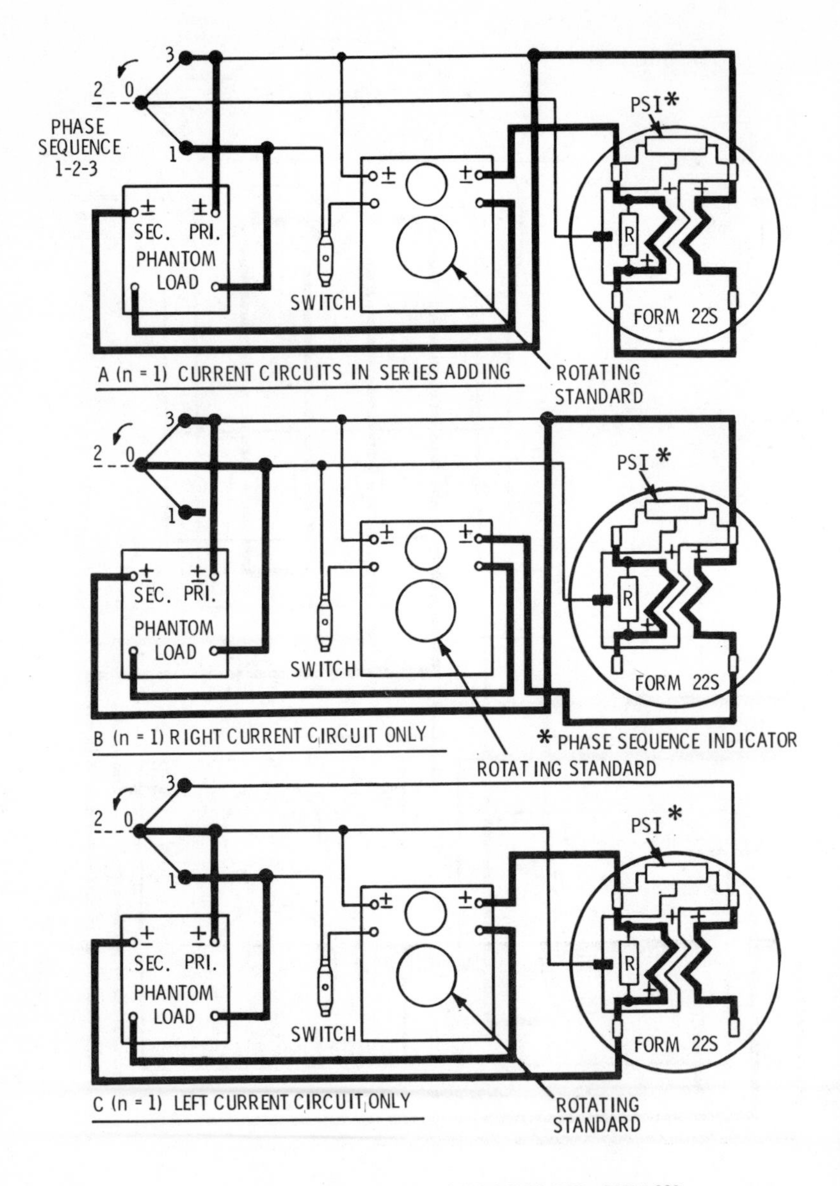

SINGLE STATOR NETWORK, SELF-CONTAINED, FORM 22S

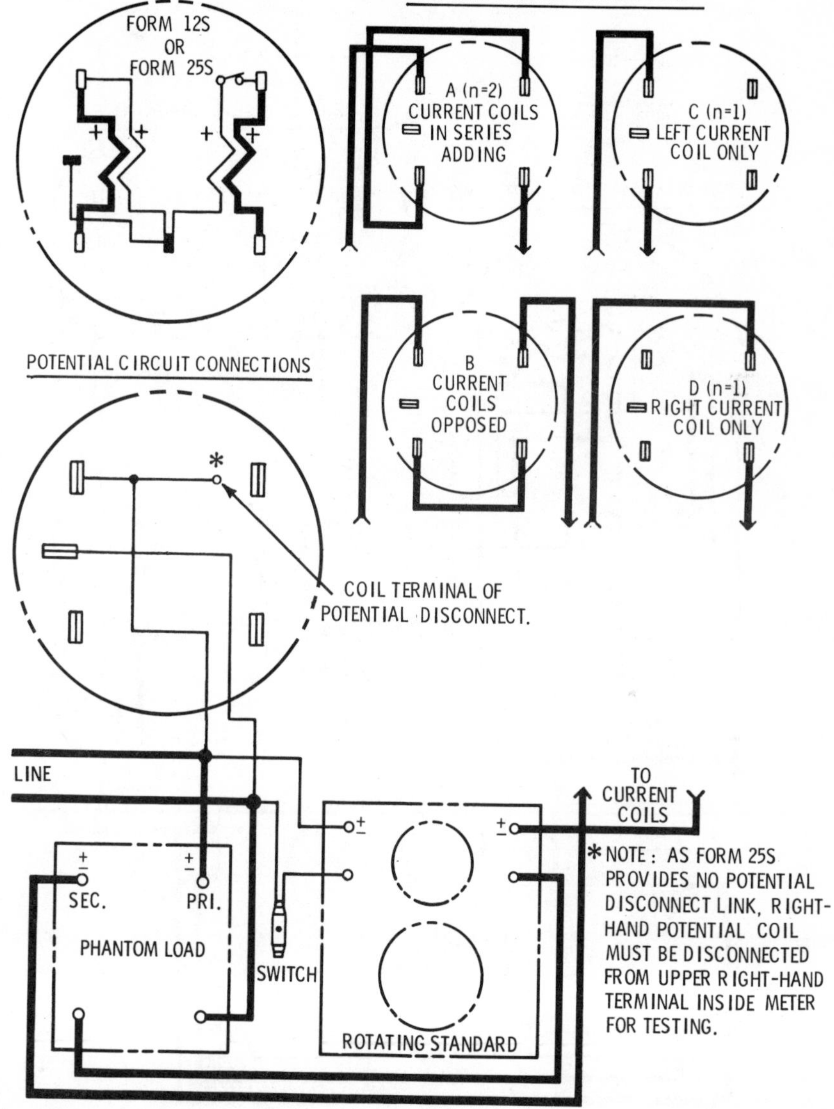

2 STATOR, 3 W NETWORK, SELF-CONTAINED, FORM 12S OR 25S

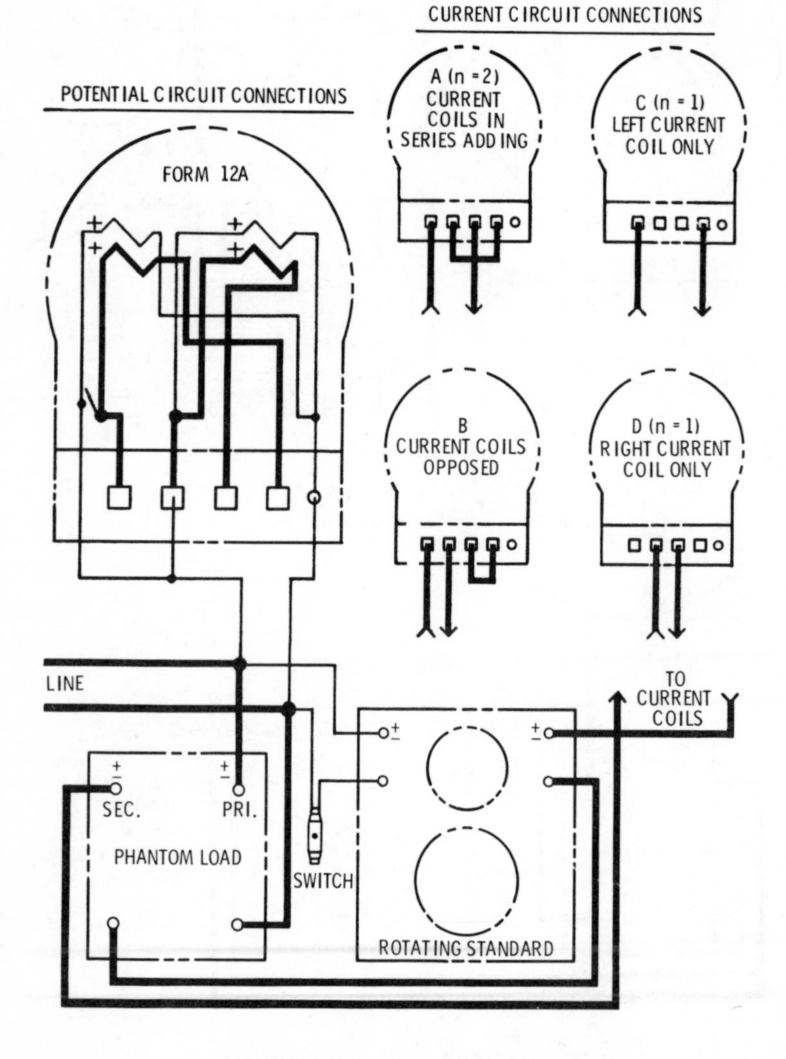

2 STATOR NETWORK, SELF-CONTAINED, FORM 12A

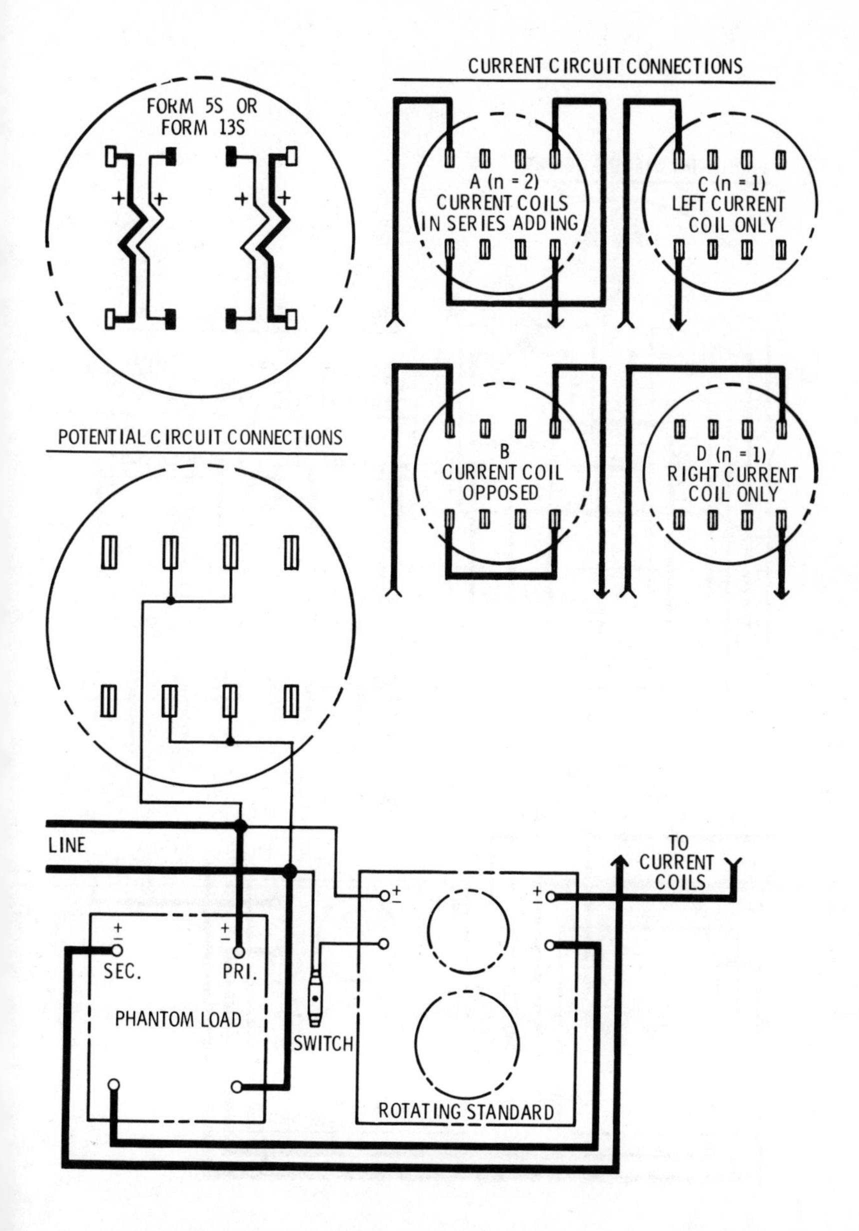

2 STATOR, 3ϕ, 3 W, SELF-CONTAINED OR TRANSFORMER RATED, FORM 5S OR 13S

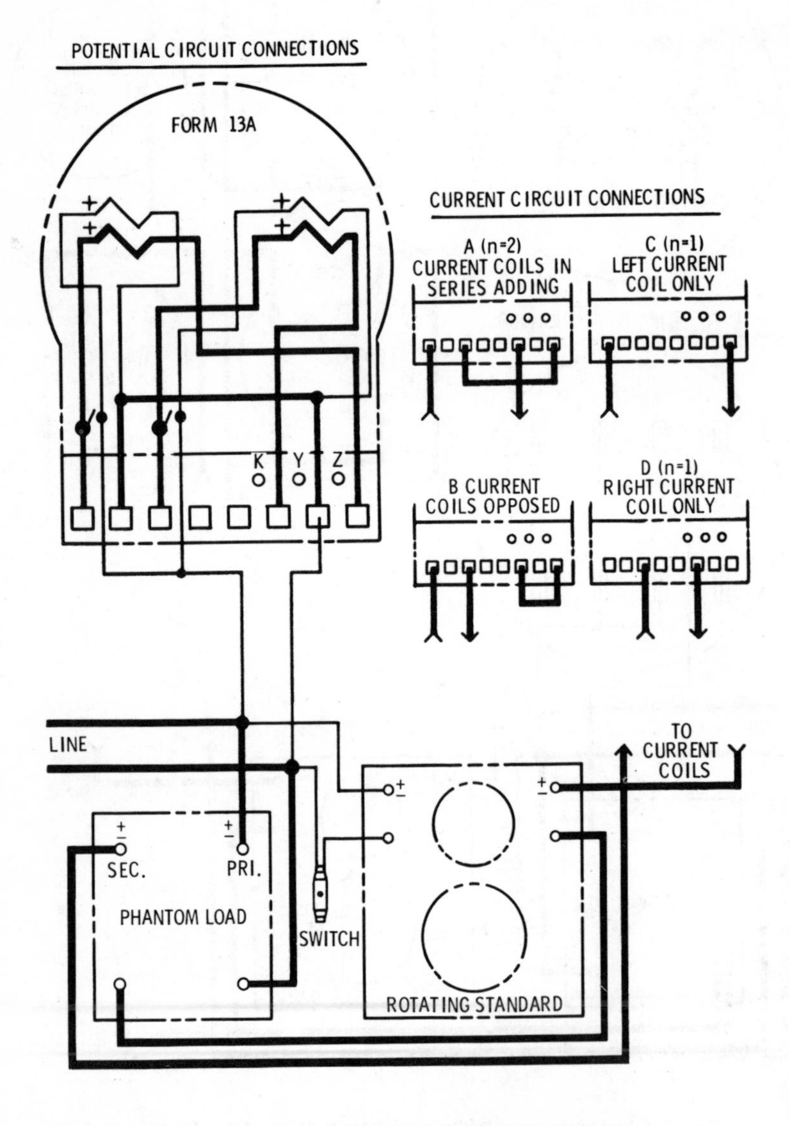

2 STATOR, 3ϕ, 3 W, SELF-CONTAINED, FORM 13A

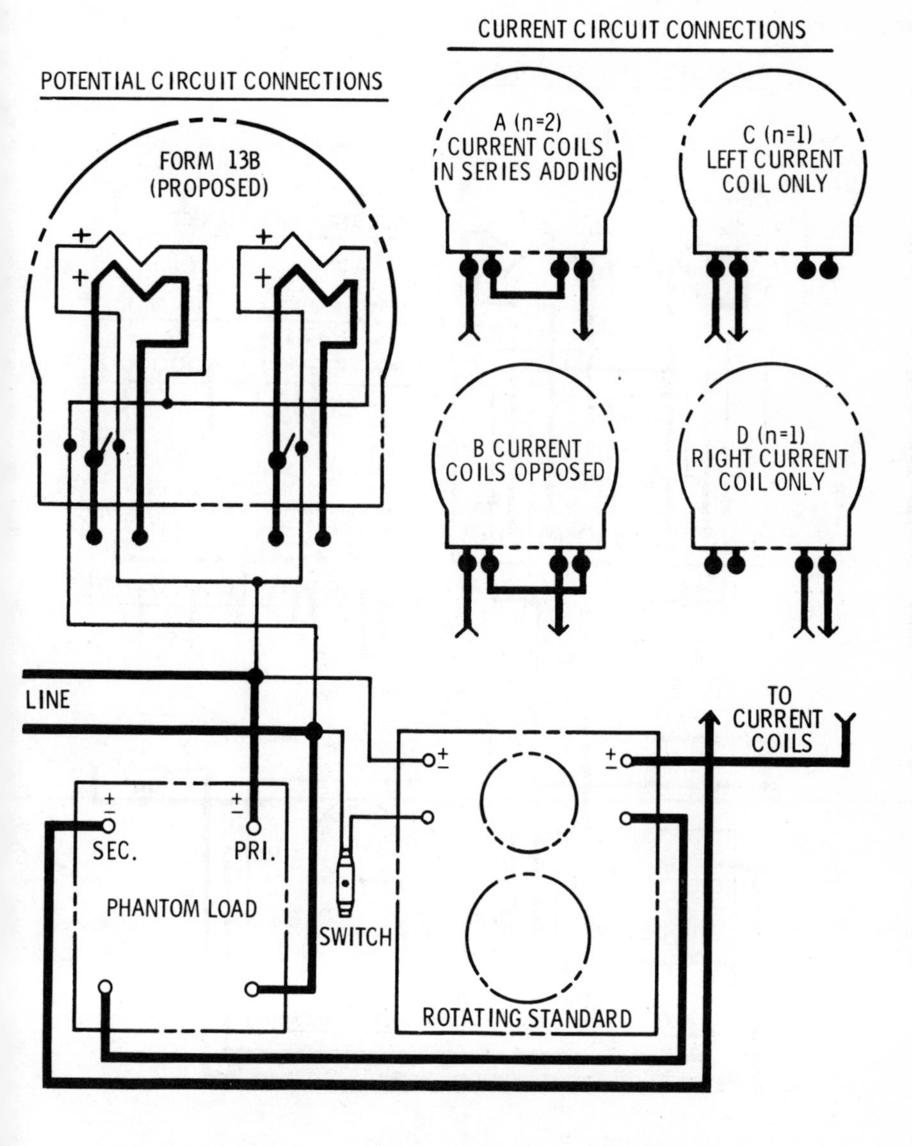

2 STATOR, 3 ϕ, 3 W, SELF-CONTAINED, FORM 13B

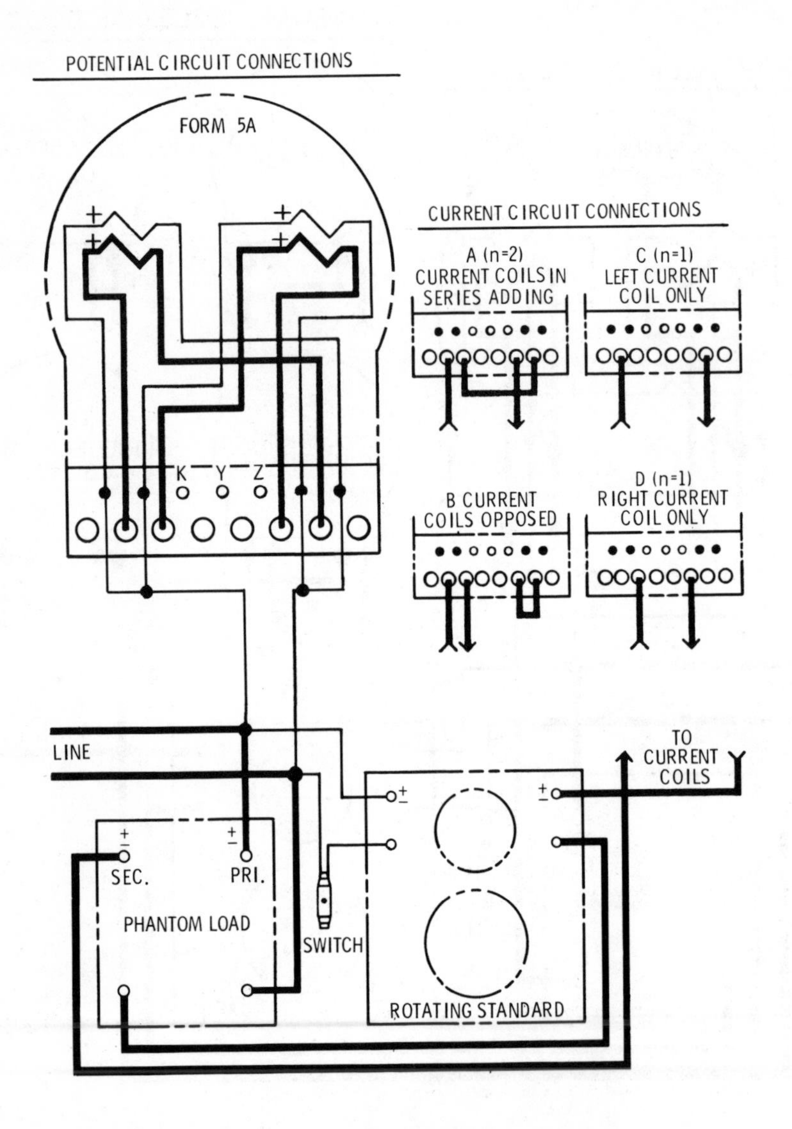

2 STATOR, 3ϕ, 3 W, TRANSFORMER RATED, FORM 5A

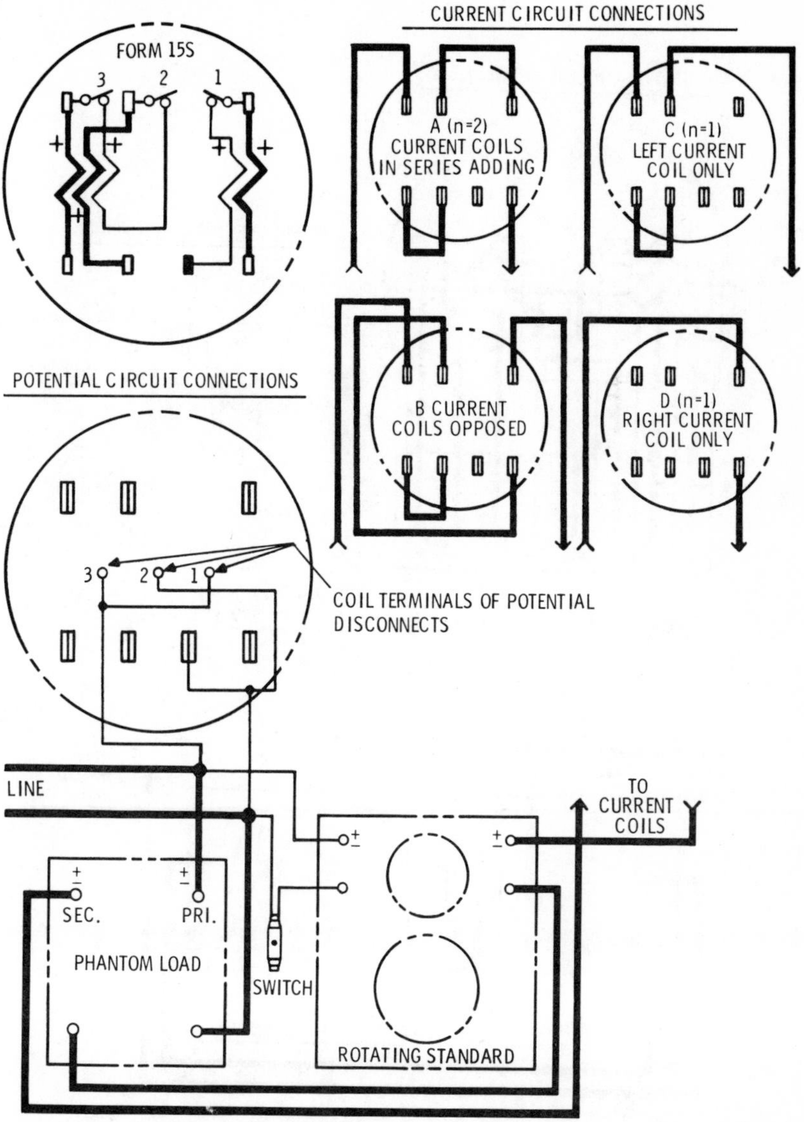

CURRENT CIRCUIT CONNECTIONS

FORM 15S

3 2 1

A (n=2) CURRENT COILS IN SERIES ADDING

C (n=1) LEFT CURRENT COIL ONLY

B CURRENT COILS OPPOSED

D (n=1) RIGHT CURRENT COIL ONLY

POTENTIAL CIRCUIT CONNECTIONS

COIL TERMINALS OF POTENTIAL DISCONNECTS

LINE

TO CURRENT COILS

SEC.

PRI.

PHANTOM LOAD

SWITCH

ROTATING STANDARD

2 STATOR, 3ϕ, 4 W. Δ, SELF-CONTAINED, FORM 15S

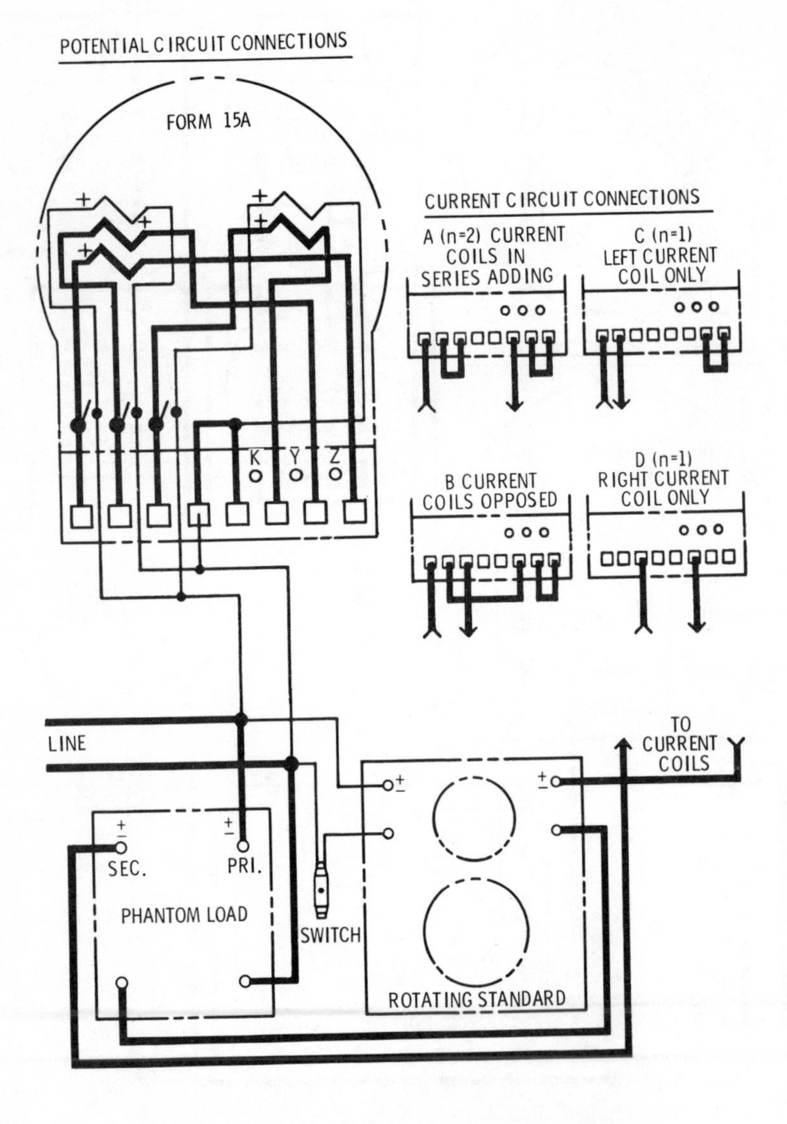

2 STATOR, 3 ϕ, 4 W, Δ, SELF-CONTAINED, FORM 15A

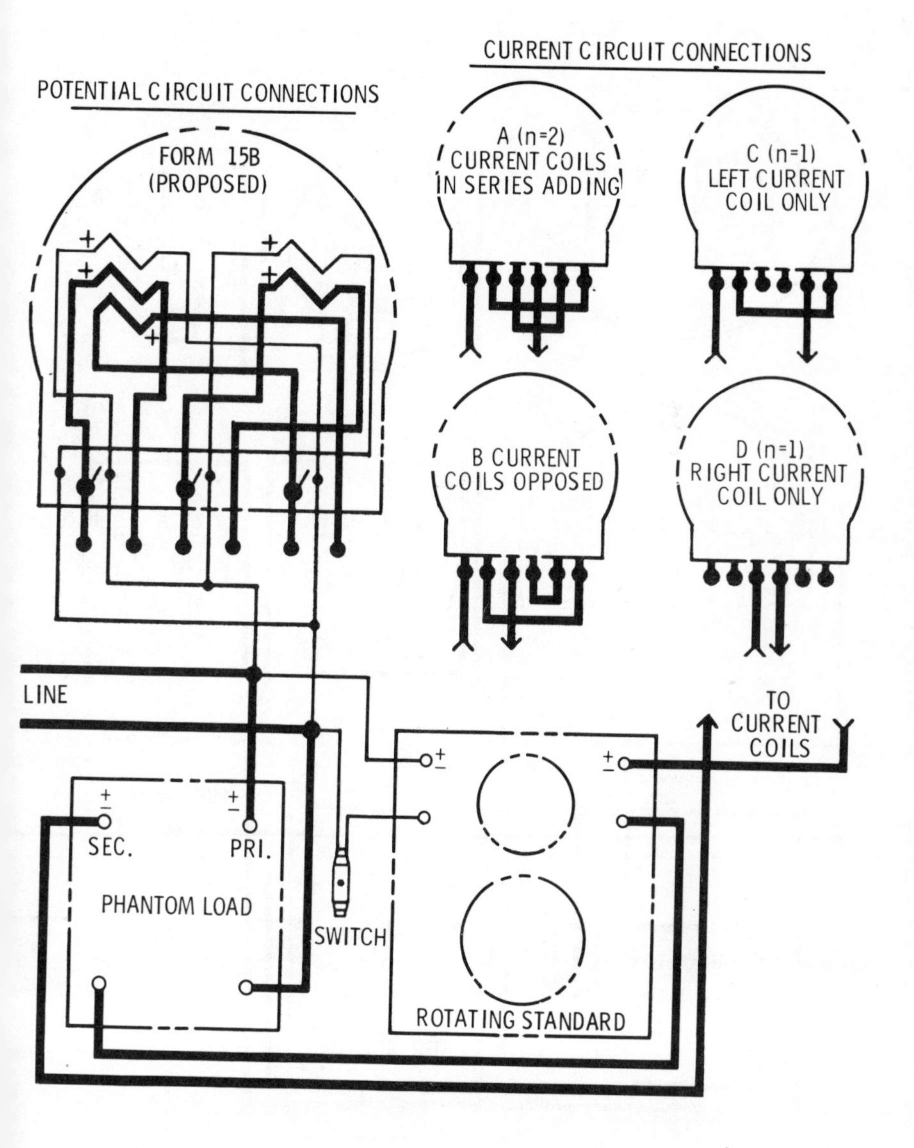

2 STATOR, 3ϕ, 4 W.Δ, SELF-CONTAINED, FORM 15B

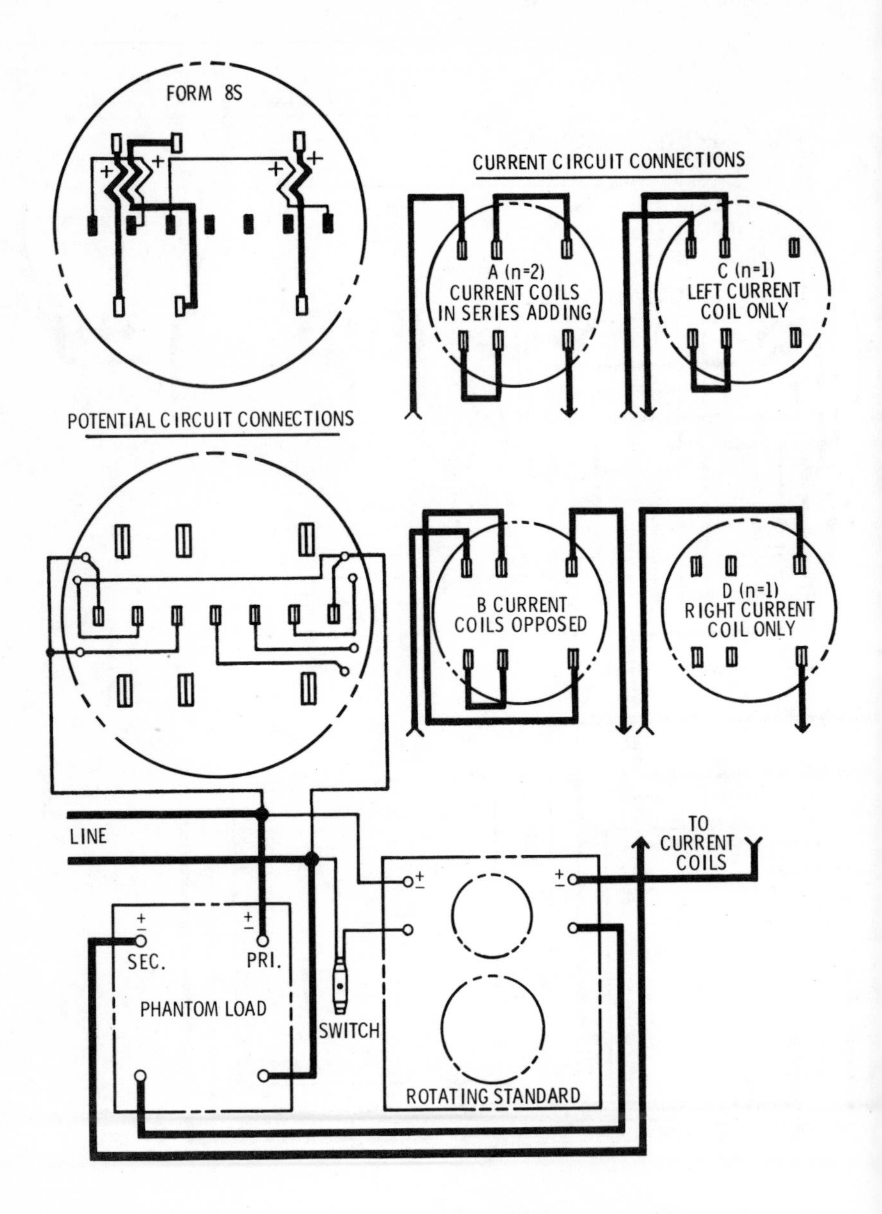

2 STATOR, 3ϕ, 4 W, Δ, TRANSFORMER RATED, FORM 8S

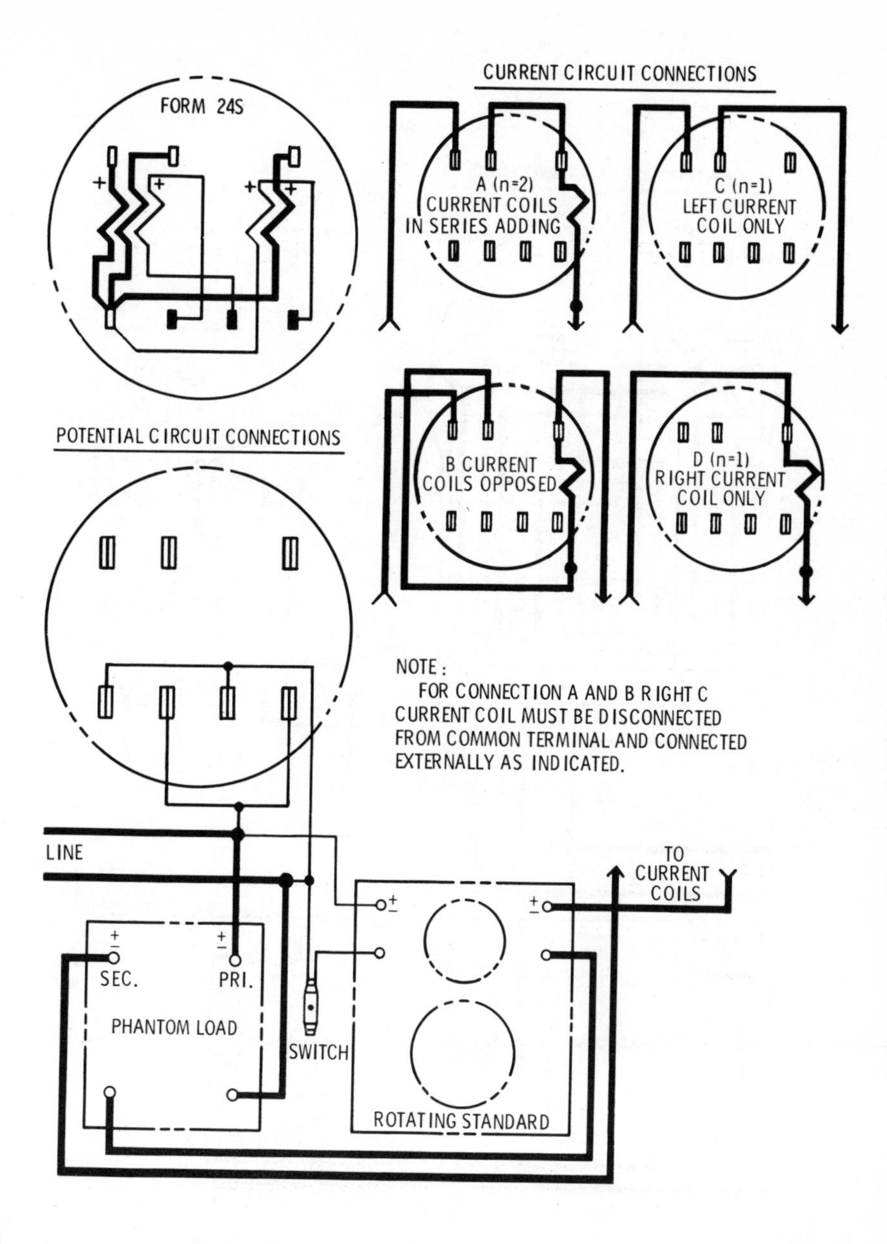

2 STATOR, 3 ϕ, 4 W, Δ, TRANSFORMER RATED, FORM 24S

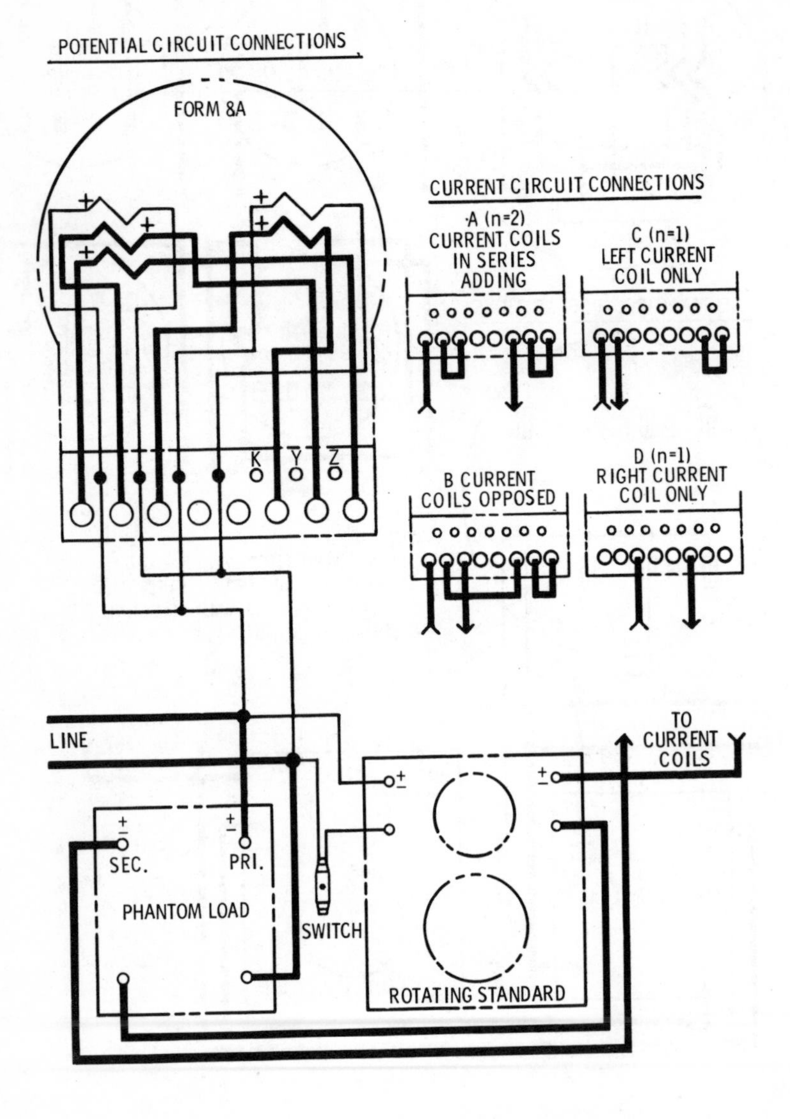

2 STATOR, 3ϕ, 4 W, Δ, TRANSFORMER RATED, FORM 8A

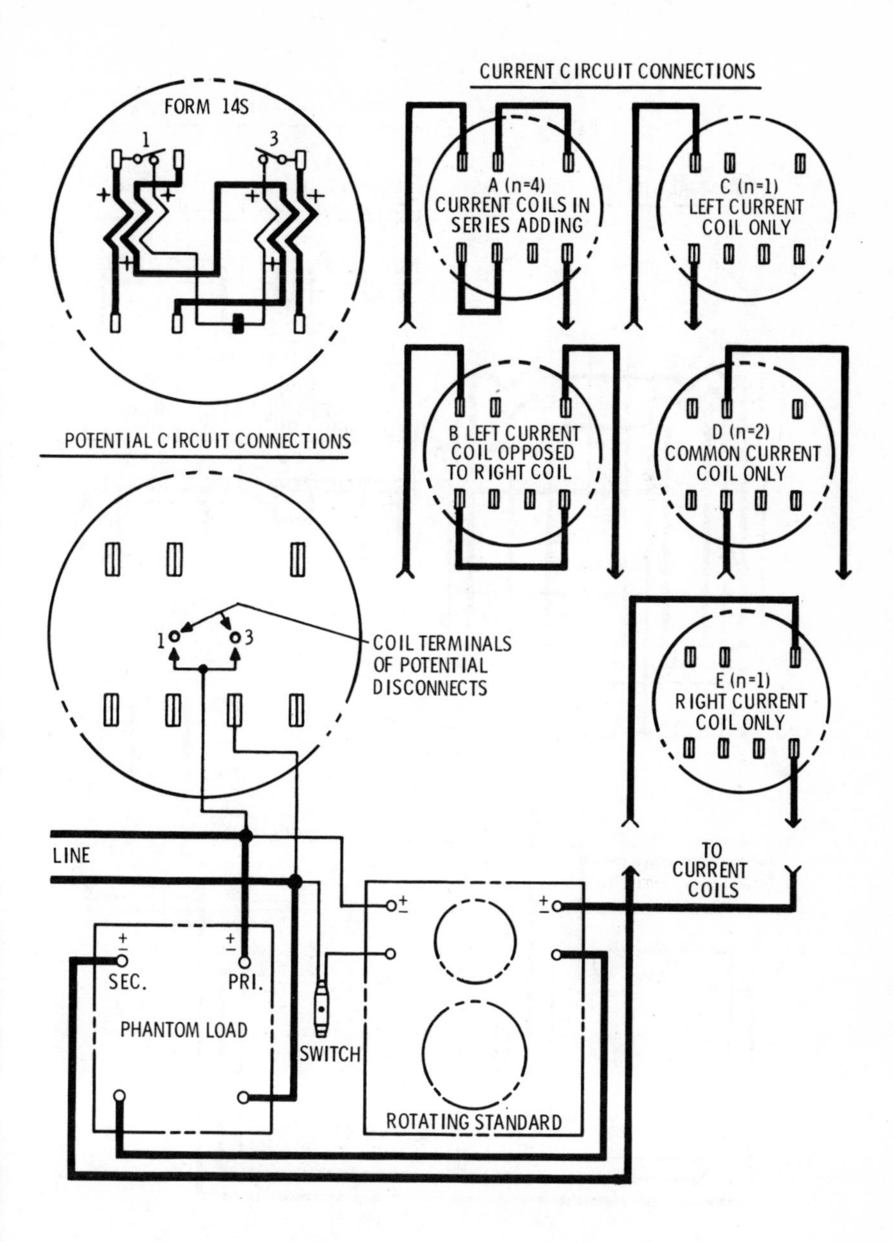

2 STATOR, 3 ϕ, 4 W, Y, SELF-CONTAINED, FORM 14S

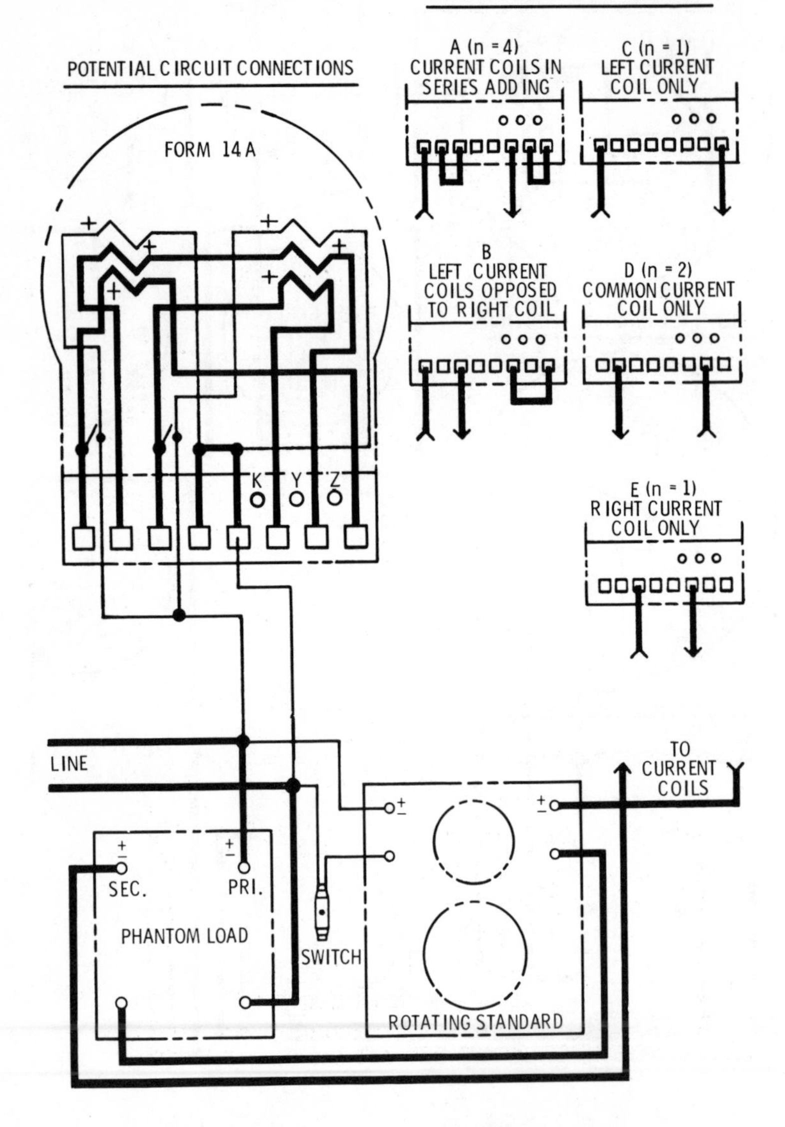

2 STATOR, 3ϕ, 4 W, Y, SELF-CONTAINED, FORM 14A

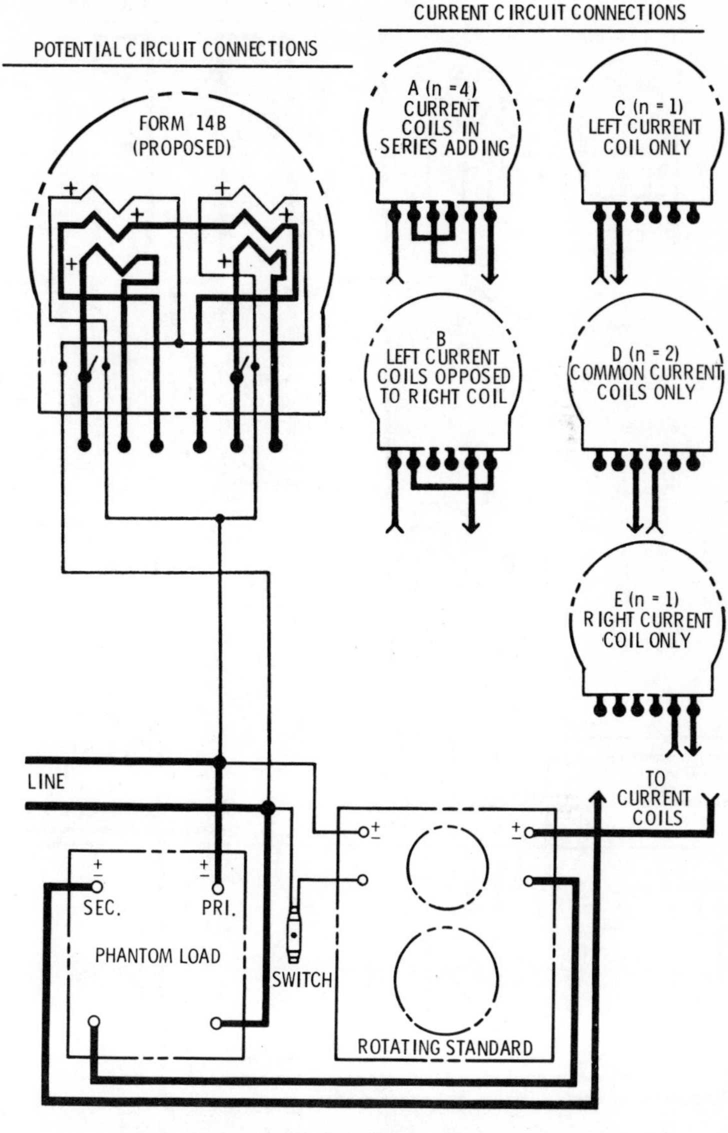

POTENTIAL CIRCUIT CONNECTIONS

FORM 14B (PROPOSED)

CURRENT CIRCUIT CONNECTIONS

A (n = 4) CURRENT COILS IN SERIES ADDING

C (n = 1) LEFT CURRENT COIL ONLY

B LEFT CURRENT COILS OPPOSED TO RIGHT COIL

D (n = 2) COMMON CURRENT COILS ONLY

E (n = 1) RIGHT CURRENT COIL ONLY

LINE

TO CURRENT COILS

SEC.

PRI.

PHANTOM LOAD

SWITCH

ROTATING STANDARD

2 STATOR, 3 ϕ, 4 W, Y, SELF-CONTAINED, FORM 14B

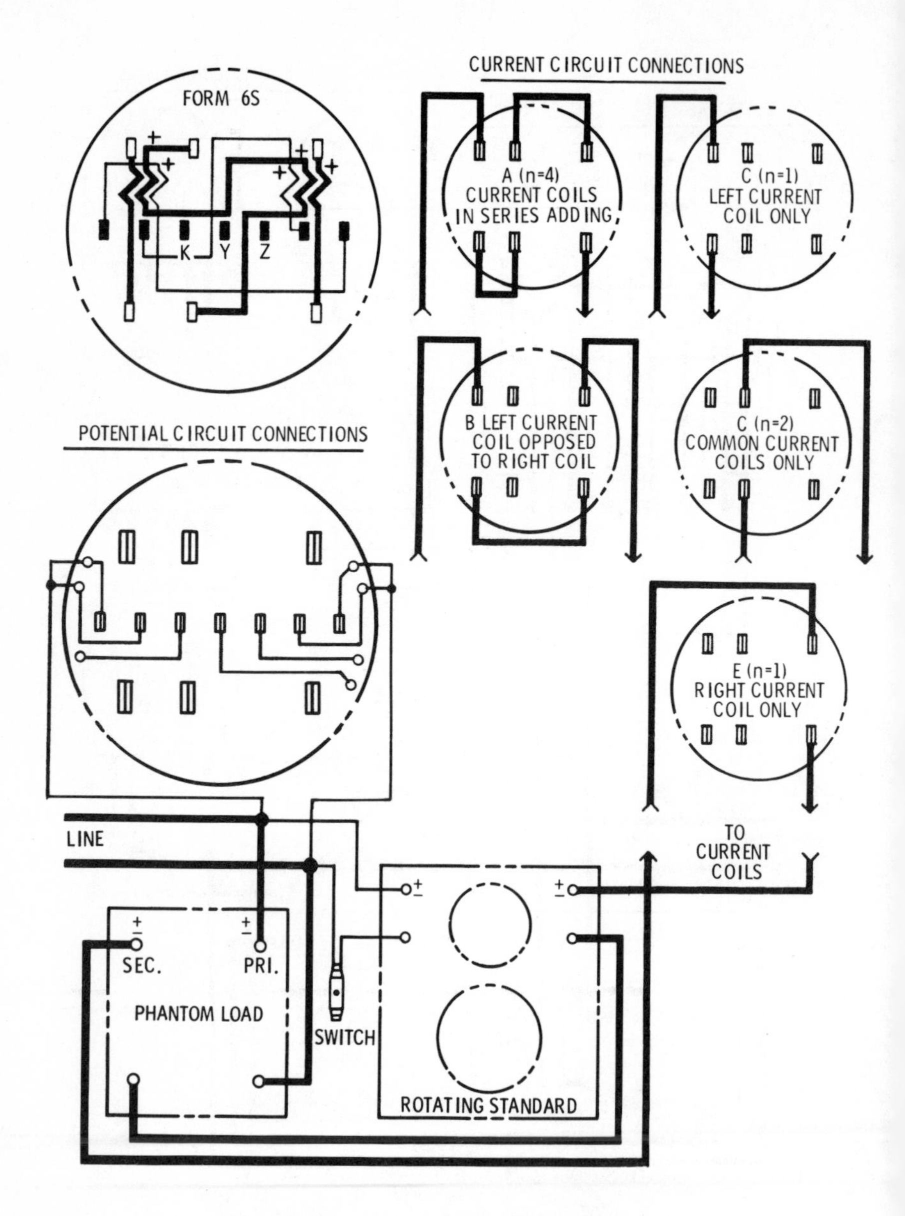

2 STATOR, 3φ, 4 W, Y, TRANSFORMER RATED, FORM 6S

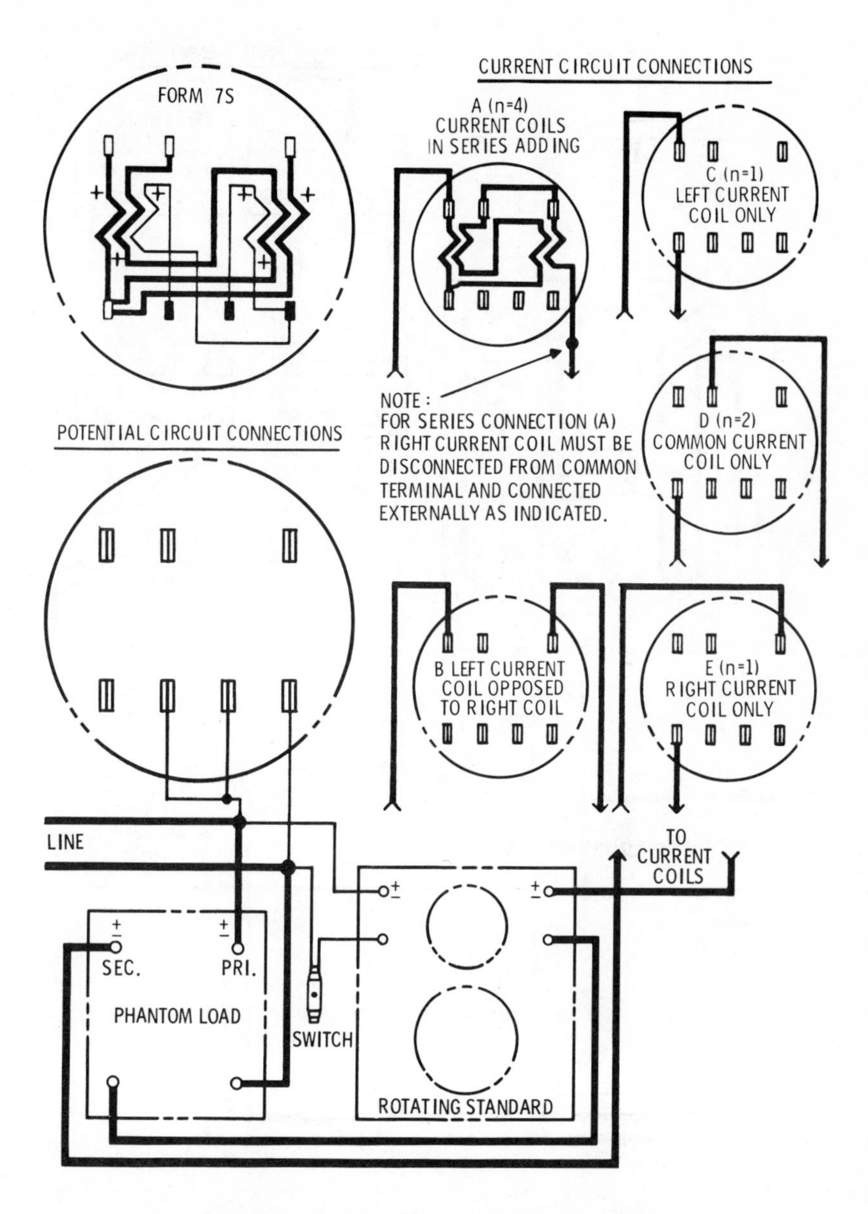

2 STATOR, 3 ϕ, 4 W, Y, TRANSFORMER RATED, FORM 7S

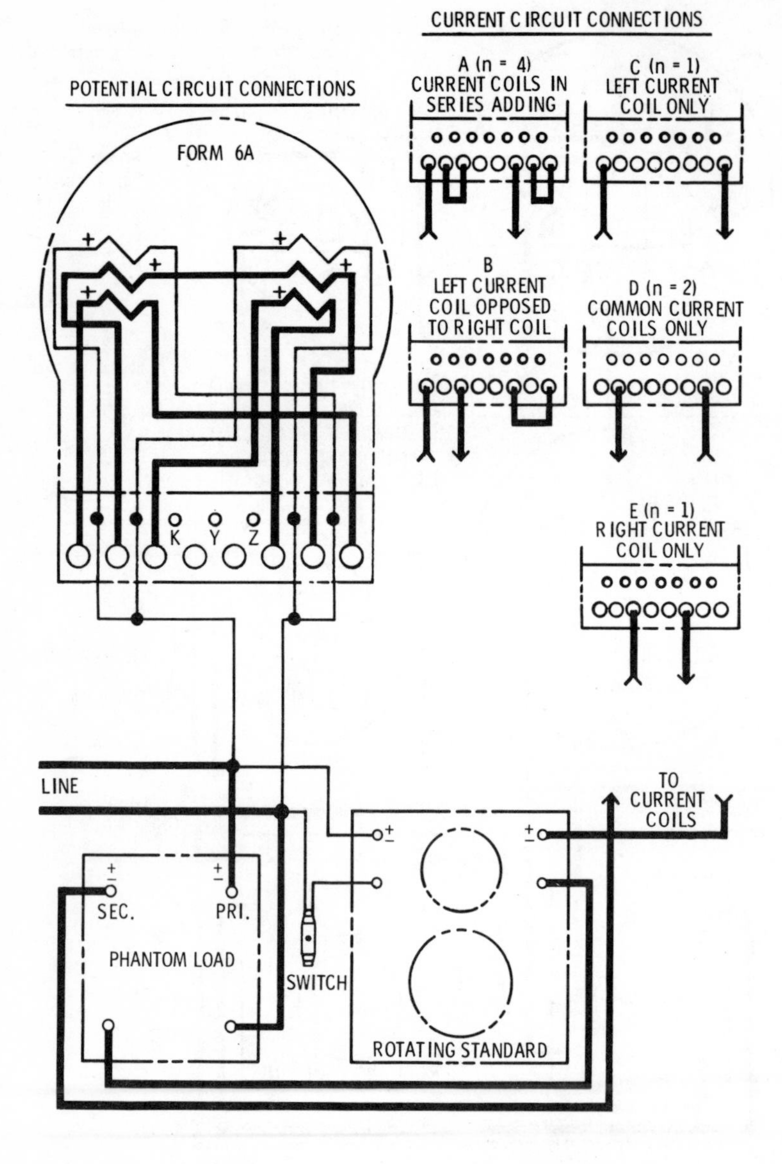

2 STATOR, 3ϕ, 4 W, Y, TRANSFORMER RATED, FORM 6A

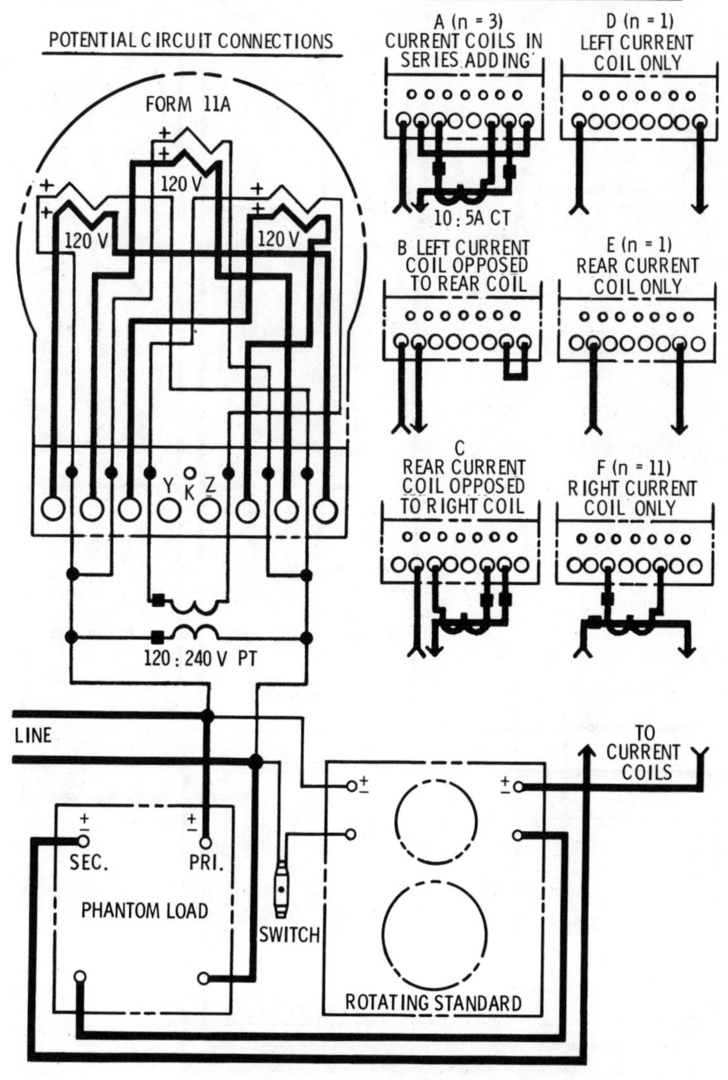

3 STATOR, 3ϕ, 4 W, Δ, TRANSFORMER RATED, FORM 11A

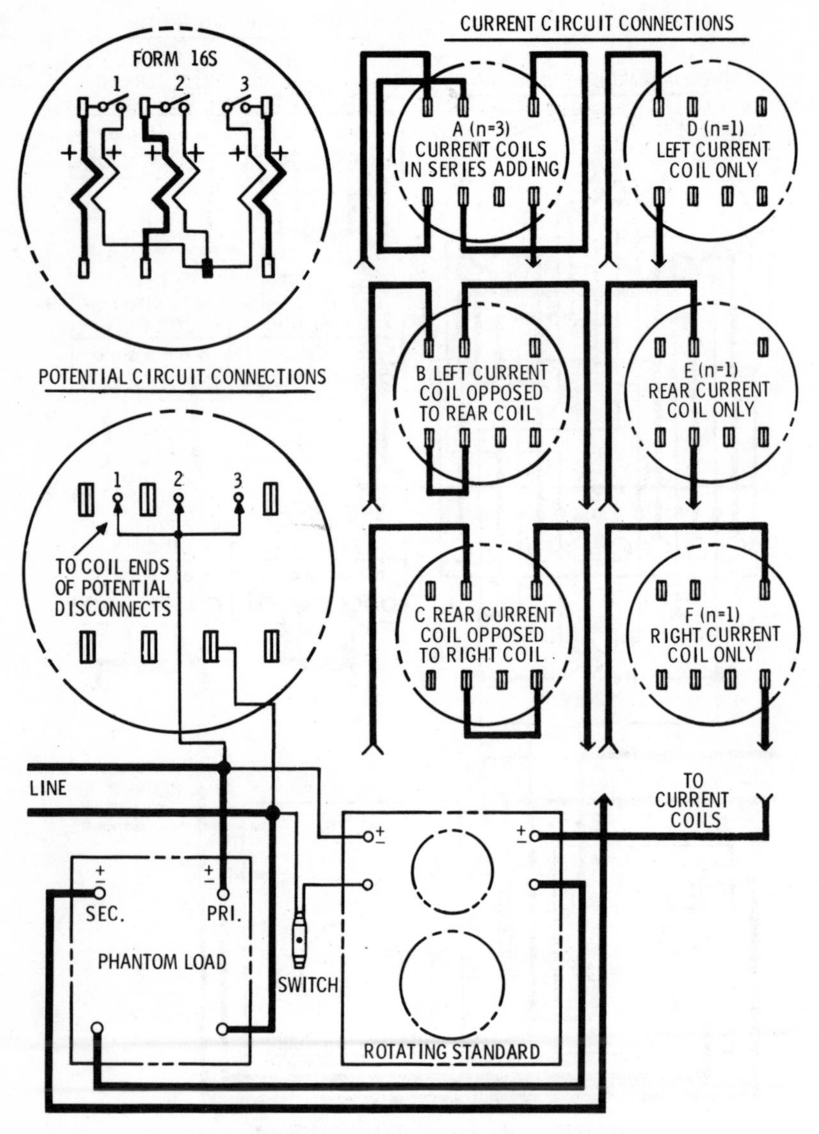

3 STATOR, 3ϕ, 4 W, Y, SELF-CONTAINED, FORM 16S

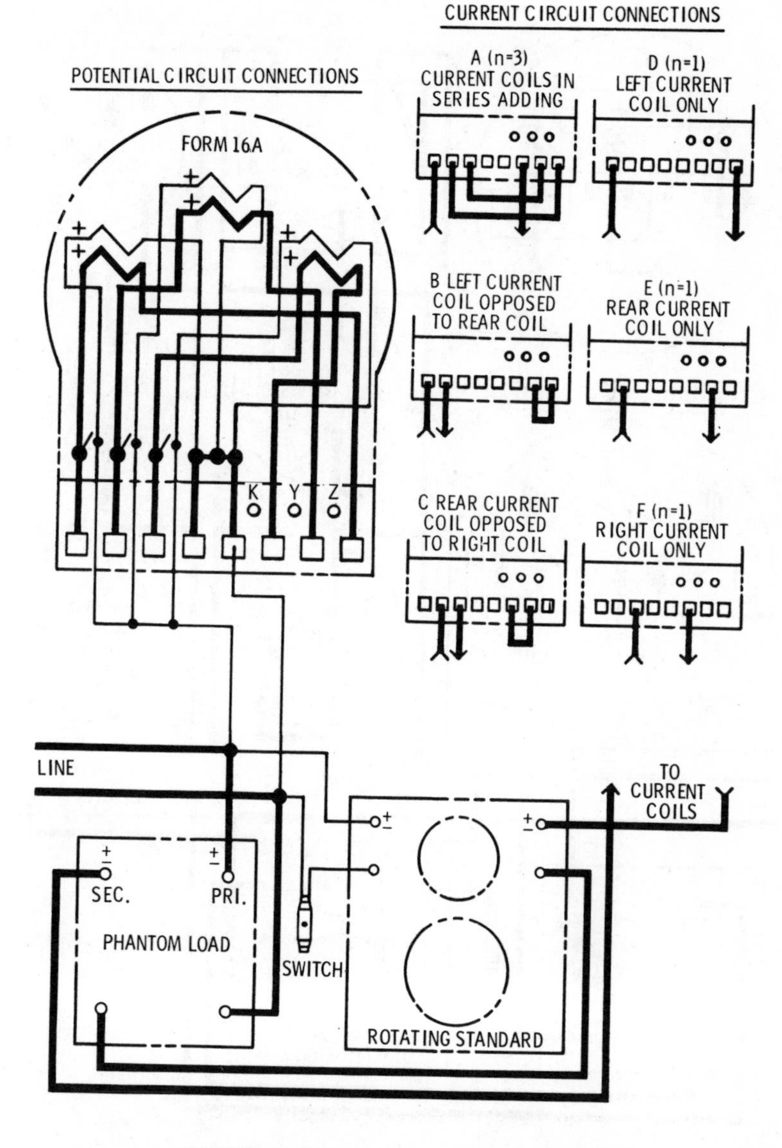

3 STATOR, 3ϕ, 4 W, Y, SELF-CONTAINED, FORM 16A

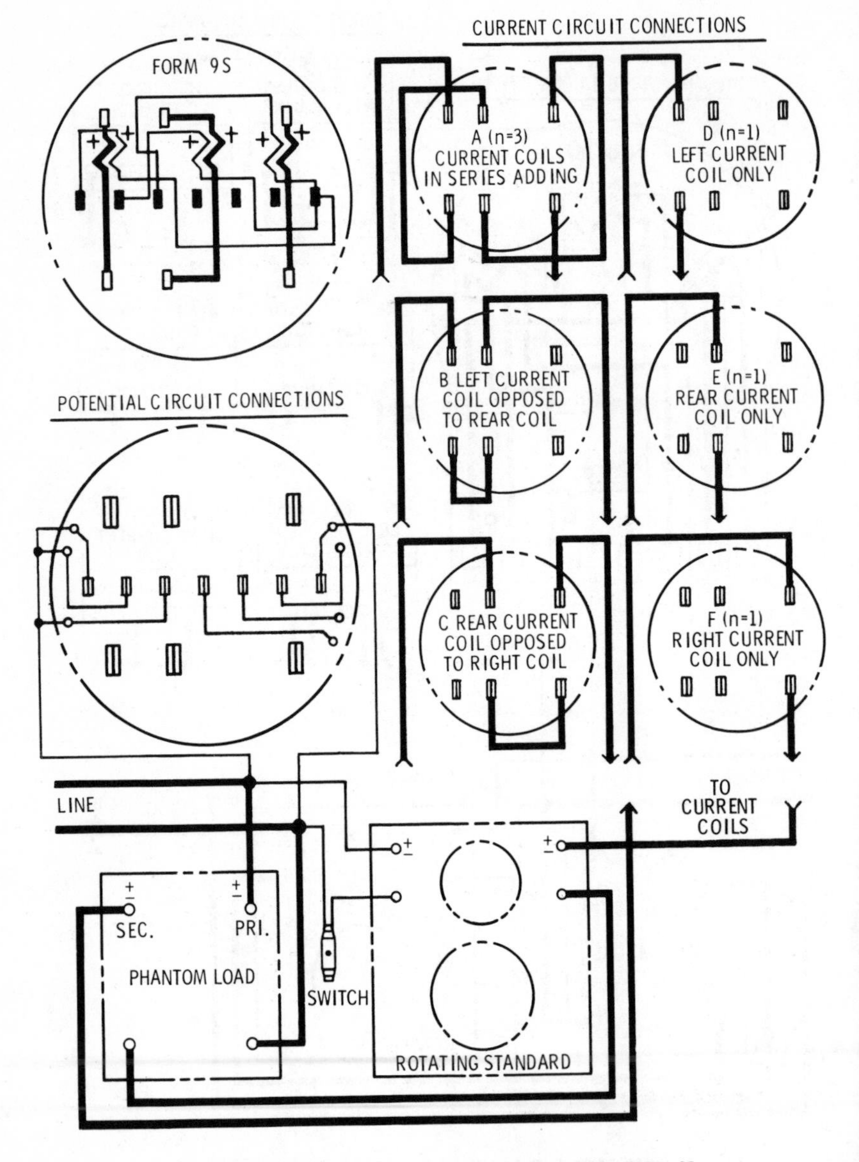

3 STATOR, 3 ϕ, 4 W, Y, TRANSFORMER RATED, FORM 9S

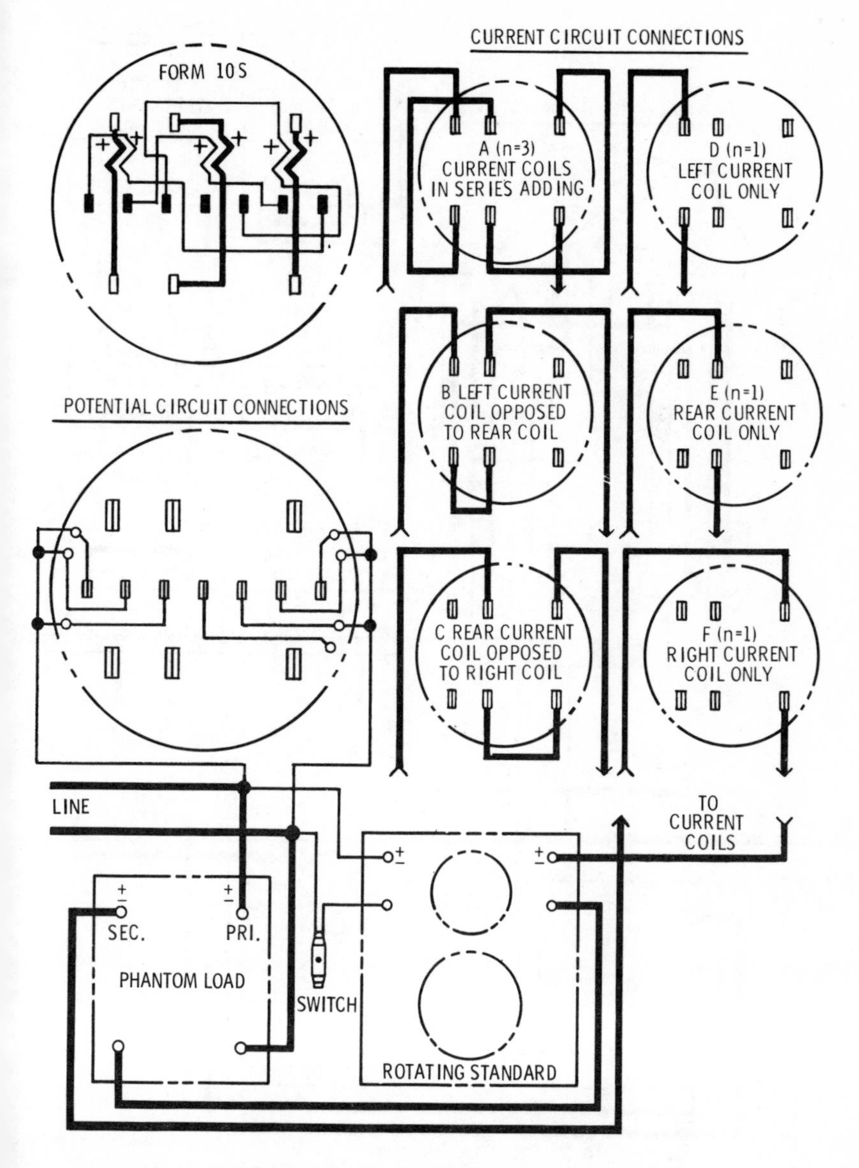

3 STATOR, 3ϕ, 4 W, Y, TRANSFORMER RATED, FORM 10S

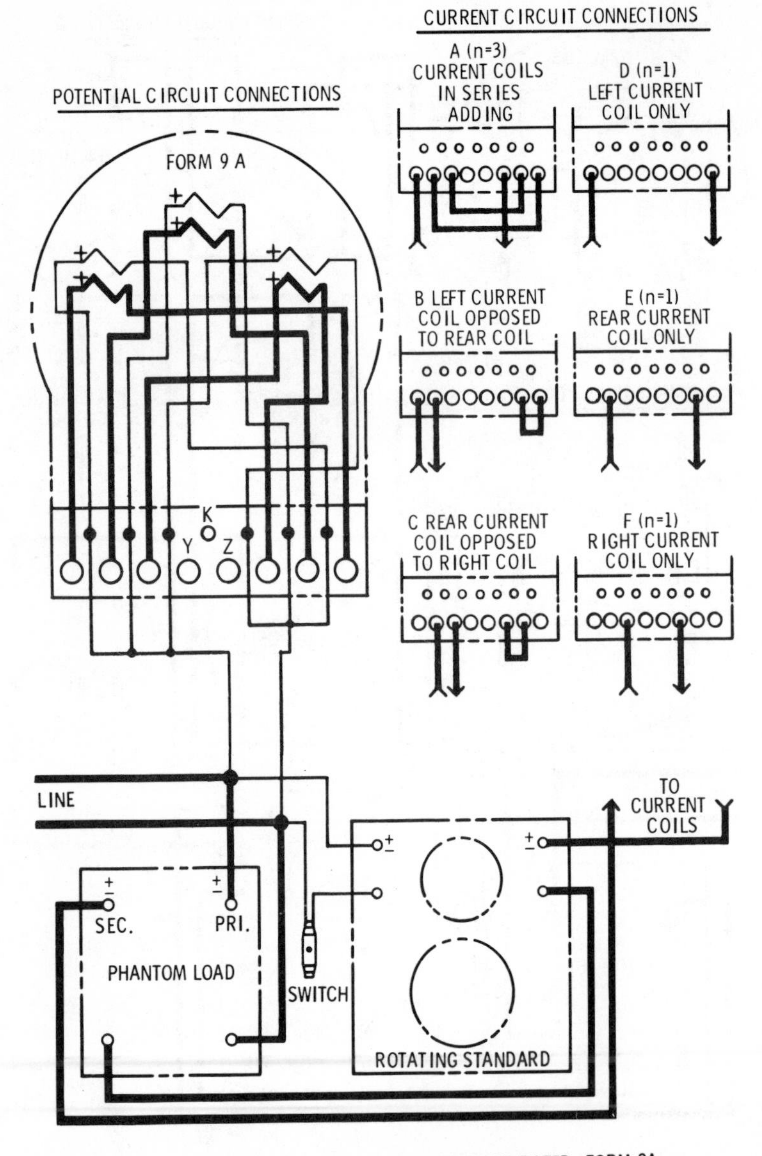

3 STATOR, 3ϕ, 4 W, Y, TRANSFORMER RATED, FORM 9A

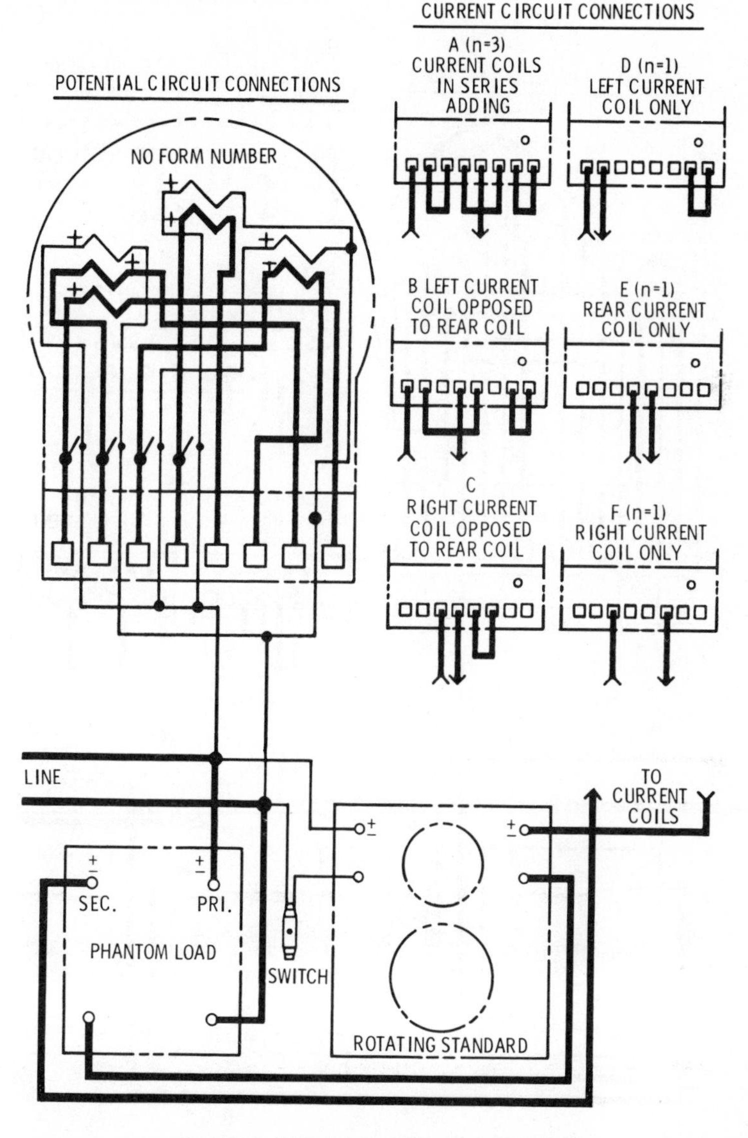

3 STATOR, TOTALIZING METER, SELF-CONTAINED

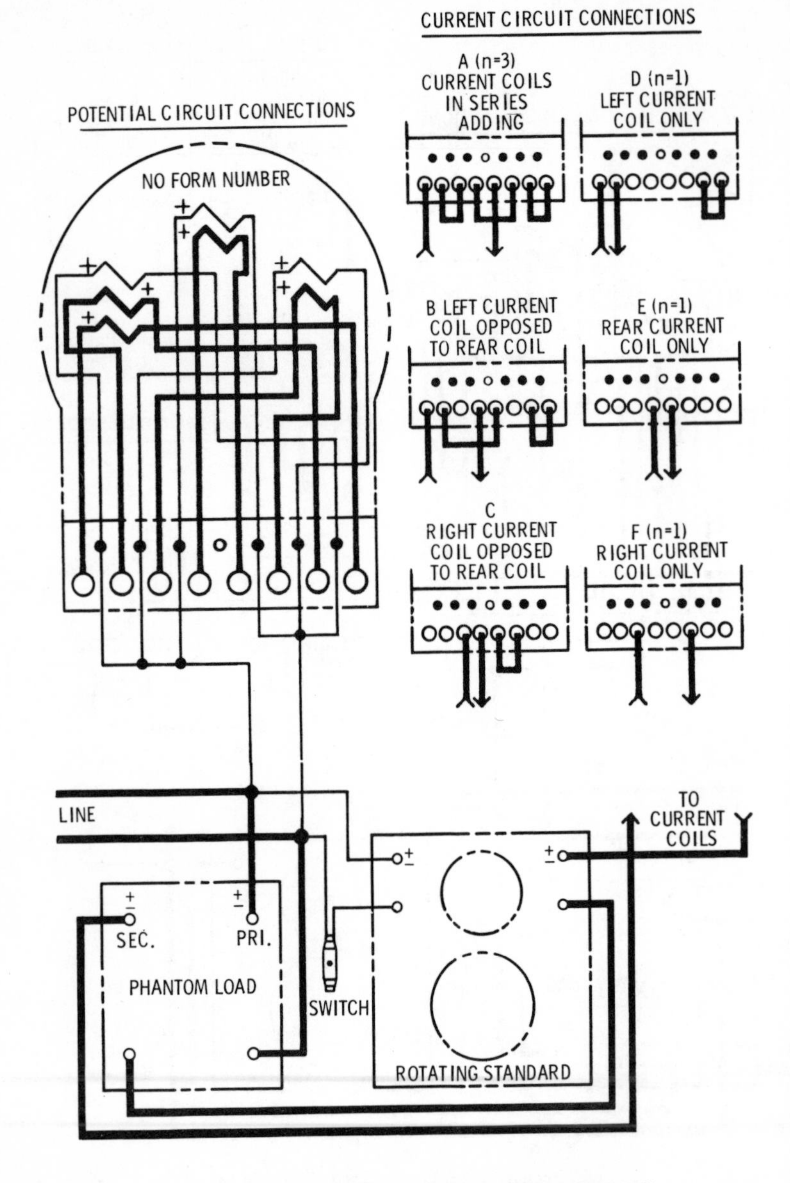

3 STATOR, TOTALIZING METER, TRANSFORMER RATED

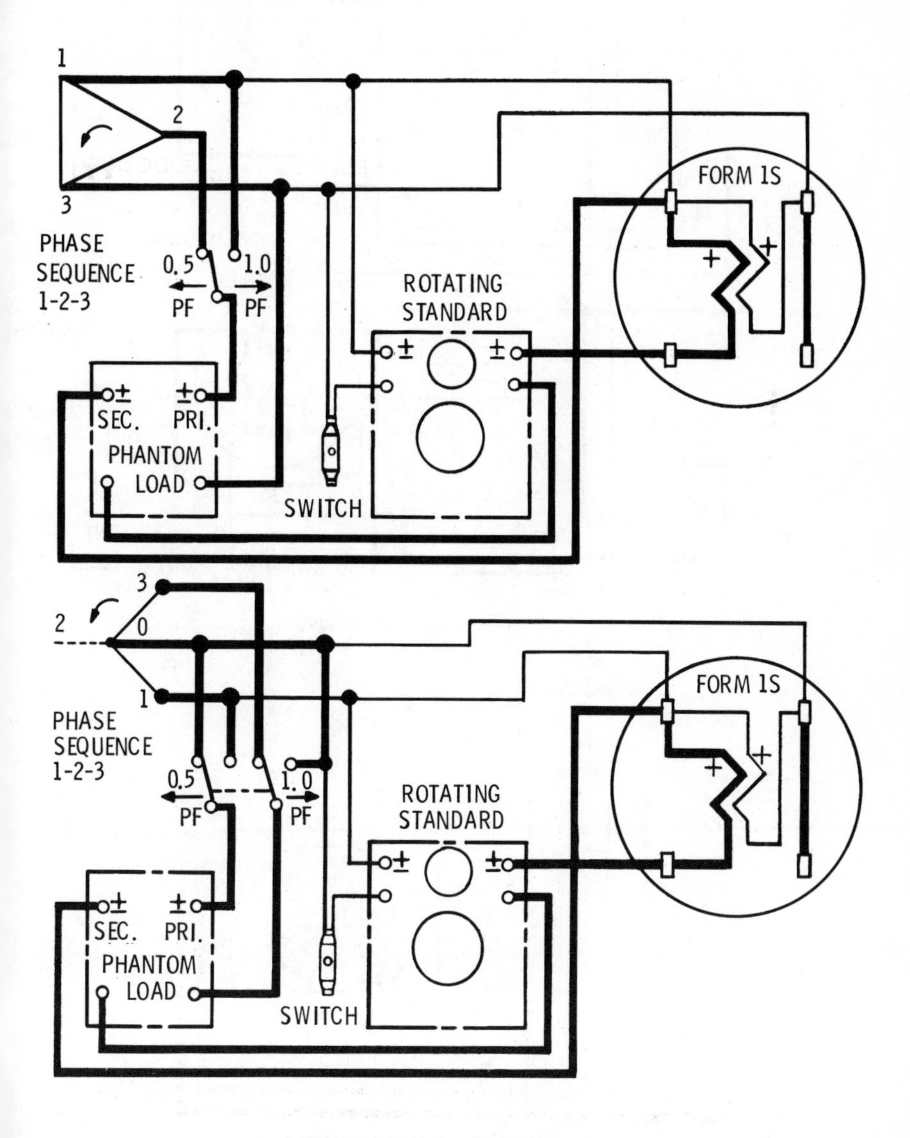

TEST AT 0.5 P.F., LAGGING

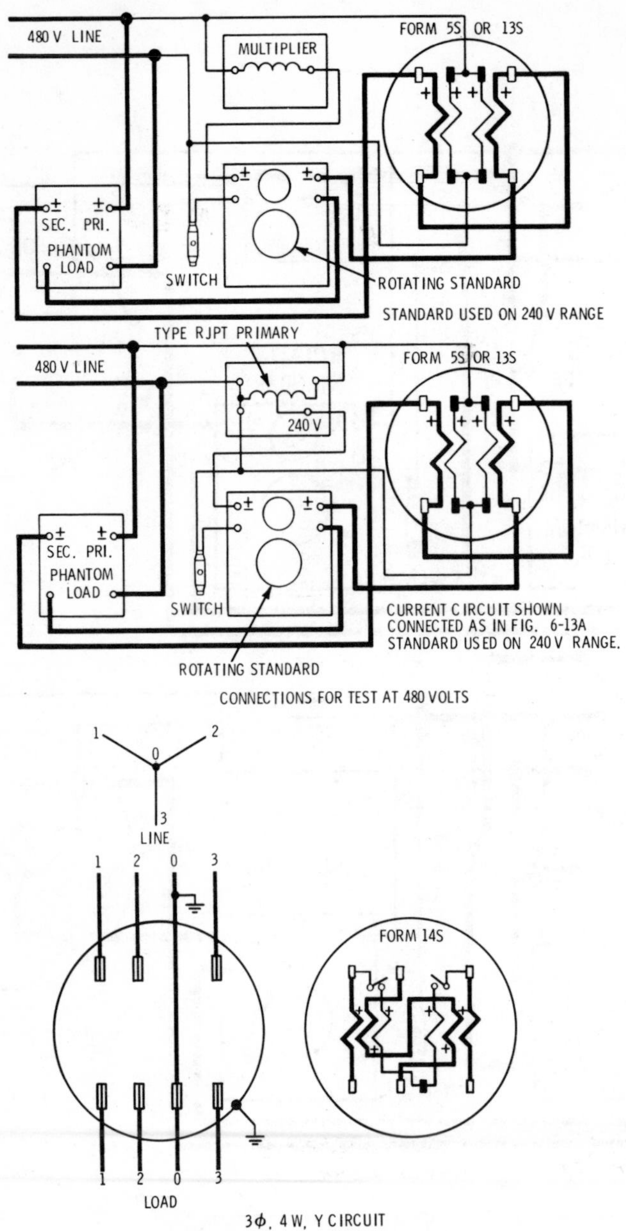

CONNECTIONS FOR TEST AT 480 VOLTS

3ϕ, 4 W, Y CIRCUIT

2 STATOR, 3 ϕ, 4 W, Y, TYPE P2SY, P20SY OR S5S METER, SELF-CONTAINED

RELAYS AND INSTRUMENT CONNECTIONS

OVERLOAD RELAY

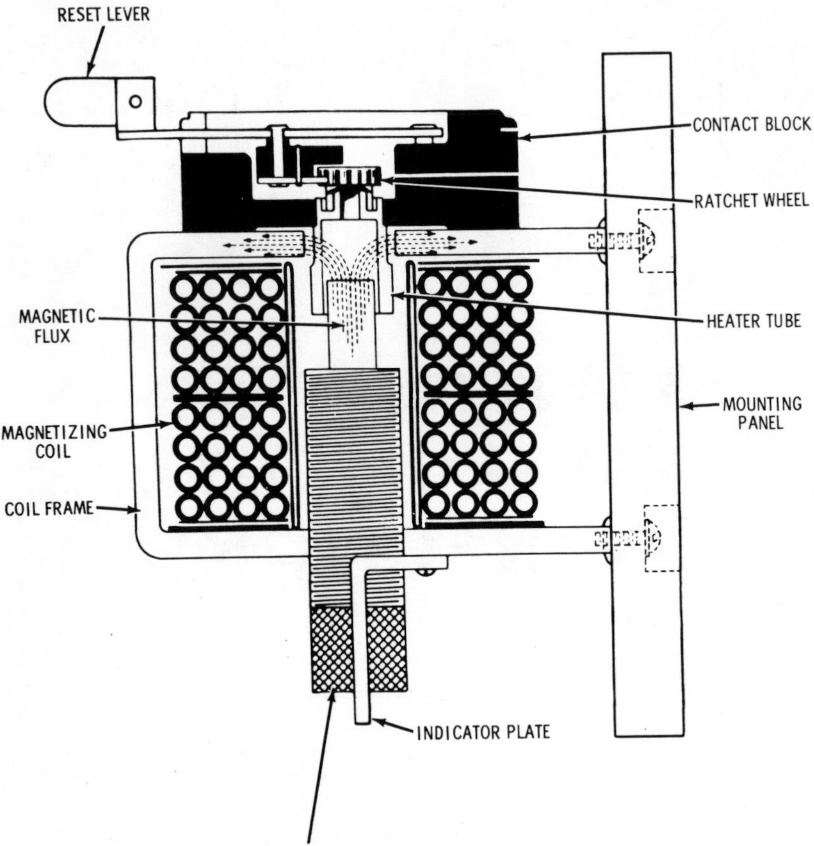

Operation—The load current flowing through the magnetizing coil sets up a magnetic flux in the core. A current directly proportional to the load current is induced in the heater tube. Under overload conditions this current generates enough heat to melt the solder holding the ratchet wheel in place. When the ratchet wheel is released, a spring trips the relay and disconnects the load from the line. The relay may be reset by pushing the reset lever after allowing a short period of time for the solder to harden.

The tripping current is set by turning the threaded core to raise or lower the position of the core within the coil. A current-indicator plate indicates the correct current adjustment.

METHODS OF OVERLOAD PROTECTION WITH INDUCTION-TYPE OVERLOAD RELAYS

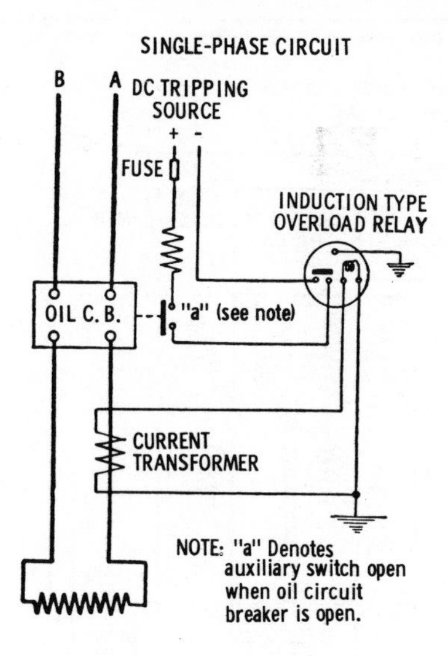

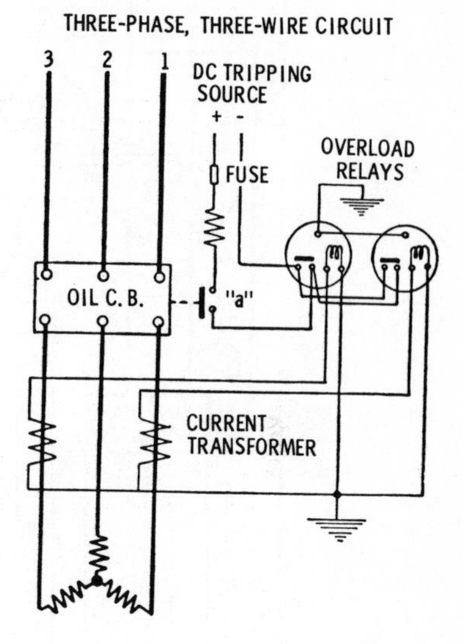

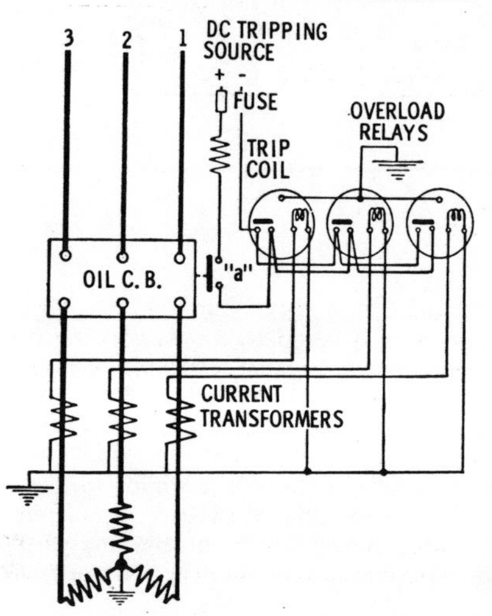

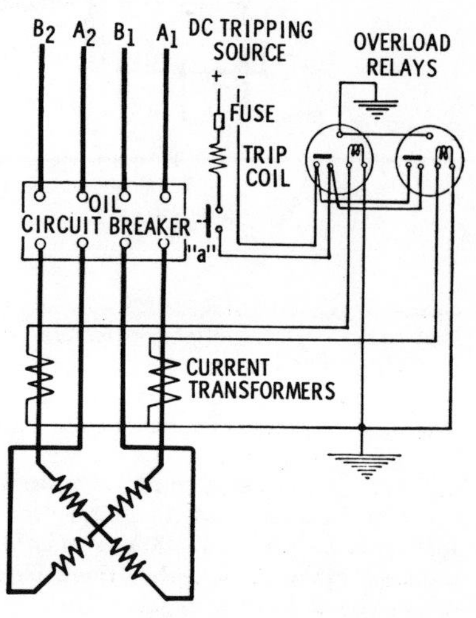

DIAGRAM OF CONNECTION FOR DIFFERENTIAL PROTECTION OF POWER TRANSFORMERS

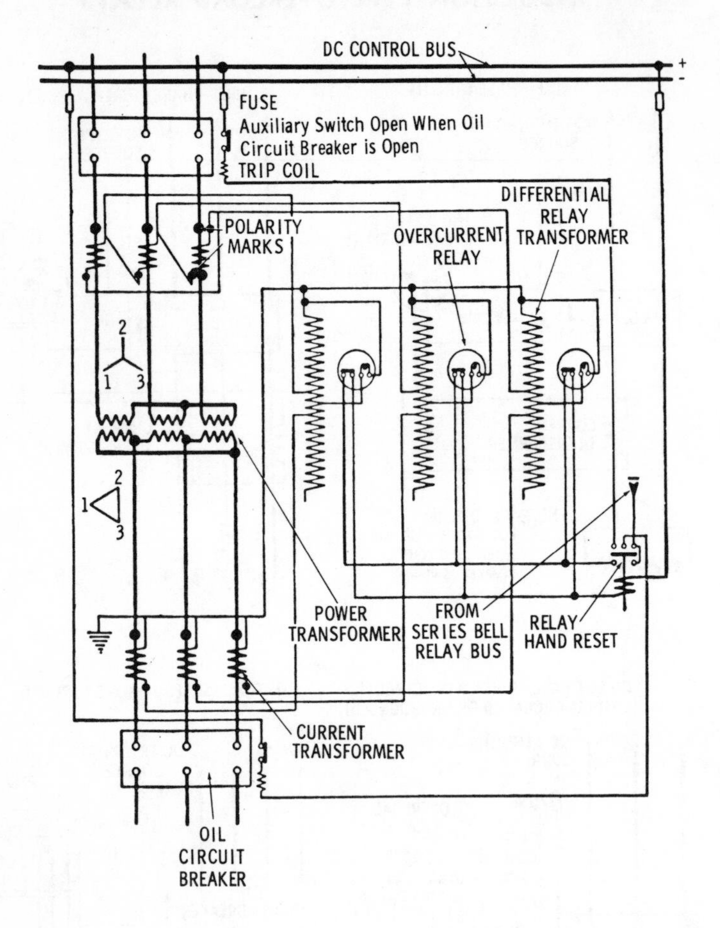

Operation—Differential protective equipment is used with power transformers most frequently when two or more are operated in parallel. Thus, when this system of protection is utilized both automatic and simultaneous tripping of the high- and low-voltage breaker is obtained in case of internal breakdown in the transformers.

It is important that current transformers be selected of proper ratios to give equal secondary currents on the high- and low-voltage side. However, most frequently the ratio of transformation is such that this is difficult to obtain, in which case taps are restorted to, which may be changed from time to time. (For operation of relays see following page.)

CONNECTIONS FOR DIFFERENTIAL PROTECTION OF TRANSFORMER WITH A TERTIARY WINDING USING INDUCTION-TYPE OVERCURRENT RELAYS

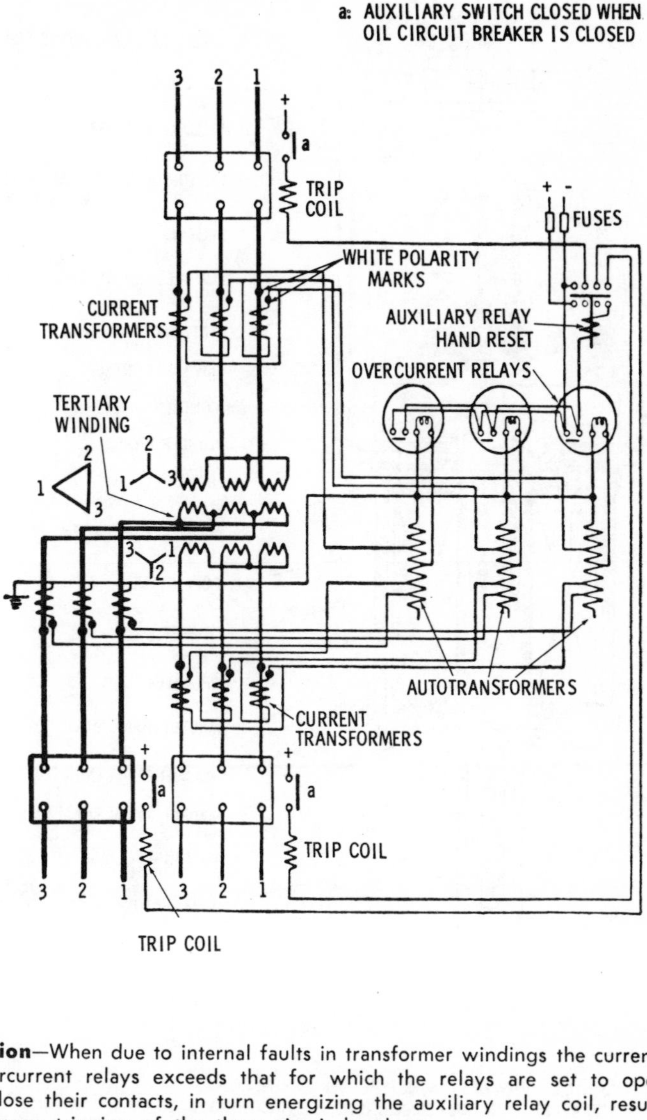

Operation—When due to internal faults in transformer windings the current through the overcurrent relays exceeds that for which the relays are set to operate, the relays close their contacts, in turn energizing the auxiliary relay coil, resulting in a simultaneous tripping of the three circuit breakers.

OVERCURRENT RELAY

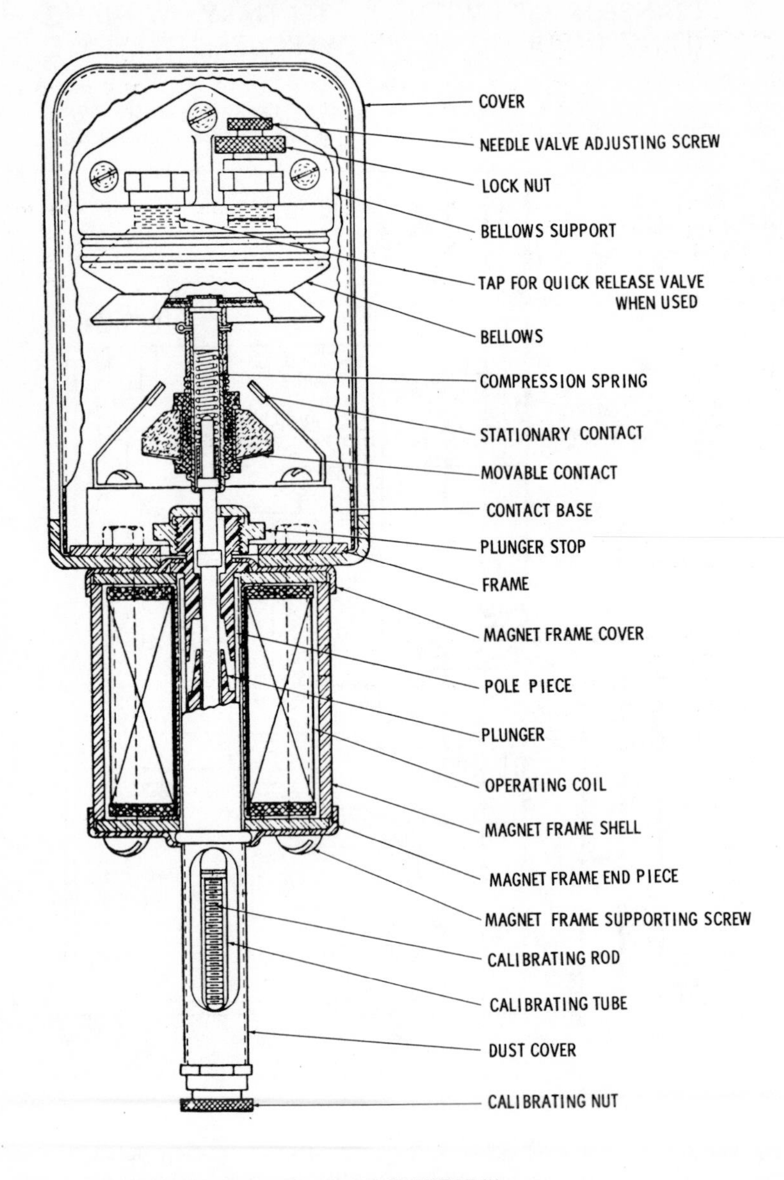

SECTIONAL VIEW OF PLUNGER TYPE OVERCURRENT RELAY
FOR APPLICATION AND CONNECTIONS, SEE FOLLOWING PAGES

PLUNGER-TYPE OVERLOAD RELAY

Operation Principles—When due to certain conditions in the circuit to be protected, the current exceeds the value at which the relay is set to operate, the plunger raises and carries up with it the movable cone contact, or it strikes against the center of the toggle mechanisms (depending upon the type of contacts in the relay) thus causing the contacts to function.

Generally, when a relay functions to open its contacts it is referred to as a **circuit-opening** type, and when it functions to close its contacts, it is referred to as the **circuit-closing** type. In this manner the function of the contacts of a relay is most frequently used as a means of identification, a relay being **circuit-closing** or **circuit-opening,** or **circuit-opening** and **circuit-closing.**

Timing Features—In regard to speed of operation a relay may be referred to as instantaneous, or time delay. The word instantaneous is a general qualifying term applied to any relay, indicating that no delayed action has been purposely introduced.

The time relays are similar in construction to the instantaneous type, except for the addition of an air bellows which limits the rate of travel of the relay plunger, and in this way introduces an interval of time to the opening or closing of the relay contacts.

This time delay may be regulated to suit the special service desired, which is accomplished by means of a needle valve located in the head of the bellows as shown on page 198. This valve controls the rate of air flow from the bellows under various operating conditions.

METHODS OF OVERLOAD PROTECTION WITH PLUNGER-TYPE, CIRCUIT-CLOSING RELAYS

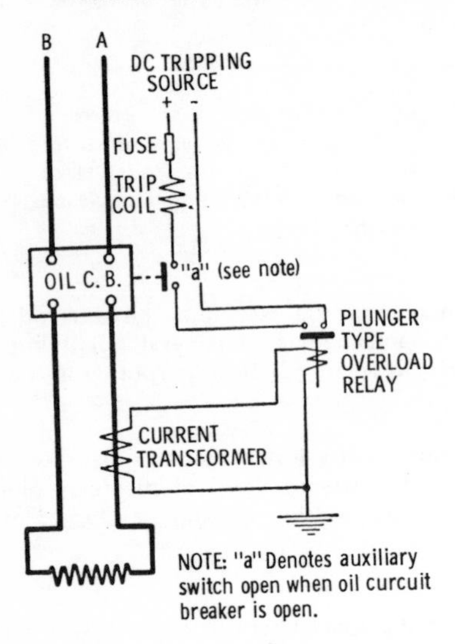

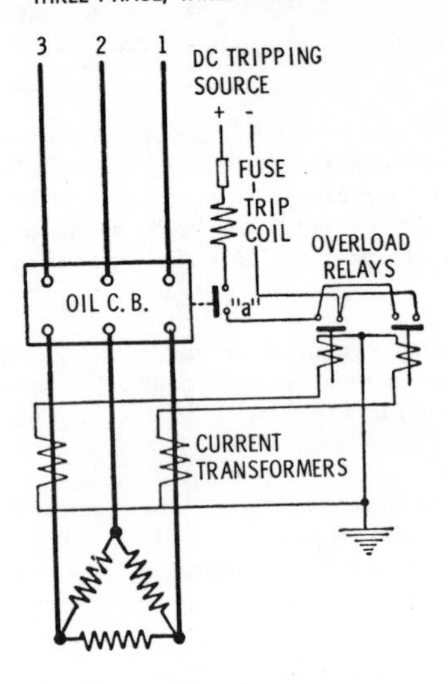

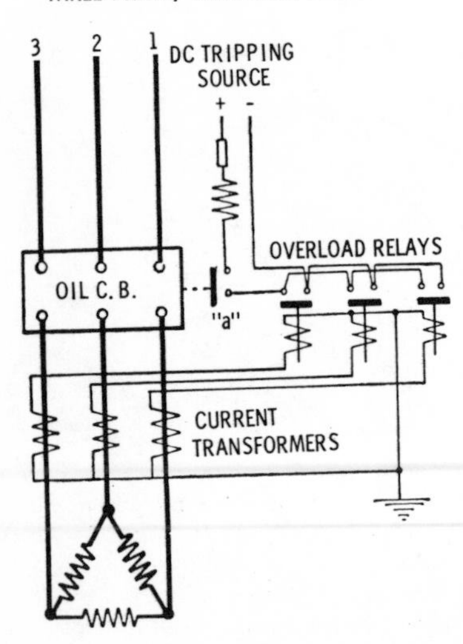

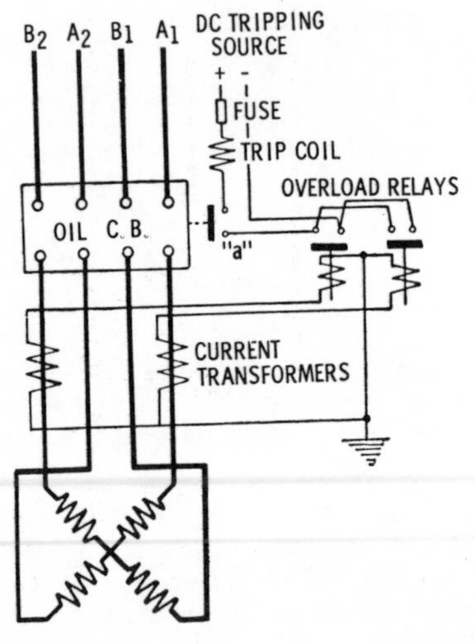

OVERLOAD PROTECTION WITH PLUNGER-TYPE, CIRCUIT-OPENING RELAYS

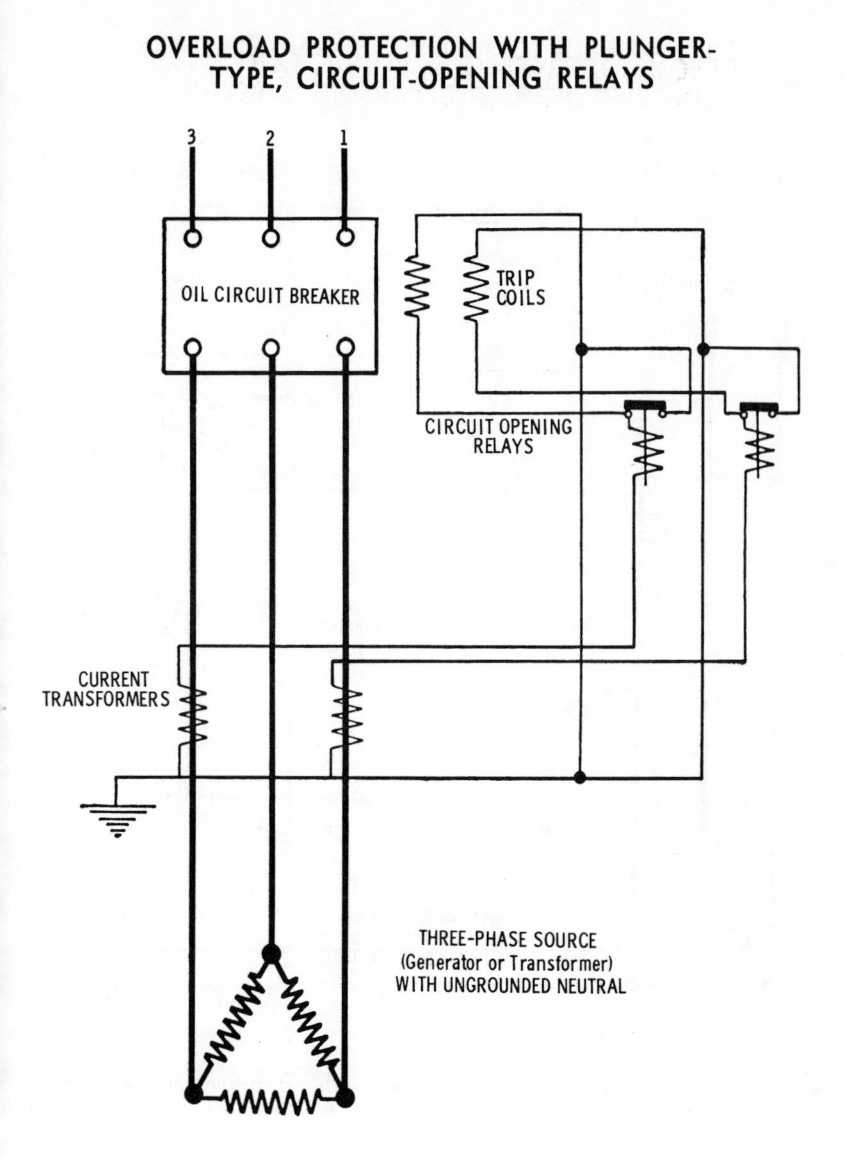

Operation—In this circuit overload protection is accomplished by means of a set of current transformers, with its associated relays and trip coils. The relay contacts are normally closed. When the overload through the trip coils exceeds that for which the relays are set to operate, the contacts open, placing the trip coils in series with the relay coils, causing the trip coils to trip the oil circuit breaker.

OVERLOAD PROTECTION WITH PLUNGER-TYPE, CIRCUIT-CLOSING RELAYS

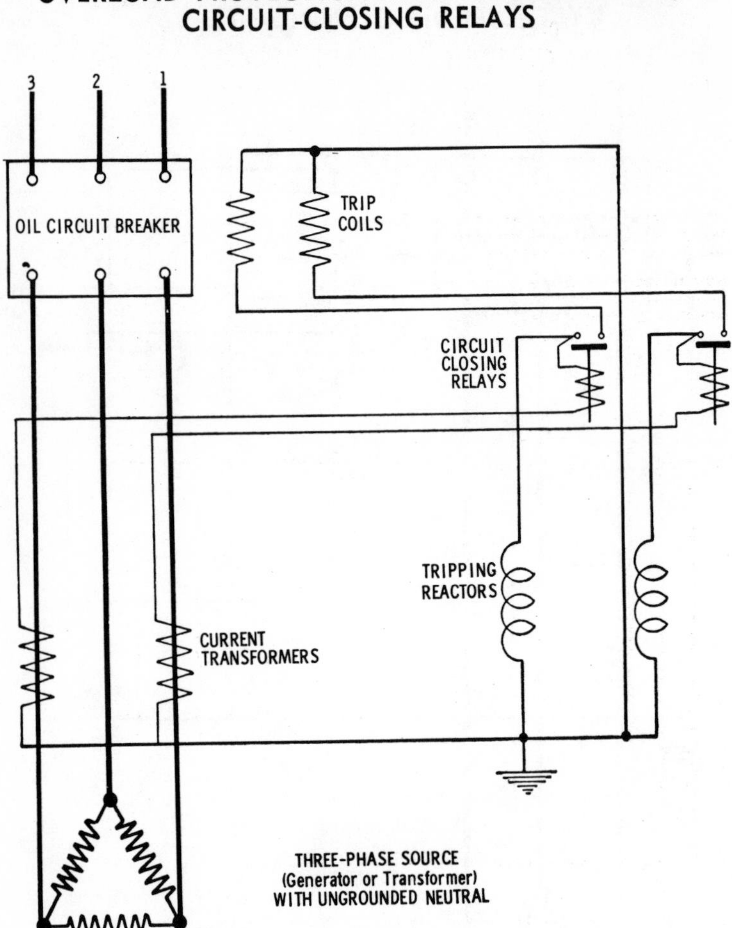

Operation—When tripping reactors are used as in overcurrent and other types of relays, instrument and meters should be connected from an extra set of current transformers.

Tripping reactors are frequently employed when a direct current or reliable alternating current is not available as a tripping source for the relays.

Normally the trip coil circuit is open and the reactor forms the closed circuit of the current transformer secondary. When the overload is of a sufficiently high value to cause operation of the relay, it closes the trip coil circuit in shunt with the reactor, causing sufficient current to be passed through the coils to trip the breaker.

TRIPPING OF TWO OIL CIRCUIT BREAKERS USING TRIPPING REACTORS AND CIRCUIT-CLOSING RELAYS

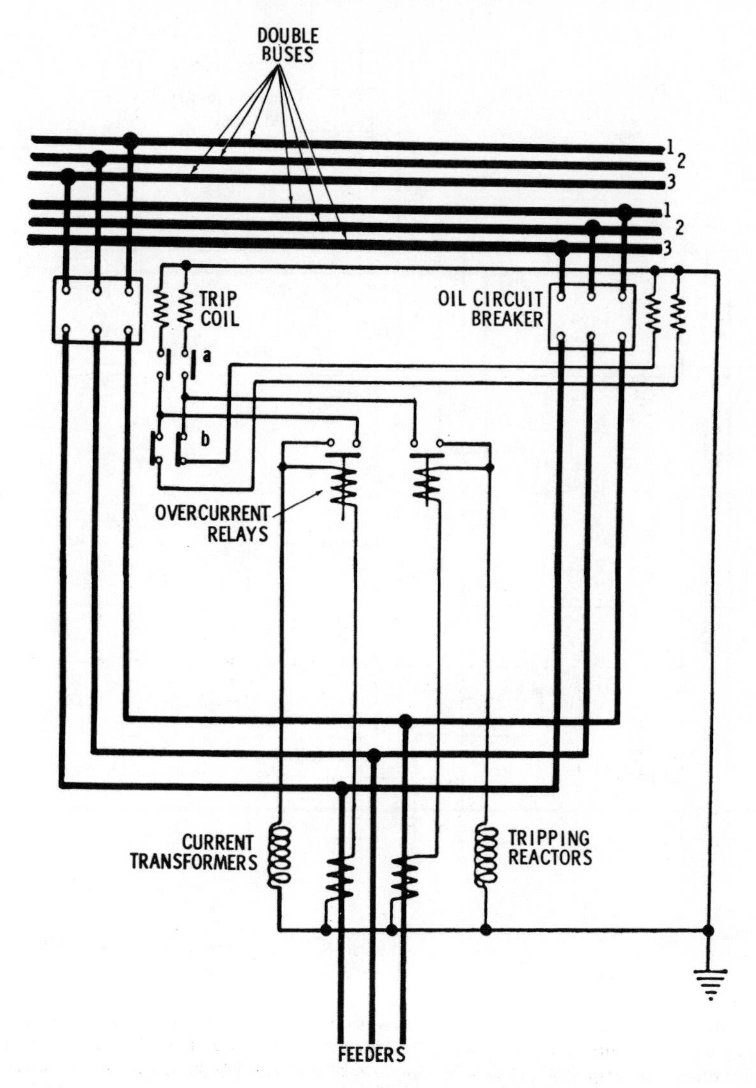

Note—Auxiliary switch "a" is open when oil circuit breaker is open, and auxiliary switch "b" is closed when the oil circuit breaker is open.

APPLICATION OF LOCKING RELAYS TO FEEDER CIRCUITS

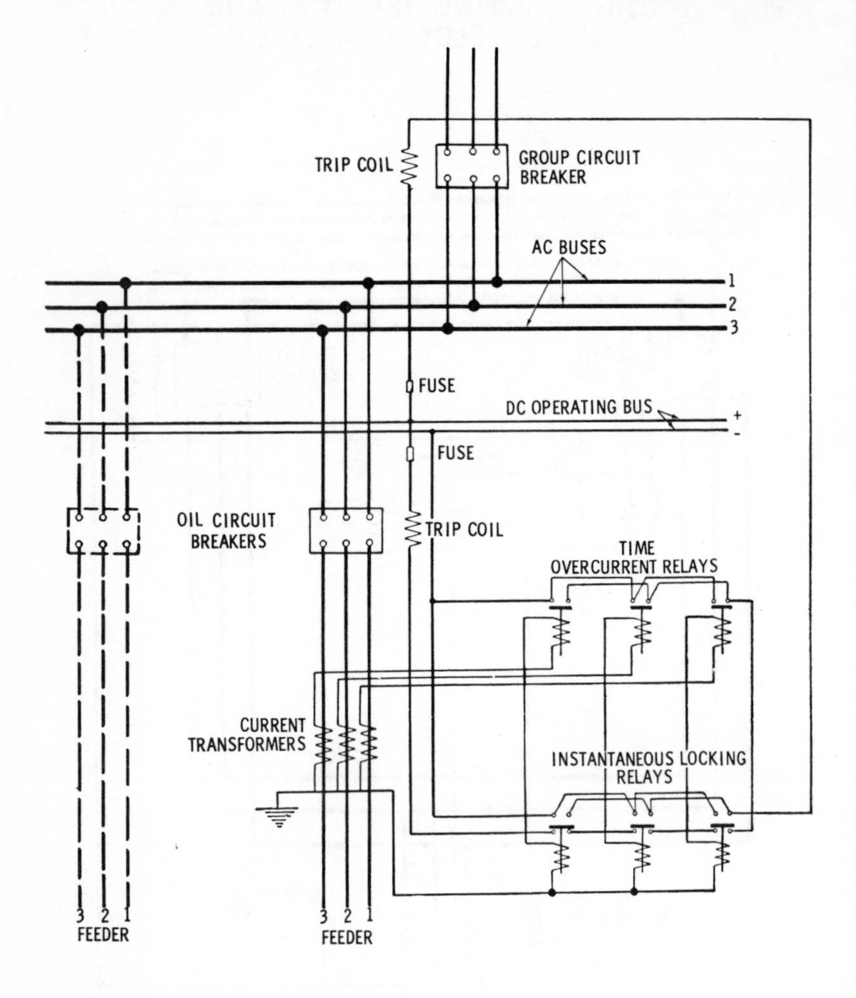

Operation—In this system each feeder is equipped with a complement of time overload relays adjusted to trip the feeder breaker on simple overcurrent, and a set of instantaneous locking relays with high current coil setting, adjusted not to function as long as the primary current does not exceed the capacity of the feeder breaker, but to function instantaneously in case the current exceeds this value.

The operation of the locking relays opens the tripping circuit of the feeder of the heavy duty group circuit breaker.

APPLICATION OF LOCKING RELAYS TO FEEDER CIRCUITS

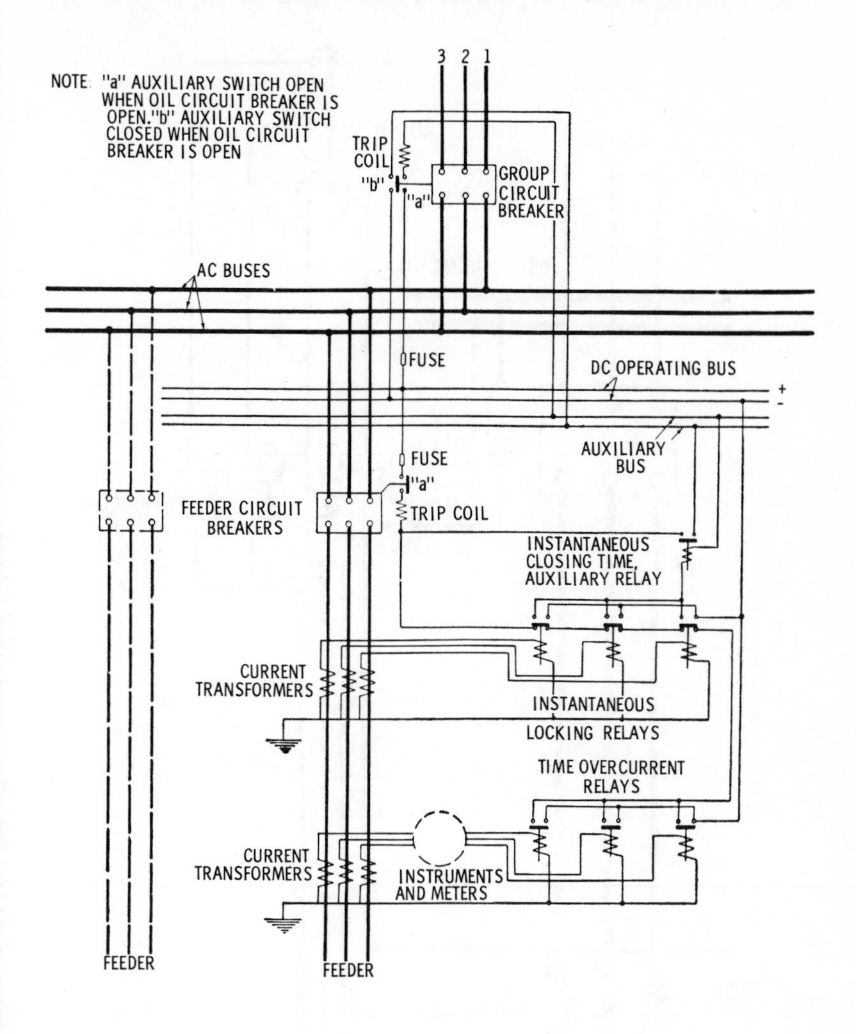

Operation—In this, as in the system shown on the previous page, the locking relays operate only upon excessive overcurrent, in which case the locking relays close the feeder breaker and open the group breaker.

An additional relay equipped with a direct current coil arranged to close instantaneously and reset itself (open) in a definite time is used as an auxiliary relay to work in conjunction with a circuit closing auxiliary switch on the group breaker to open the feeder breaker after the group breaker has been opened.

OVERLOAD-RELAY CONNECTION

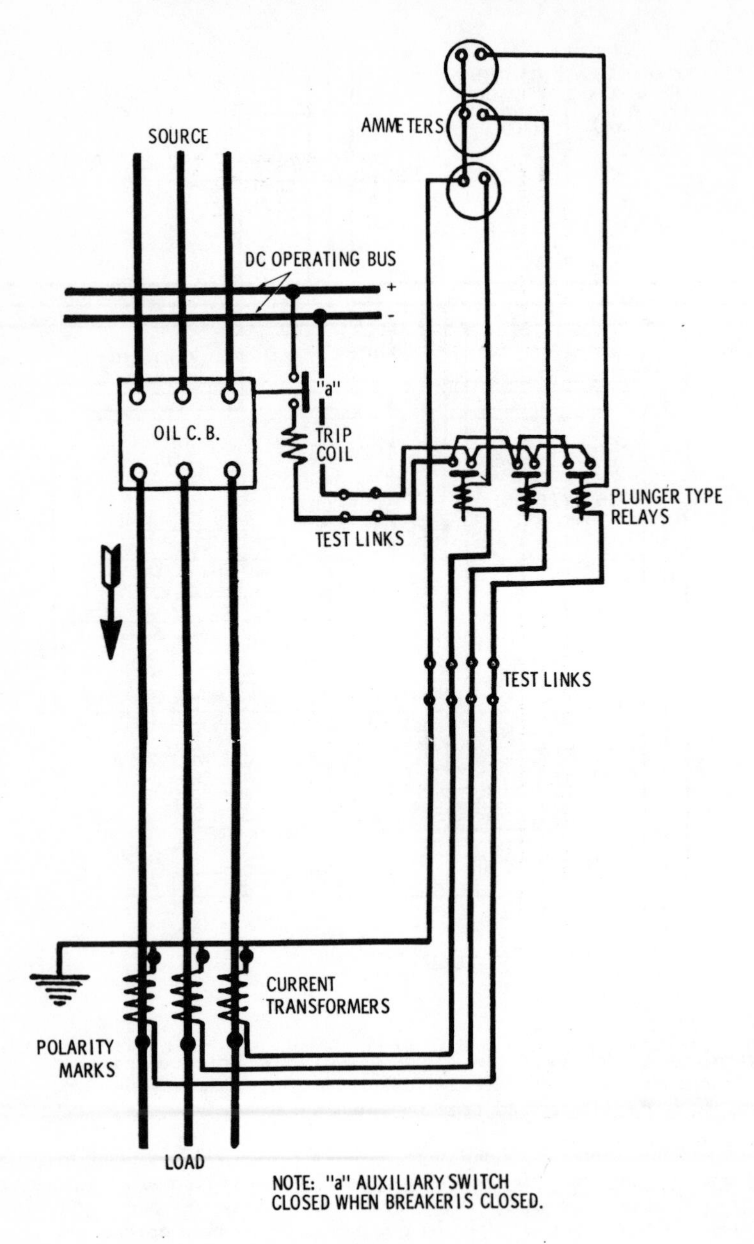

OVERCURRENT PROTECTION

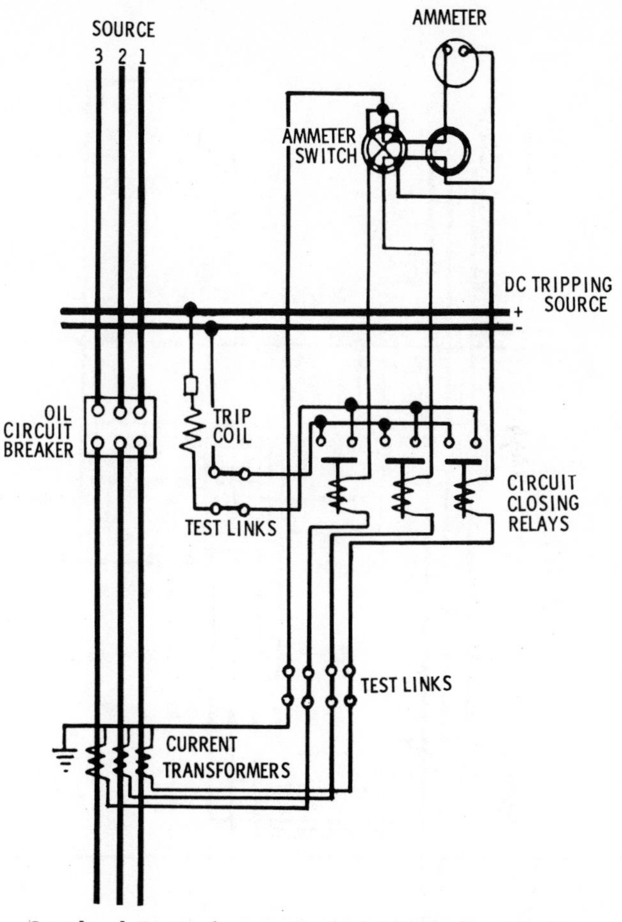

Overload Protection on Typical AC Feeder Circuit.

Operation—When current exceeds the setting of the relays, the relays will close their contacts, energizing the trip coil, which trips the oil circuit breaker.

The test links shown are optional but will, if used, facilitate the testing and calibration of instruments.

The current in each phase is measured by means of an ammeter and a three-way switch.

OVERCURRENT PROTECTION

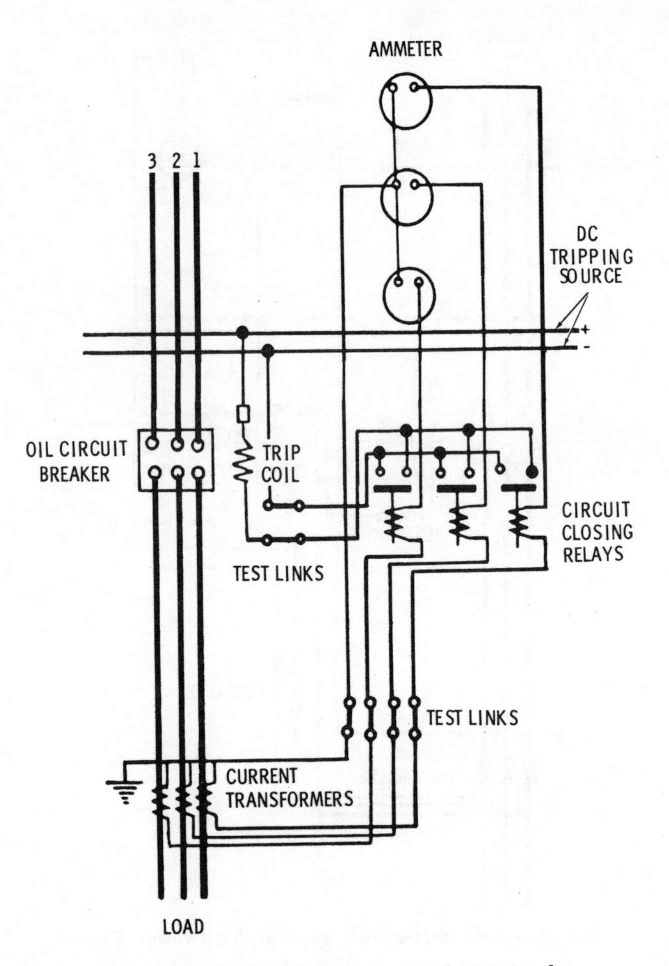

Overload Protection on Typical AC Feeder.

Operation—When current exceeds the setting of the relays, the relays will close their contacts, energizing the trip coil, which trips the oil circuit breaker.

The test links are optional but will, if used, facilitate testing or calibration of instruments.

The current in each phase is measured by individual ammeters.

OVERCURRENT PROTECTION

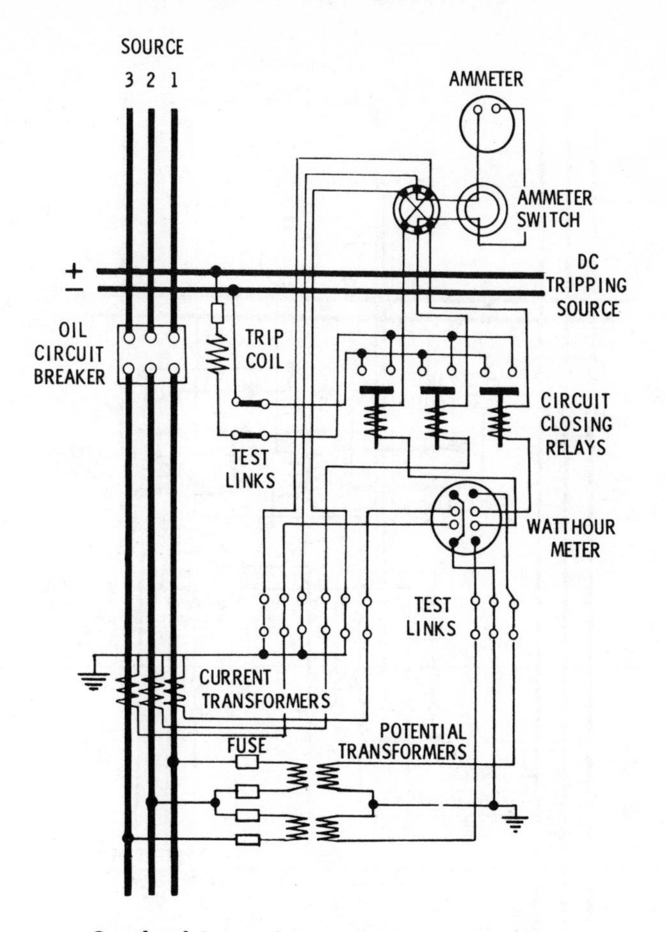

Overload Protection on Typical AC Feeder.

Operation—When current exceeds the setting of the relays, the relays will close their contacts, energizing the trip coil which trips the oil circuit breaker.

The energy is measured by means of a watthour meter; the current in each phase is measured by ammeter and three-way switch.

Test links shown are optional but will, if used, facilitate the testing of relays and instruments.

OVERLOAD PROTECTION

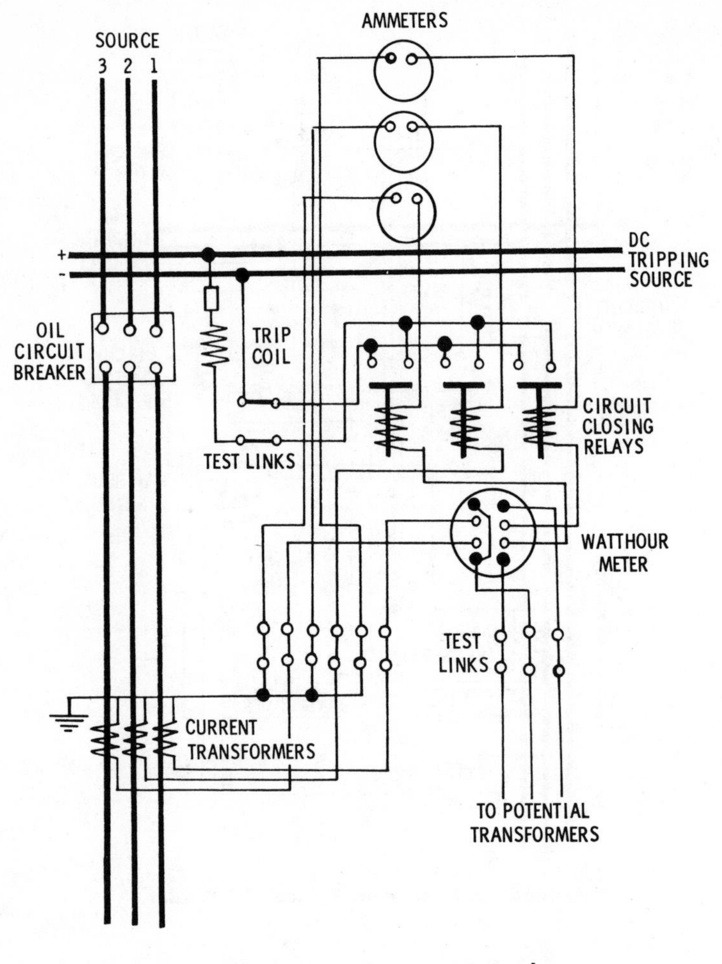

Overload Protection on AC Feeder.

Operation—When the current exceeds the setting of the relays, the relays will close their contacts, energizing the trip coil which trips the breaker.

The energy is measured by means of a watthour meter, and the current in each phase is measured by individual ammeters. Test links are optional but will, if used, facilitate the testing or calibration of relays and instruments.

OVERCURRENT PROTECTION

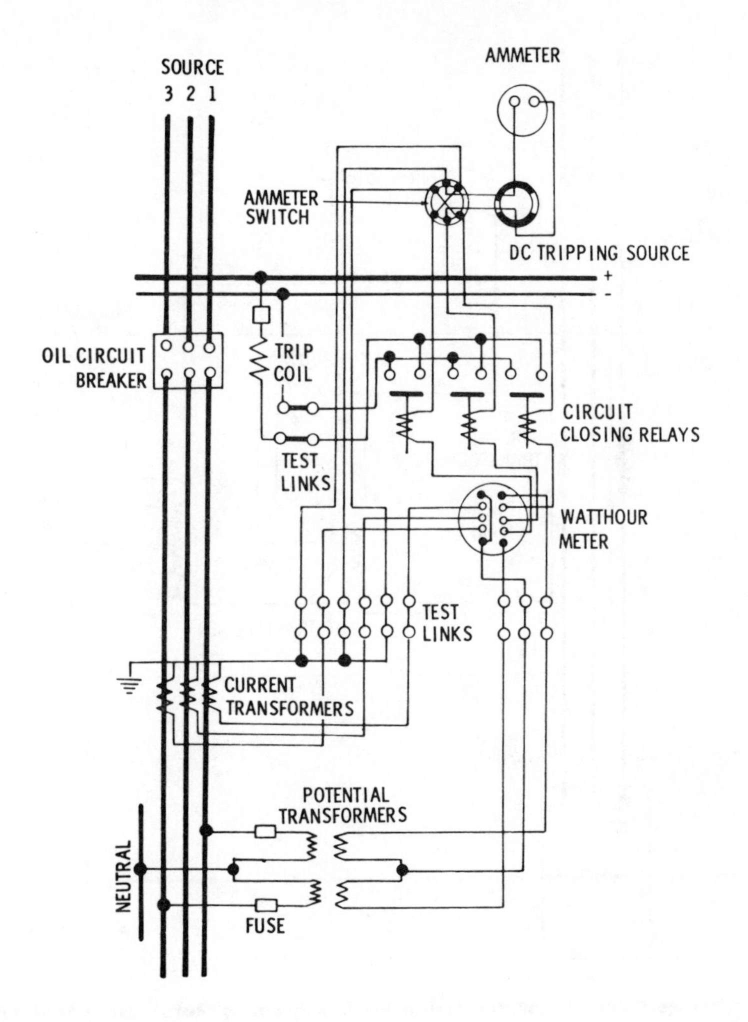

Operation—When current exceeds the setting of the relays, the relays will close their contacts, energizing the trip coil, which trips the oil circuit breaker. The energy is measured by a watthour meter, and the current in each phase is measured by ammeter and three-way switch.

Test links are optional but will, if used, facilitate the testing and calibration of relays and instrument.

TEMPERATURE OVERCURRENT PROTECTION

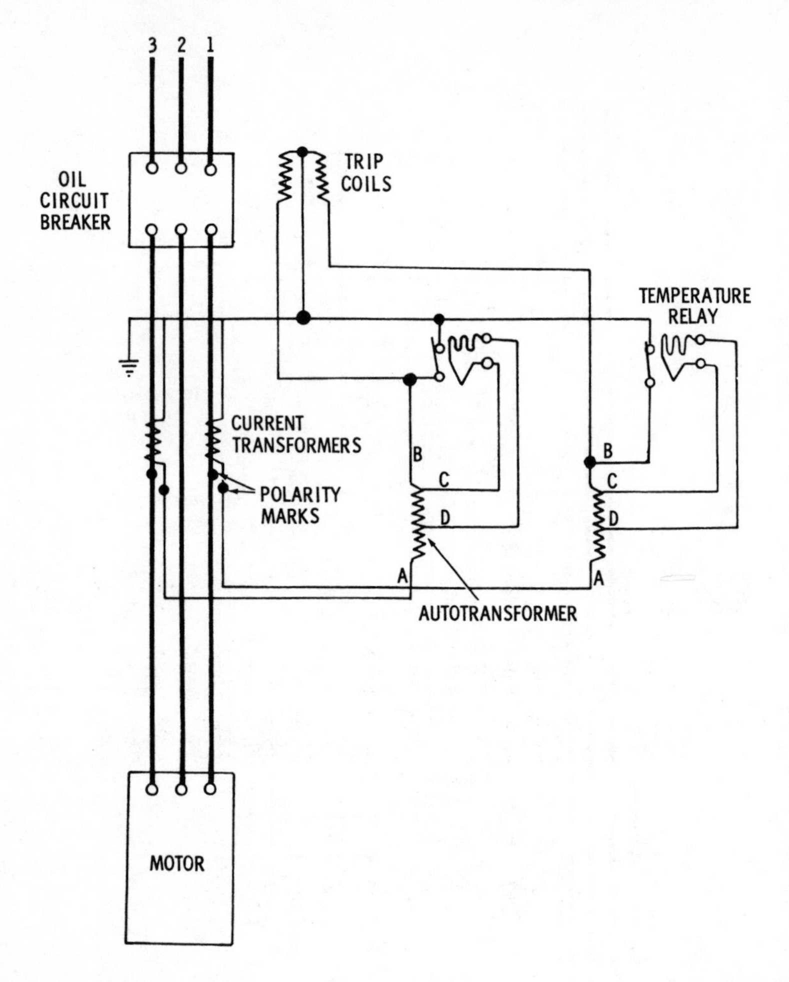

Temperature Overcurrent Protection for Synchronous Motor Using Temperature Relays.

Operation—When the overcurrent exceeds the rating at which the relays are set to operate, the heating effect of the current passing through the relays will cause the relay contacts to close and energize the trip coils, which trips the oil circuit breaker. The relay operating characteristics are usually inverse-time, in that the time to operate the relay varies inversely with the overcurrent applied.

AC MOTOR CONTROL STARTER CONNECTIONS PUSH BUTTONS AND SPECIAL SWITCHES

DEVICE		SYMBOL
COILS	RELAY AND SWITCH COILS	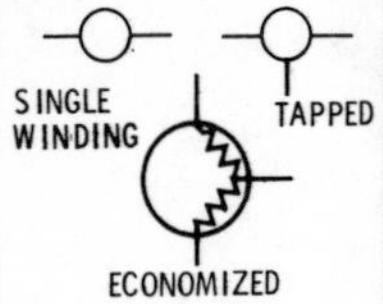 SINGLE WINDING TAPPED ECONOMIZED
CONTACTS	NORMALLY CLOSED (NC)	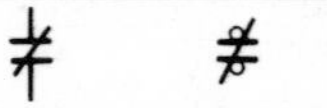 MAIN AUXILLIARY
	NORMALLY OPENED (NO)	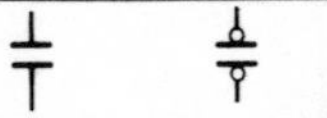 MAIN AUXILLIARY
	TIME CLOSING	 T.C.
	TIME OPENING	 T.O.
CONTACTORS	AC SOLENOID TYPE	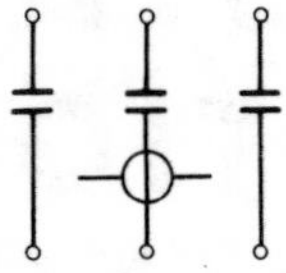
	MANUALLY OPERATED	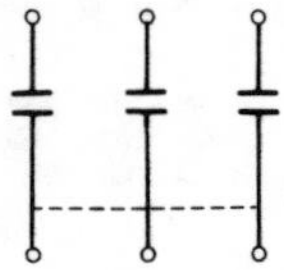

DEVICE		SYMBOL
FUSE	GENERAL	
INDICATING LAMPS	GENERAL	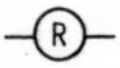 R A - AMBER B - BLUE C - CLEAR G - GREEN R - RED W - WHITE
MOTORS	3-PHASE SQUIRREL CAGE INDUCTION	
	SINGLE PHASE	
	2 PHASE 4 WIRE	
RECTIFIER	FULL WAVE WITH COLOR CODE	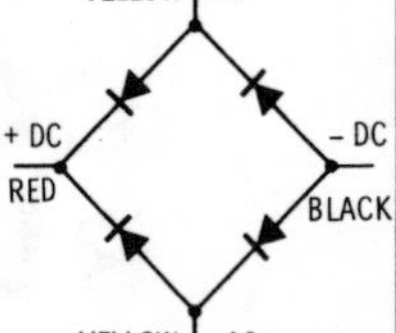 YELLOW AC + DC RED – DC BLACK YELLOW AC 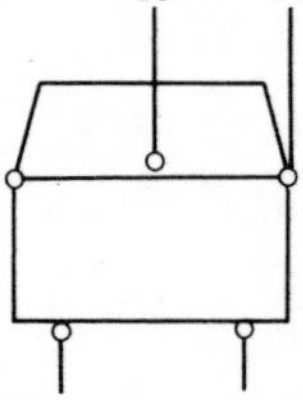 DC DC AC AC

<table>
<tr><th colspan="2">DEVICE</th><th>SYMBOL</th></tr>
<tr><td rowspan="5">RELAYS</td><td>CONTROL TYPE BR</td><td></td></tr>
<tr><td>CONTROL TYPE B</td><td> 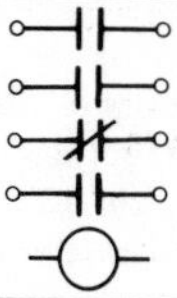REAR
FRONT
SINGLE DECK</td></tr>
<tr><td>THERMAL OVERLOAD</td><td> 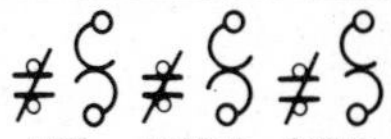LEFT MIDDLE RIGHT</td></tr>
<tr><td>TIMING (PNEUMATIC) ON-DELAY</td><td> 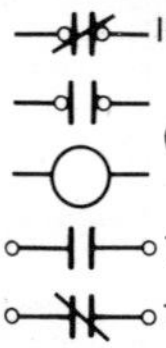INSTANTANEOUS AUXILIARY CONTACTS (WHEN USED)
T.C.
T.O.</td></tr>
<tr><td>3-WIRE THERMOSTAT</td><td> 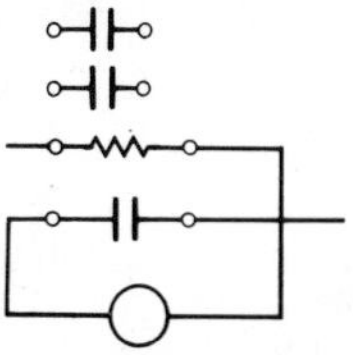</td></tr>
<tr><td>SWITCHES</td><td>FLOAT SWITCH</td><td> 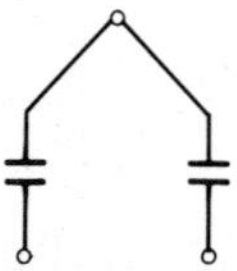WHEN THIS SYMBOL IS USED ON A DRAWING, THE SYMBOL SHOULD BE IDENTIFIED AS A "FLOAT" SWITCH</td></tr>
<tr><td></td><td></td><td></td></tr>
</table>

<table>
<tr><th colspan="2">DEVICE</th><th>SYMBOL</th></tr>
<tr><td rowspan="8">SWITCHES</td><td>LIMIT SWITCH</td><td> 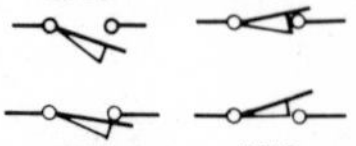NORMALLY OPEN   NORMALLY CLOSED
HELD CLOSED   HELD OPEN</td></tr>
<tr><td>PRESSURE OR TEMPERATURE</td><td> 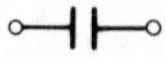WHEN THIS SYMBOL IS USED ON A DRAWING, THE SYMBOL SHOULD BE IDENTIFIED AS A "PRESSURE SWITCH" OR A "TEMPERATURE SWITCH"</td></tr>
<tr><td>PUSH BUTTON STANDARD</td><td> 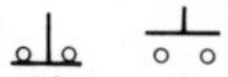NC   NO</td></tr>
<tr><td>PUSH BUTTON HEAVY DUTY, OIL TIGHT</td><td> 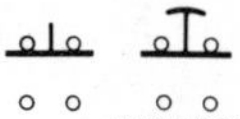MUSH ROOM HEAD</td></tr>
<tr><td>PUSH BUTTON & JOG ATTACHMENT</td><td> 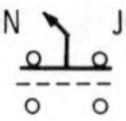PIN   JOG</td></tr>
<tr><td>STANDARD DUTY SELECTOR SWITCH</td><td> 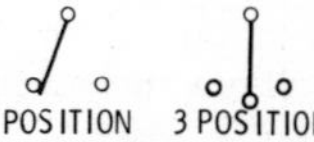2 POSITION   3 POSITION</td></tr>
<tr><td>HEAVY DUTY SELECTOR 2-POSITION</td><td> 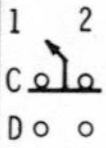1 2
C
D<table><tr><th>LETTER SYM</th><th colspan="2">POSITION</th></tr><tr><td></td><td>1</td><td>2</td></tr><tr><td>C</td><td>X</td><td></td></tr><tr><td>D</td><td></td><td>X</td></tr></table></td></tr>
<tr><td>HEAVY DUTY SELECTOR 3-POSITION</td><td>1 2 3
C
D<table><tr><th>LETTER SYM</th><th colspan="3">POSITION</th></tr><tr><td></td><td>1</td><td>2</td><td>3</td></tr><tr><td>C</td><td></td><td></td><td>X</td></tr><tr><td>D</td><td>X</td><td></td><td></td></tr></table></td></tr>
<tr><td rowspan="2">TRANS-FORMER</td><td>POTENTIAL</td><td>L1 #1 #2 #3 #4 L2
X2 X1</td></tr>
<tr><td>CURRENT</td><td></td></tr>
</table>

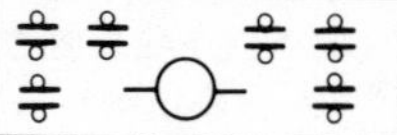

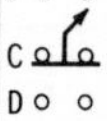

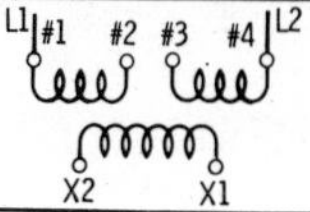

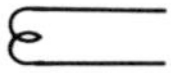

There are some common and important terms used with motor controls. They are standard "motor control terms" and are used repeatedly.

NO-VOLTAGE RELEASE

Also called: Low voltage release
two-wire control

NO-VOLTAGE PROTECTION

Also called: Low-voltage Protection
three-wire control.

The "no-voltage release" means that the motor will drop out of circuit upon a voltage failure and will pick-up again after the voltage is restored to the circuit. The diagram below will illustrate. The actuating device remains unaffected by loss of voltage, so the moment the voltage returns to the circuit the starter is again actuated and the motor starts.

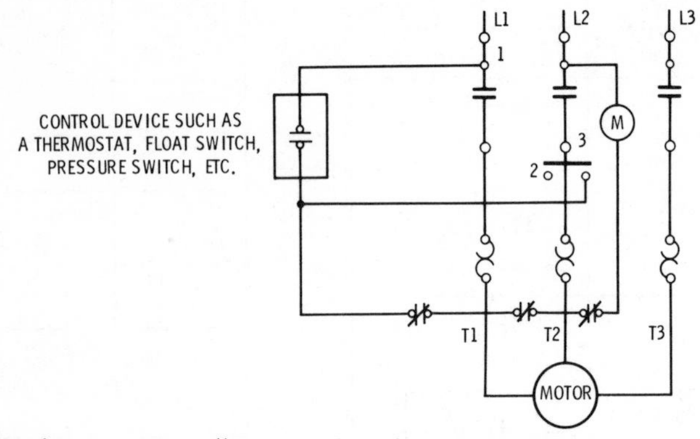

The "no-voltage protection" means when there is a voltage failure the starter will drop out of circuit and the motor will stop. When the voltage returns, the circuit will not automatically pickup and start the motor.

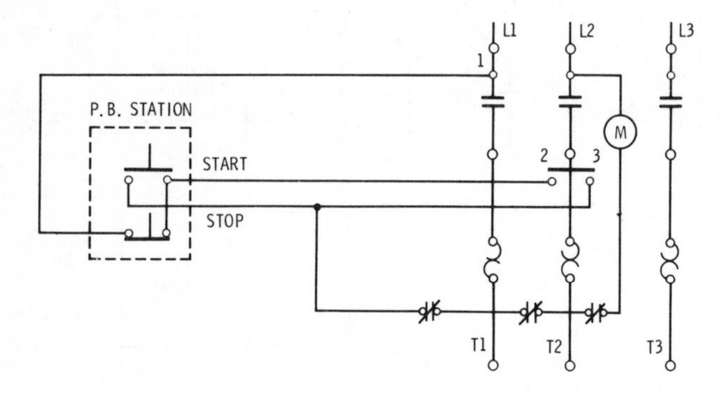

MOTOR-CONTROL METHODS

The Functions of Motor Control—The elementary functions of any motor control are starting, stopping and reversing the motor. These functions however, are only a few of the many contributions which the control renders to efficient operation of modern electric drives. Among the common functions of motor control required in motor installations are:

1. Motor Starting—This includes the basic function of starting and stopping the motor, including protection from overload and under voltage when required.

2. Motor Starting and Reversing—This includes the basic function of starting the motor where the direction of rotation of the motor is changed at will of the operator.

3. Motor Starting and Speed Regulation—Basic function of starting and stopping the motor involving speed regulation or adjustment of motor speed as by rheostat control.

4. Motor Starting, Reversing and Speed Regulation—Basic function of starting, stopping or reversing the motor involving speed regulation as by rheostat control.

5. Motor Starting and Speed Selection—Basic functions of starting and stopping the motor, involving selection of one of several basic speeds as by pole changing.

6. Motor Starting, Reversing and Speed Selection—Basic functions of starting, stopping and reversing, including selection of one of several basic speeds.

With respect to the general construction and method of starting and control required in motor installations as:

7. Across the Line Starter—This consists of a line switch (with protection as may be required) for connecting the motor directly across the supply line. This method of starting is also referred to as "full-voltage starting."

8. Starting Rheostat—Also referred to as the face-panel type. This is a type of rheostat whose stationary contacts are mounted upon the face of an insulating panel whose surface is a plane, the contacts being arranged in the form of an arc (or arcs) of a circle and the moveable contact (or contacts) being mounted upon a pivoted switch arm (or arms).

9. Primary Resistor Starter—A starter which provides reduced voltage to the primary of the motor by inserting resistance in the primary circuit during acceleration. The device includes the necessary switching mechanism, which may be manually or magnetically operated.

10. Secondary Resistor Starter—A starter which reduces the primary starting current by inserting resistance in the secondary circuit, usually the rotor

of a wound-rotor motor during acceleration. The device includes the necessary switching mechanism, which may be manually or magnetically operated.

11. Compensator or Autotransformer Starter—A starter which provides for reduced voltage starting by means of a compensator or autotransformer from which a predetermined fractional part of the winding is tapped off to produce voltage reduction to suit the particular starting load. The device includes the necessary switching mechanism to switch from the tap to full voltage and also to open the circuit of the compensator winding.

12. Star-Delta Starter—A star-delta starter is one which is applicable for starting of motors which have their windings arranged for full rated operation, with windings connected in delta, and arranged for starting at reduced voltage with windings connected in star.

13. Magnetic Controller—One wherein the main circuits are made and broken by magnetically operated switches controlled by a master switch located either within the controller or at any desired distance from the main controller.

With respect to the operation of the control circuit, magnetic controllers may be subdivided as follows:

14. Nonautomatic—Where the operator is responsible for all control functions of starting, stopping and accelerating the motor.

15. Semiautomatic—Where the rate of acceleration after starting by the operator, is dependent upon accelerating contactors, which are adjusted to function under predetermined conditions of currents, voltages and time.

16. Full Automatic—Where all basic functions including starting or stopping of the motor, are performed without the necessity of manual direction in any degree after being initially energized.

With respect to their type for operating magnetic controllers, the sub-classification of master switches may be made as follows:

17. Drum Switch—In general for nonautomatic types of controllers.

18. Push Button—In general for semiautomatic types of controllers.

19. Automatic Switch—Operated by float, pressure, etc., for full automatic controllers.

20. Emergency-Run Feature—Provides a means of temporarily rendering the overload device inoperative during an emergency.

With respect to the proximity of the master switch, magnetic controllers may be subclassified as follows:

21. Distant Motor Control—Where the master switch is mounted apart from the main control panel.

MOTOR-CONTROL METHODS

22. Local Motor Control—Where the master switch may be combined with the main control panel.

The service classification of control resistors with reference to the duty period are:

23. Continuous Rating—Where the load is required to be carried for an unlimited period.

24. Periodic Rating—Where the load can be carried for alternate periods of load and rest.

25. Standard Periodic Rating—Where the starting and intermittent duty may be standardized as light, heavy, or extra heavy starting, or intermittent duty classifications.

The kind of protection to be provided required may be termed as:

26. Low- or Under-Voltage Protection—Operates to cause and maintain the interruption of the main power or reduction or failure of voltage.

27. Low- or Under-Voltage Release—Operates to cause the interruption of power upon reduction or failure of voltage, but not to maintain the interruption of power upon return of voltage.

28. Overload Protection—Operates to protect against excessive current to cause and maintain the interruption of current not in excess of six times the rated motor current.

29. Short-Circuit Protection—Where the overload protection does not provide for short-circuit protection, such short-circuit protection shall be provided as by fuses.

30. Single-Phase Protection or Indication—Where required, shall indicate and protect the personnel and equipment upon the failure of any part of the circuit which would cause an open phase.

MOTOR-PROTECTIVE DEVICES

The functions of motor protective devices are to protect the motor against certain abnormal conditions, such as:

1. Overloads.

2. Short circuits.

3. Under voltage, etc.

This includes devices which function on the basis of temperature, voltage, frequency, or time, and cause the switching mechanism to operate when a predetermined set of conditions exist.

MOTOR-CONTROL METHODS

Overload Relays—These devices are of two general types, namely:

1. Thermal overload relay, and
2. Dash-pot overload relay.

Thermal and dash-pot overload relays are again divided into two types as:

1. Hand reset, and
2. Automatic reset.

As the name denotes, hand reset relays must be reset by hand after having tripped (usually by pressing a button projecting through the enclosing case) whereas the automatic reset types reset themselves automatically.

Thermal Overload Relay—The thermal overload relay consists of a heater coil which is connected directly to the line of the motor, and which heats up directly in the proportion of current flowing through it. The thermostatic metal is made up of two metals rolled together and wound into a spiral. The two metals have different expansion coefficients so the spiral will unwind as it heats up. A shaft through the center of the spiral and anchored to the end of it will rotate and actuate the switching mechanism.

A Two-Pole Temperature Overload Relay—NOTE: Cover removed from unit to show thermostatic metal strip.

Because one end is solidly anchored, all of the movement occurs at the free end.

MOTOR-CONTROL METHODS

A relay of this type is quite accurate and its operating characteristics are well defined. It is possible to design a heater that will raise the temperature of the thermostatic strip at the same rate as the temperature in the motor with which it is being used, and so adjust the relay that it will trip the control contacts when the temperature of the motor has reached the allowable maximum.

Because some time is required to transmit the heat from the heater coil to the thermostatic strip, it is not affected by momentary current increases. This makes it possible to start the motor with an inrush of six to ten times normal current without tripping the relay.

Dashpot Overload Relay—The dashpot overload relay uses the mechanical retardation principle of a dashpot to retard the movement of a core in a magnetic field, produced by a solenoid coil in series with the motor leads. Such an arrangement is affected by the quality of the mechanical clearance between piston and cylinder wall, changes in viscosity of the dashpot oil caused by temperature variations and other extraneous conditions tending to upset its accuracy.

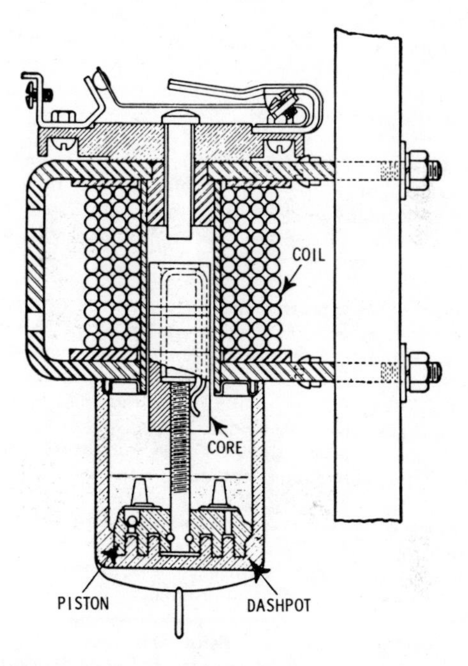

Typical Dashpot, Magnetic-Type Overload Relay, with Oil Dashpots to Give Inverse Time Tripping Characteristics.

MOTOR-CONTROL METHODS

Control Methods of Squirrel-Cage Motors—Polyphase induction motors of the squirrel-cage type may be started and controlled by several methods. They are:

1. Directly across the line.

2. By means of autotransformers.

3. By means of resistors or reactors in series with the stator winding.

4. By means of the star-delta method.

Starting Current-Torque Relationship—Before discussing the various methods which are available for reducing starting current and improving line-voltage conditions, it is important to have a thorough knowledge of the effects of reduced voltage starting on the motor, as well as on the power system.

Any method which reduces the starting current to the motor is accomplished by a reduction in starting torque. Therefore, it is essential to know something about the load-torque characteristics in determining if a given current limitation can be met. In other words, there are boundary conditions in which the permissible current to be taken from the line would not provide the needed output torque at the motor shaft, necessary for the successful acceleration of its connected load. With all starting methods, the torque of a squirrel-cage motor varies as the square of the applied voltage at the motor terminals.

Across-the-Line Method of Starting—This is generally the most economical method of starting, but on account of the large starting current required, is usually limited to motors up to 5 horsepower. With this method of starting, the motor is connected direct to full line voltage by means of a manually operated switch or a magnetic contactor.

Starting by Means of Autotransformer—In the case of the autotransformer type of starting, the current taken from the line varies as the square of the voltage applied to the motor terminals, and it is convenient to remember that the torque and line currents are reduced at the same rate. Thus, an autotransformer starter designed to apply 80% of the line voltage to the motor terminals will produce 64% of the torque that would have been developed if the motor had been started on full voltage, and will at the same time draw 64% as much current from the line as would have been required for full-voltage starting.

Starting by Means of Resistors or Reactors—With resistor or reactor starting, the starting current varies directly with the voltage at the motor terminals, because the resistor or reactor is in series with each line to the motor and must carry the same current that flows in each motor terminal.

It is evident therefore, that the resistor and reactor type of reduced voltage starting requires more line current in amperes per unit of torque in foot-pounds than does the autotransformer type.

Thus, if a motor connected to a loaded centrifugal pump is started with 80% tap on the autotransformer, the initial torque is 64%. If, on the other hand, the motor were started with a primary resistor, limiting the starting voltage to 65% of line voltage, the initial torque would only be 42.25%.

On some power systems, it is necessary to meet a restriction on the rate of current increase in starting. The rate of increase of current is determined to meet the conditions as they exist at that particular point on the system where the motor is started.

Starting By Means of the Star-Delta Method—Three-phase induction motors of the squirrel-cage type may occasionally be started by the star-delta method. This starting method is associated only with motors designed for their full power with the delta-connected, three-phase winding. There must also be provided additional leads from the motor which when regrouped will result in a star arrangement of the three-phase winding.

There will be six main leads required from the motor to accomplish the switch from start across the line (star-connection) to run across the line (delta-connection).

MANUAL COMPENSATOR WITH THERMAL OVERLOAD RELAY

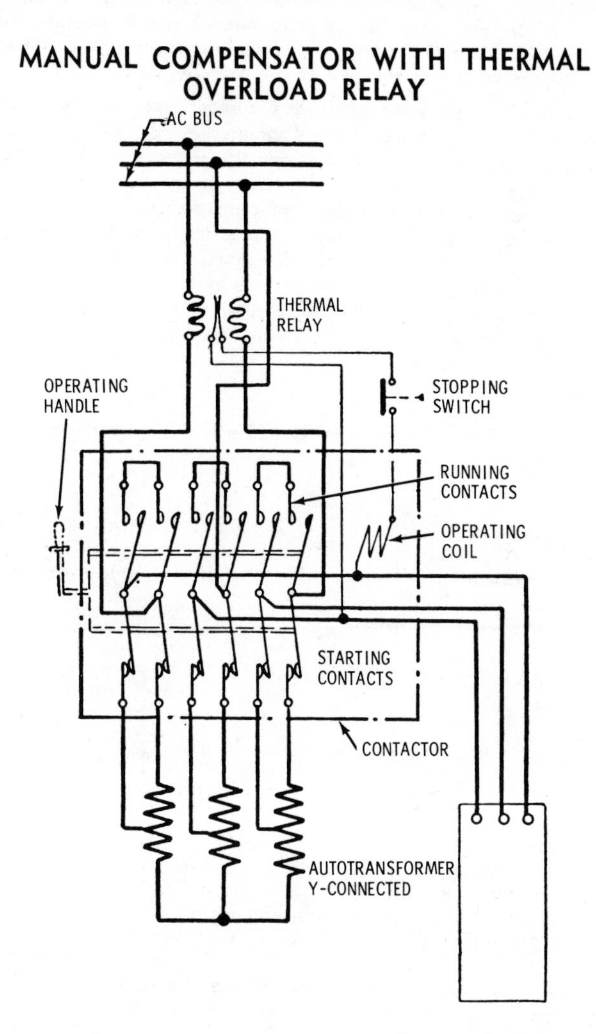

The starting connection is always star, since the voltage is $1/\sqrt{3}$ or 57.8% of the delta or line voltage.

From the foregoing it follows that this type of reduced-voltage starter (which is limited to 57.8% of line voltage at starting) can be employed only where the motor has a light starting load. In all other applications higher starting voltages are obtained with the resistor, reactors or autotransformers as previously outlined.

Operation—The starting of the motor is accomplished as follows: With the motor disconnected from the line, the operating handle is in off position.

First Step—Throw the operating handle to the start position. The motor is now connected to the line at reduced voltage through the taps of the autotransformer. **Second Step**—After the motor has reached normal speed, operate the handle to run position; at this time the starting contacts are automatically disengaged, and the motor is connected across the line at full voltage.

Starters and other information containing contacts, relays, etc., may be shown in two manners: (1) "wiring diagrams," diagrams "A" & "B" are wiring diagrams, they show the relative position of the parts and the connections. (2) "line diagrams or schematics," are what are generally used. They give the necessary information as to the sequence of operation and the location of components in the order in which they should appear in the circuit. Schematics are practically always used by electricians for troubleshooting, they readily show the effect of opening components.

Of the two types of diagrams, the schematic is by far the most used. In the following information, both will be shown in many places.

Below are schematics for the wiring diagrams for A and B. "A 1" is a schematic for "A" and "B 1" is a schematic for "B."

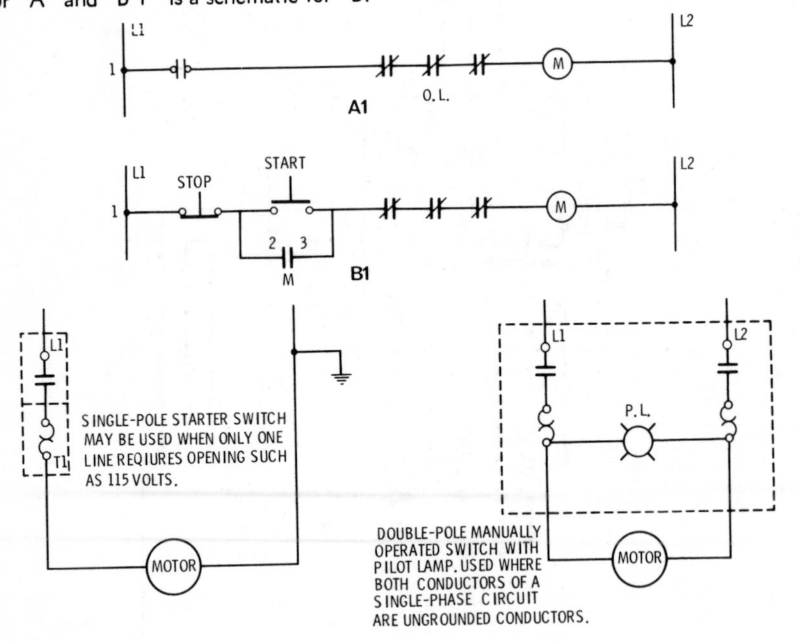

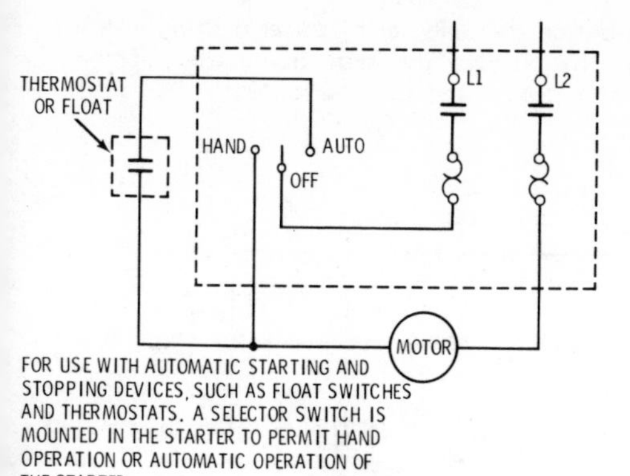

FOR USE WITH AUTOMATIC STARTING AND STOPPING DEVICES, SUCH AS FLOAT SWITCHES AND THERMOSTATS. A SELECTOR SWITCH IS MOUNTED IN THE STARTER TO PERMIT HAND OPERATION OR AUTOMATIC OPERATION OF THE STARTED.

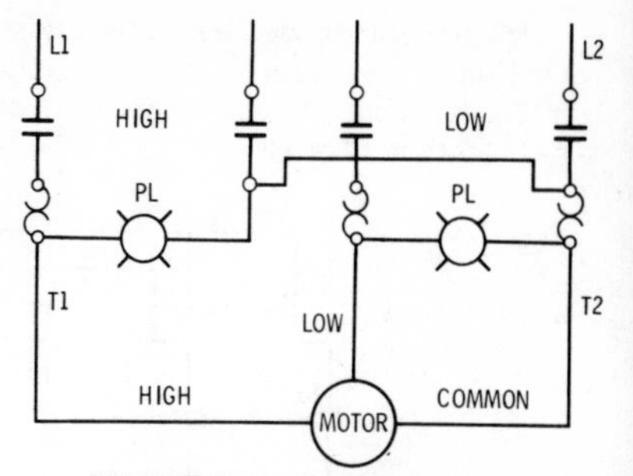

TWO SPEED SINGLE-POLE MOTORS, SUCH AS FAN MOTORS. THIS DIAGRAM CONSIST OF TWO MECHANICALLY INTERLOCKED SWITCHES.

MANUALLY OPERATED STARTERS

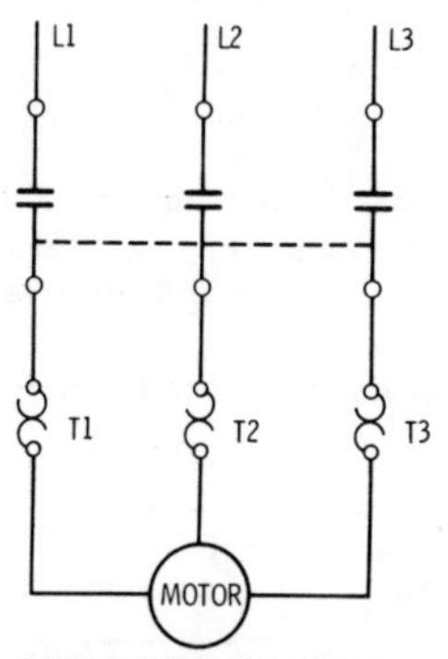

3 PHASE OR 2 PHASE, 3 WIRE (FOR 2 PHASE, 3WIRE, L2 AND L3 ARE COMMON)

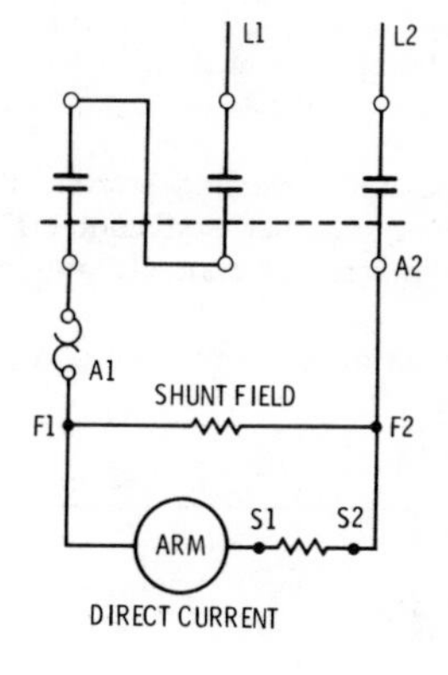

DIRECT CURRENT

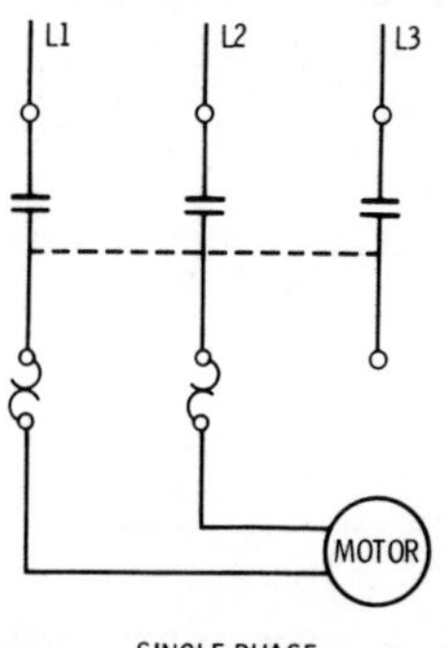

SINGLE PHASE

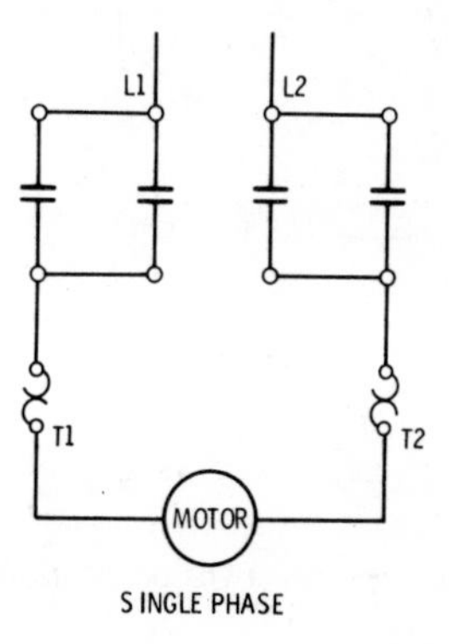

SINGLE PHASE

Motors are stopped by pushing a stop button manually, or by solder melting in the O.L. upon overloading. Motor is started by pushing the start button, but if the motor has shut off due to an overload, do not push the start button until the solder has had time to reset.

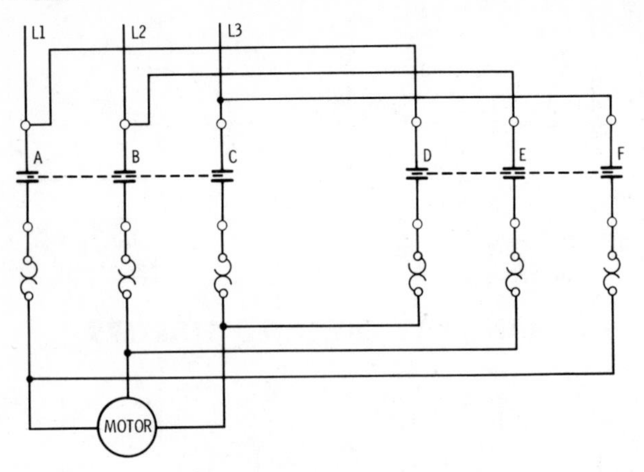

Manually operated reversing starter, three-phase. Notice contacts "A", "B" and "C" are operated together and contacts "D", "E" and "F" are operated together. The two different sets of contacts are interlocked, so that only one set may be on at a time.

Also notice that in reversing, two phases are reversed so as to cause the three-phase motor to reverse.

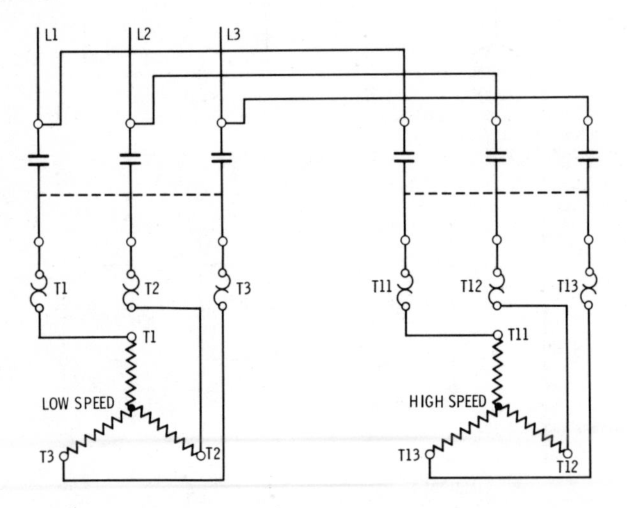

Manually operated two-speed starter. This is for a motor with separate windings only.

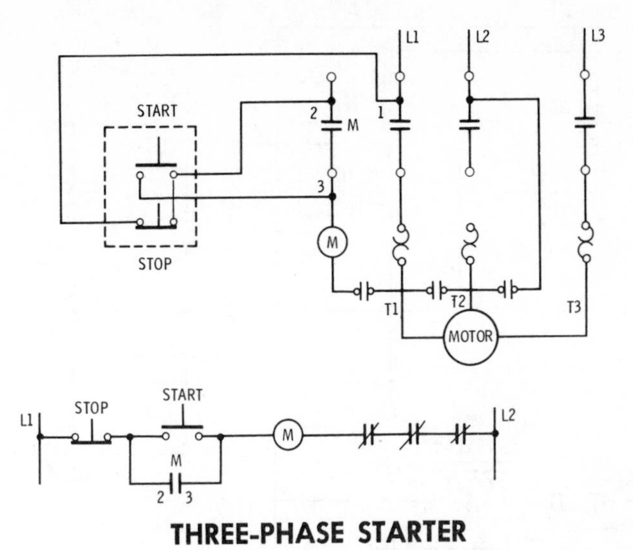

THREE-PHASE STARTER

This is a standard three-phase starter, with standard stop and start push button. The top sketch is a diagram and the bottom sketch is commonly called a schematic or line diagram.

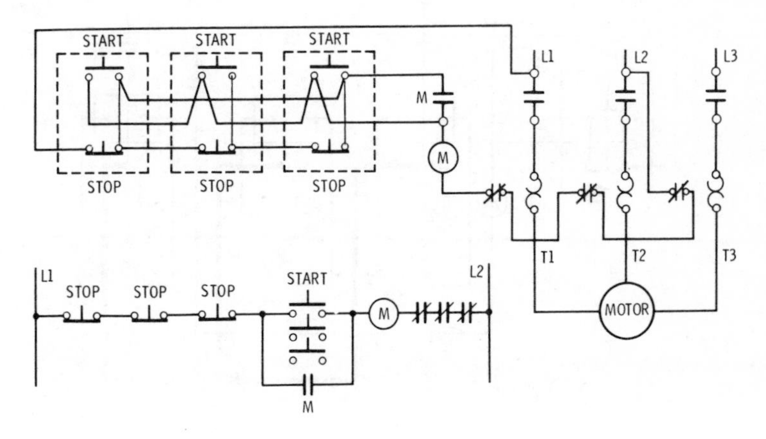

This is a three-phase starter (magnetic) with more than one stop and start push button. This motor may be started or stopped from more than one point.

Three-phase magnetic starter for a 480 volt motor, using a 480/120 volt transformer to obtain 120 volts for the control circuit. Take note that one side of the magnetic coil is connected to the grounded side of the transformer. There is no way that a ground ahead of this coil can cause the motor to start. This meets all National Electrical Code requirements.

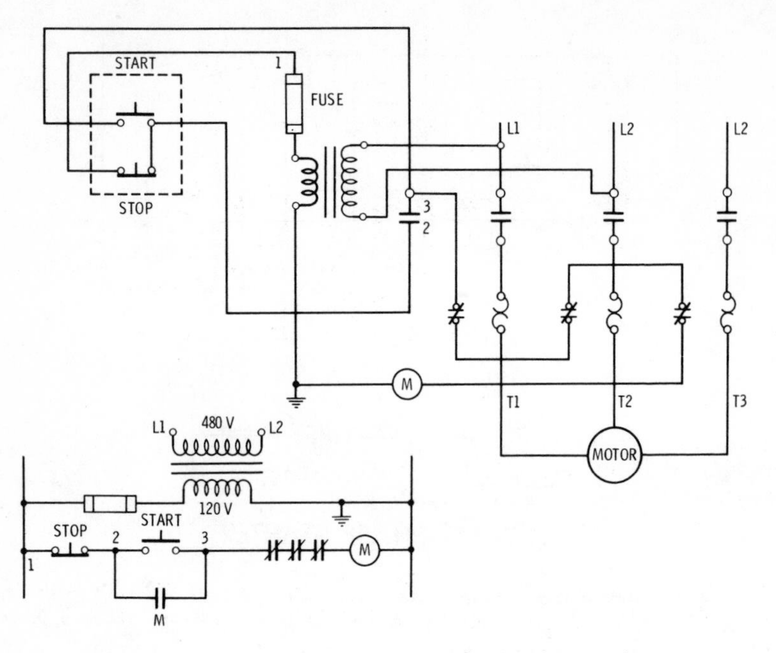

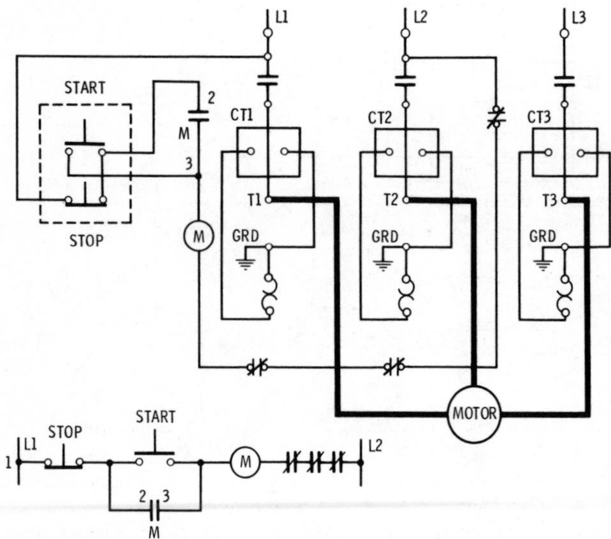

This is a standard wiring circuit using push button stop and start stations. This is a Size 5 starter, which is a very large starter, so in this case current transformers were installed so that Size 1 overload relays might be used. The ratio of the current transformers would have to be considered in sizing the overloads.

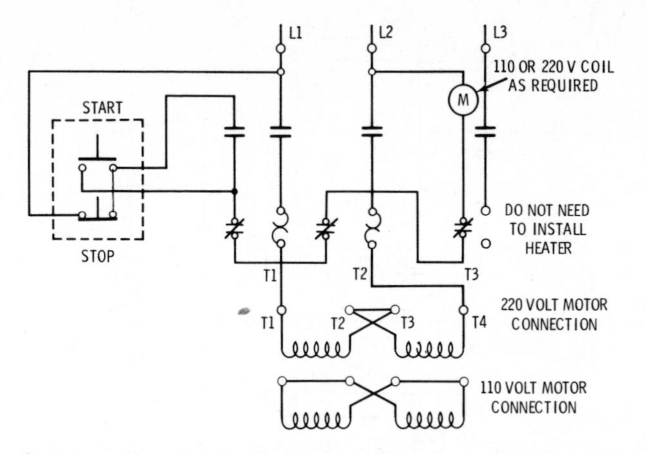

Using a 3-phase starter (standard) for a single-phase motor. If connected to 110 volts, connect the neutral to L2, to prevent accidental starting of motor due to ground in control circuit.

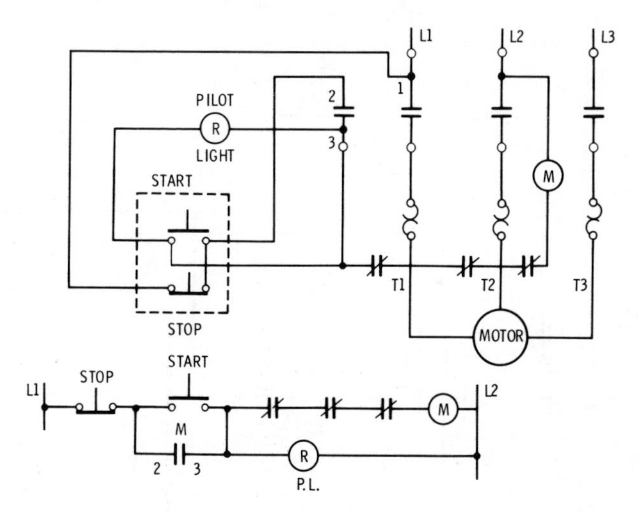

Stop and start station for 3-phase starter with pilot light to indicate when motor is running.

Stop and start station for 3-phase starter with pilot light to indicate when motor is stopped. "A" is an auxiliary contact added to the starter.

Three 3-phase starters are operated from a single "stop-start" station. An overload on any one of the motors will drop out all three starters. The wiring connection "Y" on each coil must be removed from each starter. It is necessary to disconnect the power to all lines to the starters to completely disconnect the equipment, since they are interconnected.

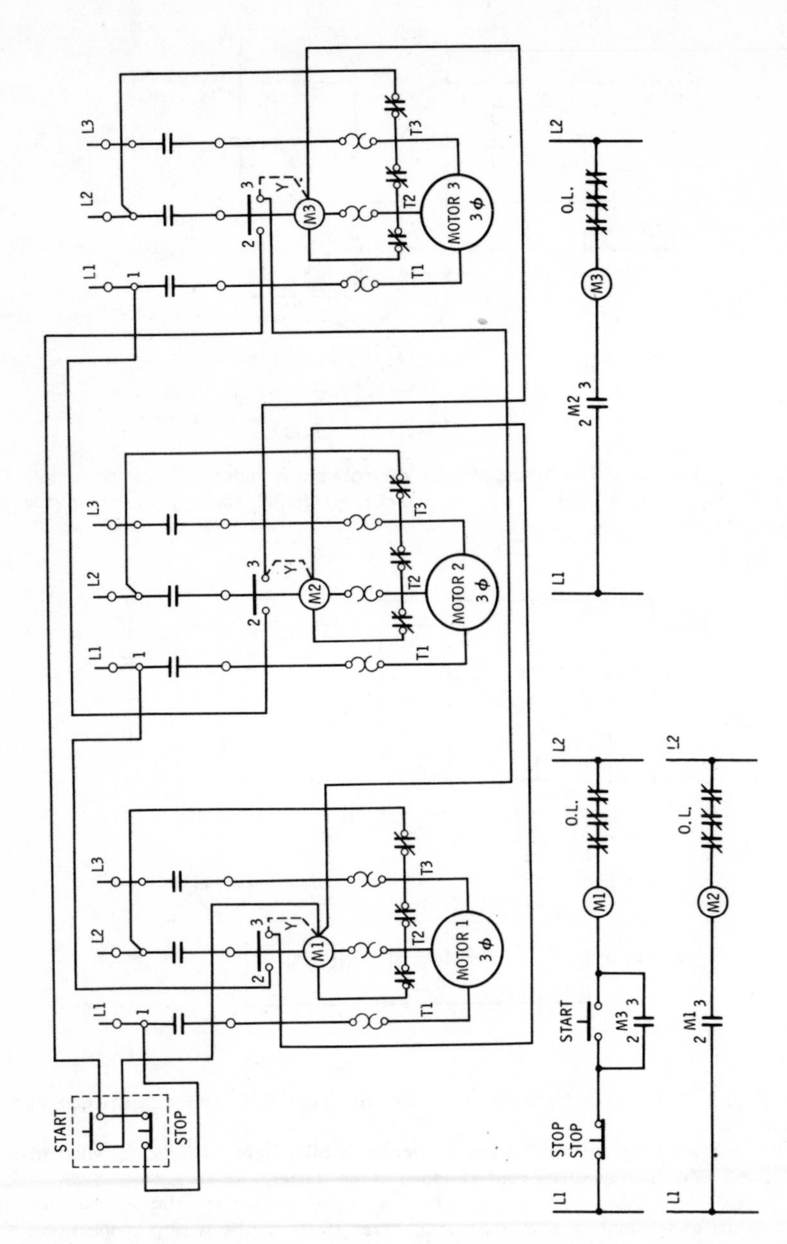

Two 3-phase starters are to be operated from a single station, but a short delay must prevent them from being energized together. TR, is a time-delay relay, which is usually adjustable for time. It is energized when the first starter is depressed,

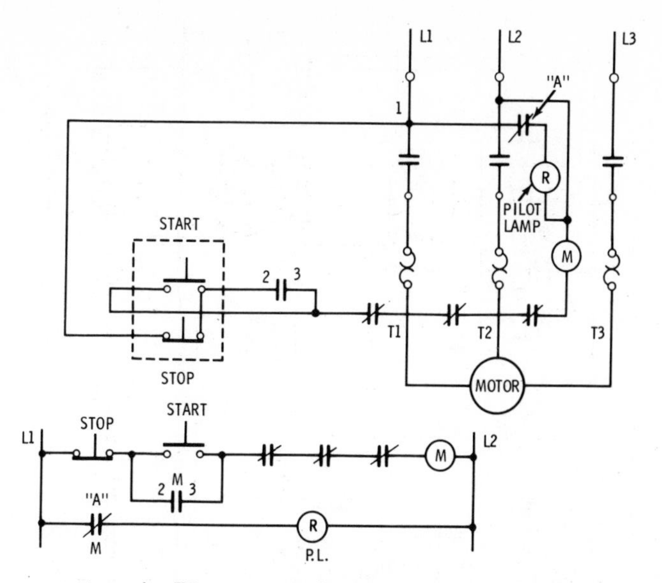

and after a set time, the TR energizes the second starter. Both starters are stopped together.

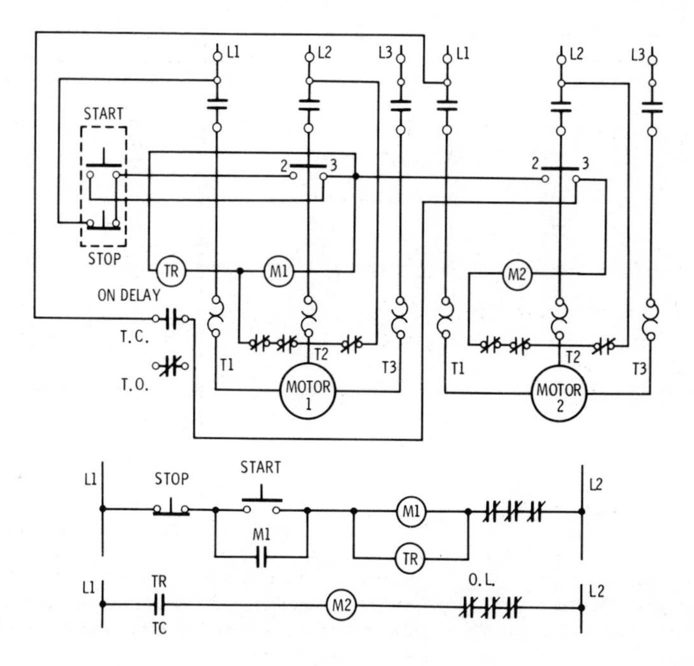

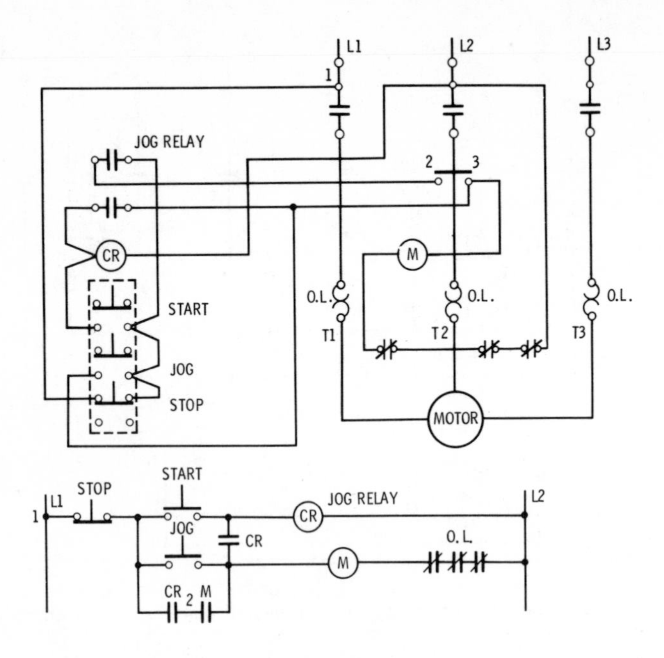

JOGGING

Jogging safety is provided by a relay. The purpose of jogging is to have the motor operate only as long as the "jog" button is held down. The motor must have

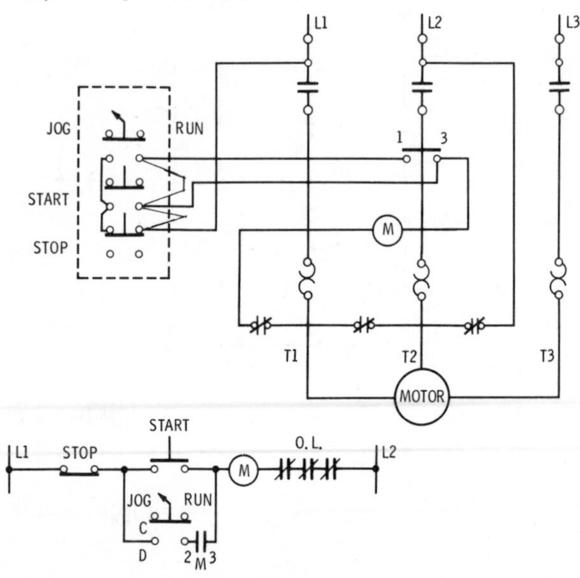

no chance to lock-in during jogging and for this reason the "jog relay" is used. Pushing the "start button" operates both the motor starter and the jog relay, causing the starter to lock-in through one of the relay contacts. When the jog button is pushed the starter operates, but this time the relay is not energized and thus the starter cannot lock-in.

A motor jogging with a 3-phase starter, a jog-run selector switch, and start-stop station. When the selector switch is in position "D" it is in run position. When the selector switch is in "C" position contacts "M" are not able to receive current on the "2" side so the start button is used to jog the motor.

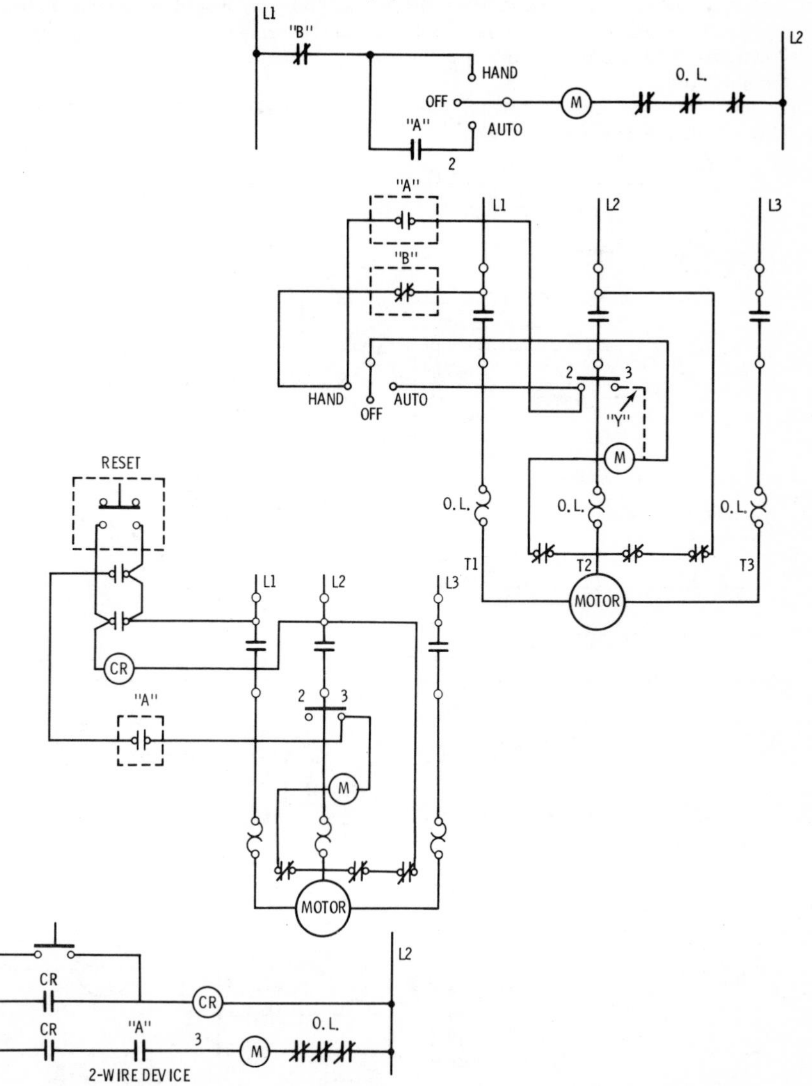

A 3-phase starter operated by a pressure switch or thermostat with a manual control provided by a selector switch. High pressure cut-off switch can be provided. "A" represents the thermostat or low-pressure switch and "B" represents high-pressure cut-out or safety switch. The connection represented by the dotted lines "Y," is to be removed.

A two-wire control with a control relay is added to provide no-voltage protection. When a 3-phase starter is being used, no-voltage protection is often necessary in a location where line disturbances are frequent. Since the two-wire pilot device is not affected by loss of voltage, the contacts will remain closed even though there is no power on the line. This means that without the protection of no-voltage protection, repeated voltage dips would cause the starter to chatter and the welding of contacts might result.

On a voltage dip or outage, the reset must be pushed before restarting. "A" represents the 2-wire pilot device.

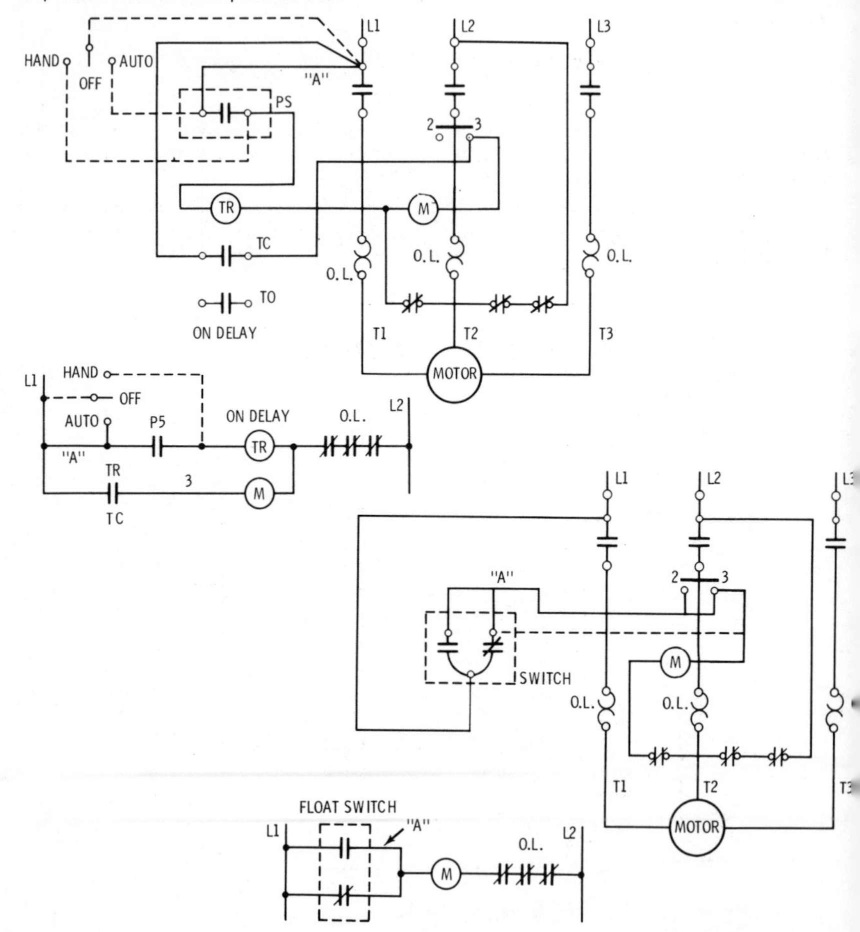

Surge protection is often necessary when the pump is turned off and the long column of water is stopped by a check valve. The force of the sudden stop may cause surges which operate the pressure switch contacts, thus subjecting the starter to "chattering."

Backspin is the name given to the backward turning of a centrifugal pump when a head of water runs back through the pump just after it has been turned off. Obviously starting the pump during backspin might damage the pump or motor, or break the pipeline.

This system provides backspin and surge protection on stopping-time delay between the pressure switch closing and motor starting. The pressure switch energizes the timer (TR) but the motor cannot start until the time delay contact has closed. The timer can thus be set for a time long enough to allow all surges and backspin to stop.

The dotted lines show how a selector switch can be added to by-pass the pressure switch if necessary. This is often used for motor testing purposes. It does not eliminate the time delay however. If the selector switch is added, the wire "A" must be removed.

FLOAT SWITCH CONTROLS STARTER

This diagram is intended for a float switch and tank operation. When the water reaches the low level, the float switch closes and pumping will continue until the water reaches the high level.

For sump pumping, remove wire "A" and connect as per dotted line. At low level the switch stops the pump and the pump will not start again until the water reaches the high level.

SEQUENCE CONTROL

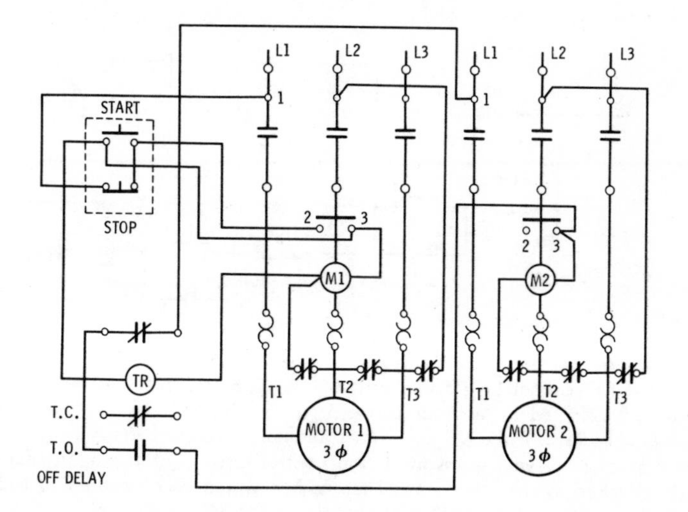

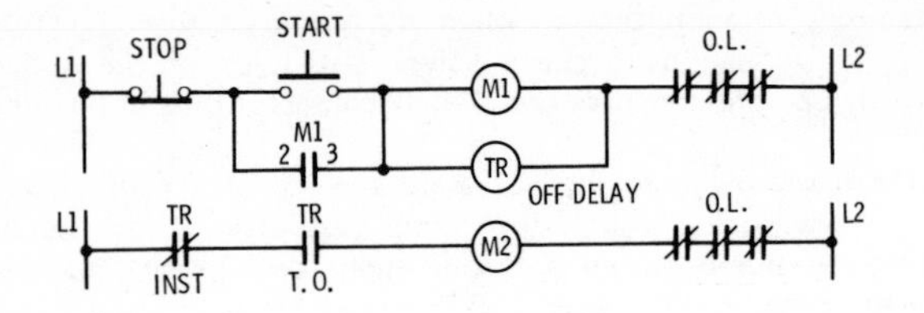

Sequence control of two motors is designed for one to start and run for a short time after the other stops. In this system, it is desired to have a second motor started automatically when the first is stopped. The second motor is to run only for a given time. Such an application might be found where the second motor is needed to run a cooling fan or a pump.

To accomplish this, an off-delay timer (TR) is used. When the start button is pressed, it energizes M1 and TR. The operation of TR closes its time-delay contact but the circuit to M2 is kept open by the opening of the instantaneous contact. As soon as the stop button is pressed, both M1 and TR are dropped out. This closes the instantaneous contact on TR and starts M2. M2 will continue to run until TR time runs out and the time-delay contact opens.

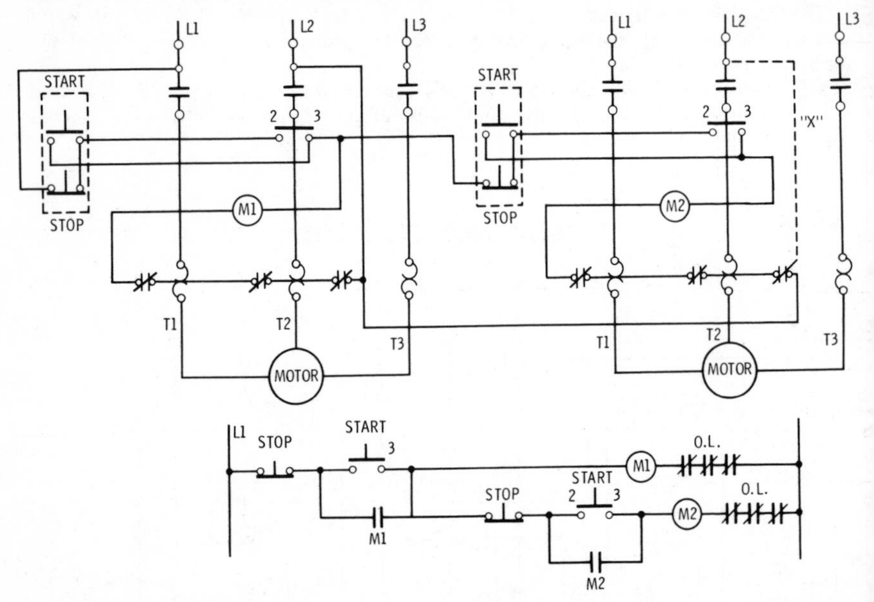

Starters Arranged for Sequence Control of a Conveyor System—The two starters are wired so that M2 cannot be started until M1 is running. This is necessary if M1 is driving a conveyor fed by another conveyor driven by M2. Material from the M2 conveyor would pile up if the M1 conveyor could not move and carry it away.

If a series of conveyors is involved, the control circuits of the additional starters can be interlocked in the same way. That is, M3 would be connected to M2 in the same "step" arrangement that M2 is now connected to M1, and so on.

The M1 stop button or an overload on M1 will stop both conveyors. The M2 stop button or an overload on M2 will stop only M2. If standard starters are used, wire "X" must be removed from M2.

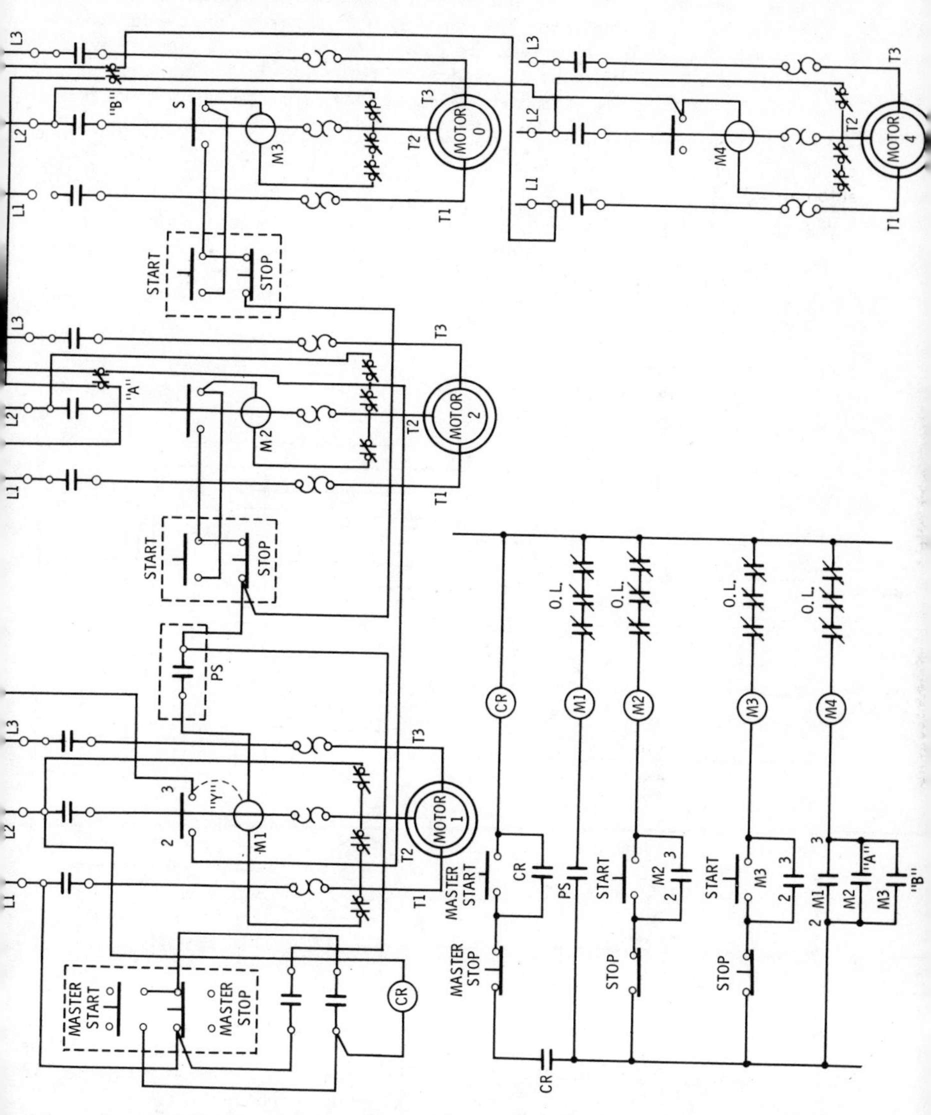

Operation of Any One of Several Starters Causes a Pump or Fan Motor to Start— Several motors are to be run independent of each other, some of the starters must be actuated by two-wire and some by three-wire pilot devices. Whenever any one of these motors is running, a pump or fan motor must also run.

A master start-stop push button station with a control relay is used to shut down the entire system in an emergency. Control relay (CR) provides "three-wire" control for M1 which is controlled by a two-wire control device such as a pressure switch. Motors M2 and M3 are controlled by standard start-stop push button stations.

Auxiliary contacts on M1, M2, and M3 control M4. These auxiliary contacts are all wired in parallel so that any one of them may start M4. Auxiliary contacts have been added to M2 and M3 for this purpose. The standard "hold-in" contact on M1 may be used as an auxiliary if wire "Y" is removed. "Hold-in" contacts are not required with a two-wire control device when used.

The auxiliary contacts are designated as "A" and "B" on the wiring diagram. These contacts are easily added. When this system is used, the phase connections on all of the starters must be the same. That is, L1 of each starter must be connected to the same incoming phase line, L2 and L3 of each starter must be similarly phased out.

REVERSING STARTERS

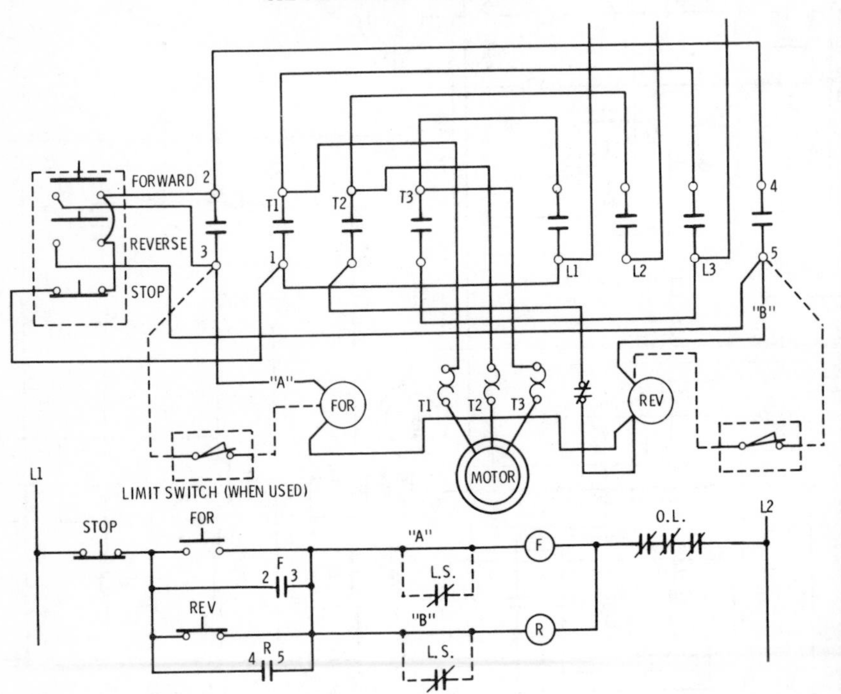

Standard wiring with "Forward-Reverse-Stop" push button station—The "stop" button must be depressed before changing directions. A mechanical interlock is provided, however electrical interlocks are not furnished on size 00 reversing starters. In the diagram, only one O.L. switch is shown for simplicity. The two others required for three phase motors, are connected in series with the one O.L. shown, as in the schematic.

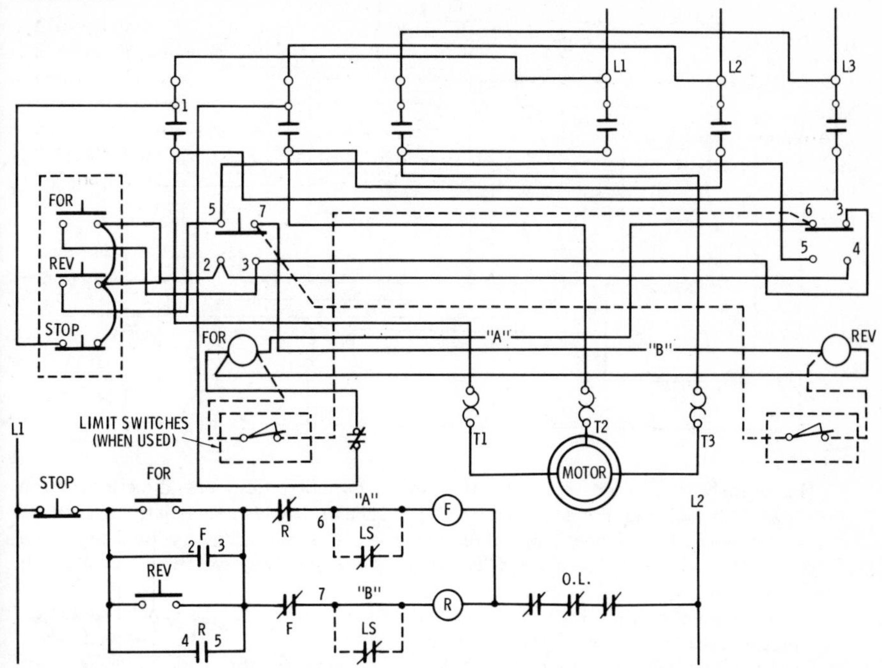

Standard wiring for "Forward-Reverse Stop" push button station—The "stop" button must be depressed before changing directions. A mechanical interlock and electrical interlocks are supplied as standard on all reversing starters size 0 and larger. Limit switches can be added to stop the motor at a certain point in either direction. Connections "A" and "B" must be removed when limit switches are used. Only one O.L. is shown in the diagram, three must be installed in series as illustrated in the schematic.

REVERSING STARTER—VARIATION

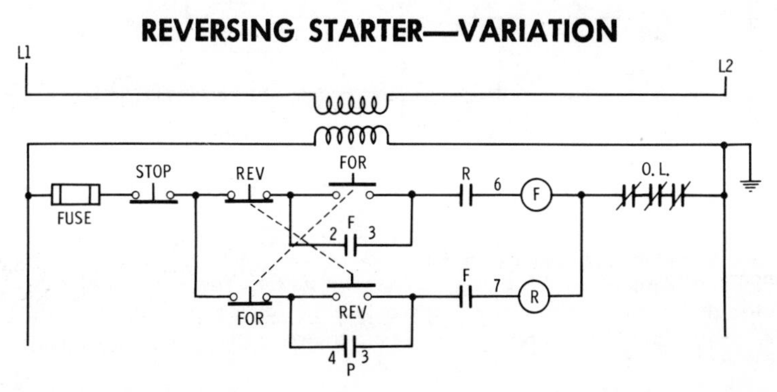

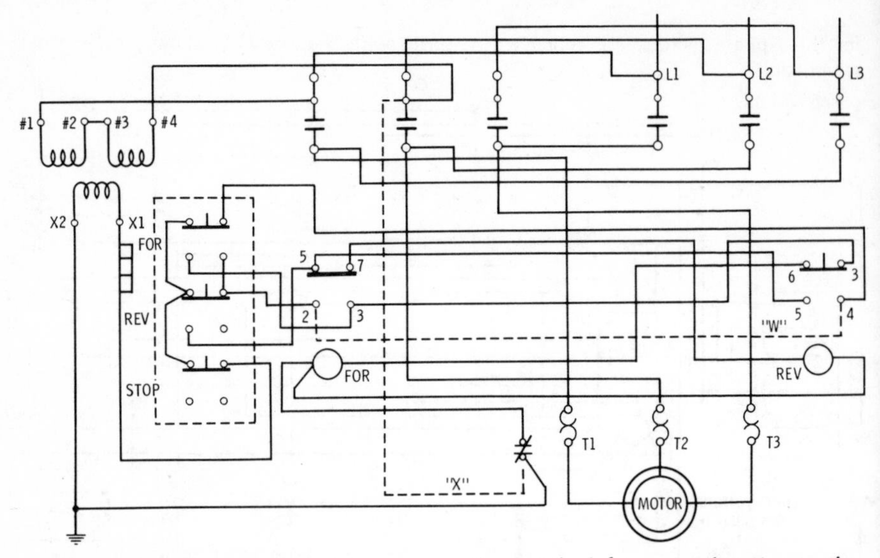

The push button wired to the starter can be switched from one direction to the other without pushing the stop button. This scheme allows immediate reversal of the motor when it is running in either direction. It is not necessary to depress the "stop" button when changing direction. A standard reversing switch can be used if wire "W" is removed.

The diagram shows the control circuit set up for reduced voltage control, although this may not be necessary at times. Notice that wire "X" must be removed when reduced voltage control is used. Only one O.L. is shown in the diagram, there must be three installed in series, as illustrated in the schematic.

JOGGING

Starting, Stopping and Jogging in Either Direction—This is one of the safest and most desirable methods of jogging a reversing starter. The motor will run steadily in either direction and also can be jogged in either direction. A bulletin 800T Type KE2A jogging unit in an oil tight enclosure are shown in the diagram.

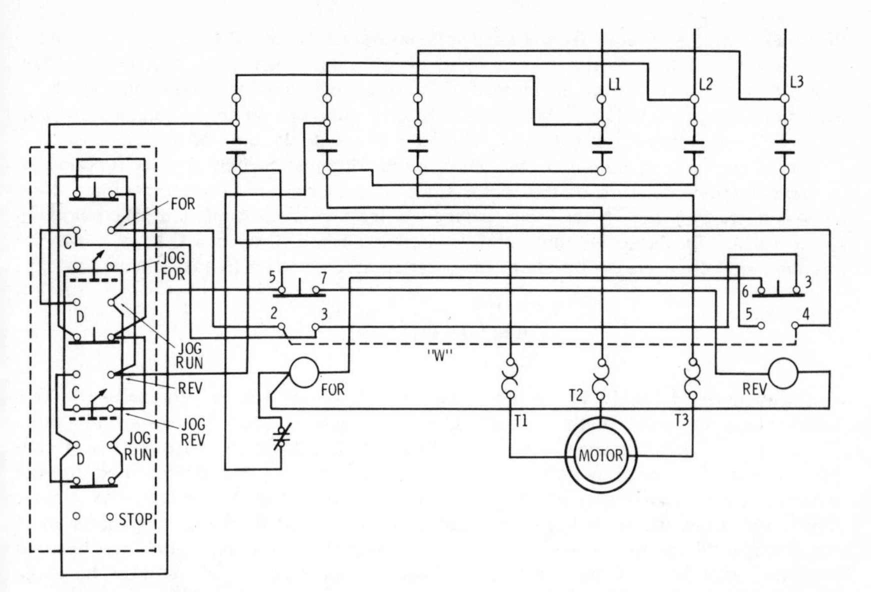

To jog, the jogging ring must be turned to "jog position". The NC contact on the jogging unit is held open thus insuring safe action by preventing either contactor from locking in during jogging. One O.L. only is shown in the diagram, three are required as shown in the schematic.

PLUGGING

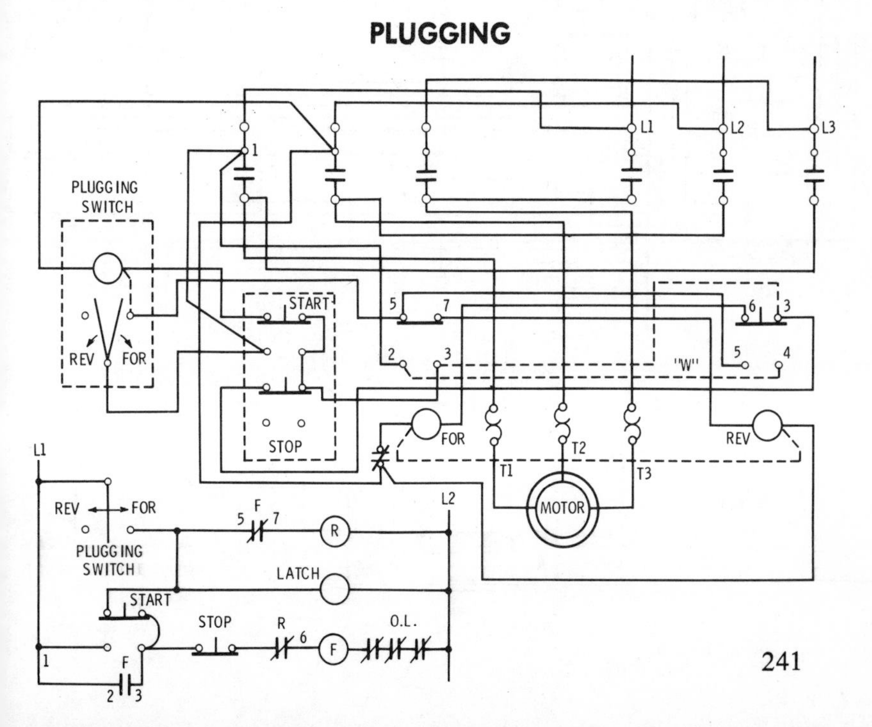

Plugging a Motor to Stop from One Direction—Safety Latch Provided—This system is for a motor that is to run in one direction only and must come to an immediate stop when the stop button is pressed. The reversing contactor reversing switch is used only for plug-stopping and not for reverse running. With a standard switch, wire "W" and all wires represented by dotted lines should be removed.

The safety latch is built into the Zero Speed Plugging Switch and its function is to prevent accidental turn of the motor shaft from closing and plugging switch contacts and starting the motor. This protective feature is optional and the plugging switch can be furnished without safety switch if desired. Only one O.L. is shown in this diagram, three are to be installed in series as illustrated in the schematic.

ANTIPLUGGING

Antiplugging—Motor Is to be Reversed, but it Must Not be Plugged— A zero speed plugging switch with normally closed contacts is used to prevent plugging. The schematic diagram shows that with the motor operating in one direction, a contact on the zero speed switch opens the control circuit of the starter used for the opposite direction. The open contact will not close until the motor has slowed down, and, thus, the reversing switch cannot be energized to change the direction of the motor until the motor is moving slowly. A standard reversing switch can be used with this application. Only one O.L. is shown in the diagram, there must be three O.L. in series as shown in the schematic.

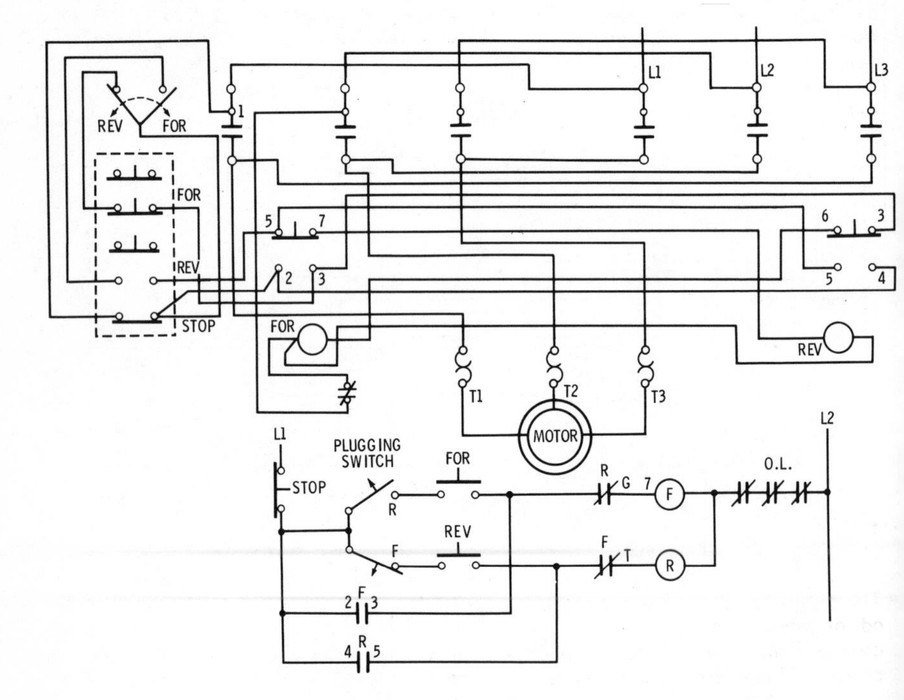

MULTISPEED MOTOR STARTER— FOR SEPARATE WINDING MOTORS

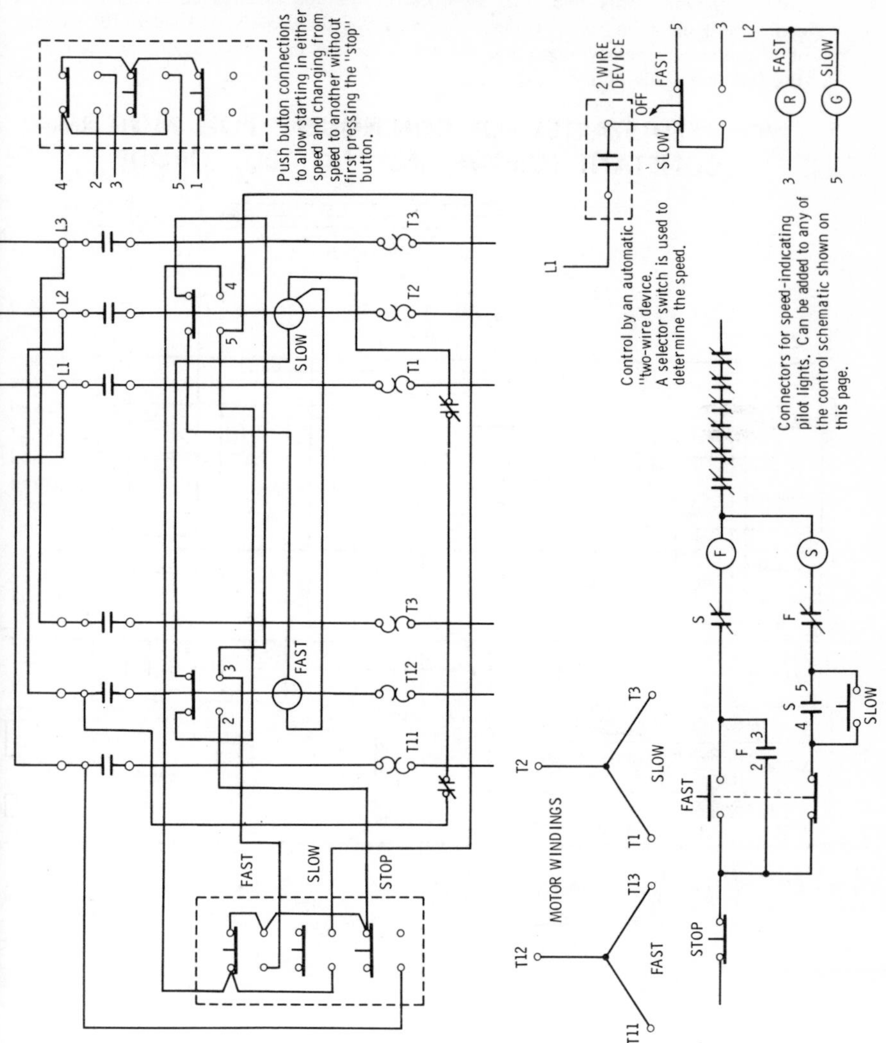

A standard connection with a two-speed separate winding motor is shown above. The wiring diagram and line diagram illustrate connections for the following method of operation: Motor can be started in either "fast" or "slow" speed. The change from slow to fast can be made without first pressing stop button. When changing from fast to slow, the stop button must be pressed between speeds. The pilot device diagrams shown on the right side panel illustrate other connections that can be made to obtain different sequences and methods of operation.

In the diagram, only two O.L. are shown. One for slow speed and one for fast speed. There must be six in series, the three for the slow speed sized for the current of the slow speed winding and the three for the fast speed sized for the current of the fast speed winding.

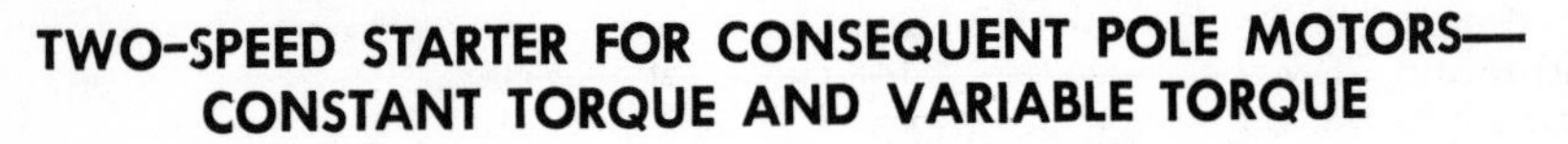

TWO-SPEED STARTER FOR CONSEQUENT POLE MOTORS—CONSTANT TORQUE AND VARIABLE TORQUE

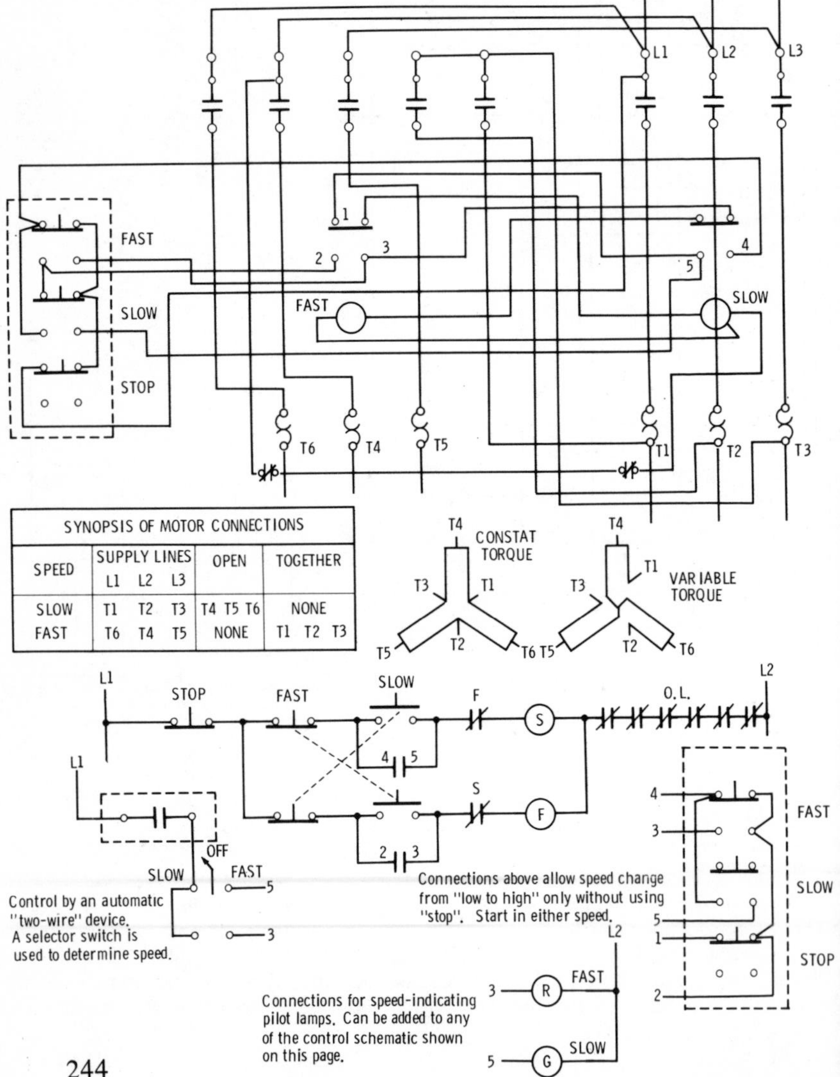

SYNOPSIS OF MOTOR CONNECTIONS					
SPEED	SUPPLY LINES L1	L2	L3	OPEN	TOGETHER
SLOW	T1	T2	T3	T4 T5 T6	NONE
FAST	T6	T4	T5	NONE	T1 T2 T3

A standard diagram used with a consequent pole constant or variable torque motor is shown above. The wiring diagram and line diagram in the above panel illustrate connections for the following method of operation: Motor can be started in either "fast" or "slow" speed. The change from slow to fast or from fast to slow can be made by first pressing the stop button. The pilot device diagrams shown in the side panel illustrate other connections that can be made to obtain different sequences and methods of operation.

Only two O.L. are shown in the diagram, but six will be required as shown in the schematic. They shall be sized for the current at slow speed for three, and at the current for fast speed for the other three O.L.'s.

STARTERS FOR CONSEQUENT POLE MOTORS— CONSTANT HORSEPOWER

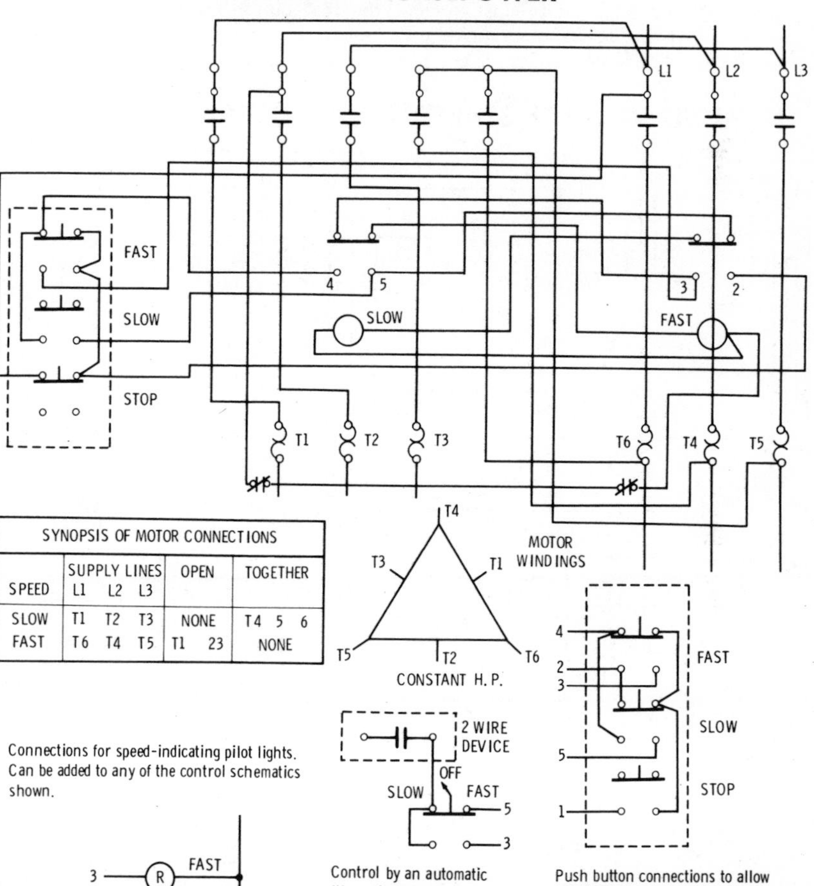

SYNOPSIS OF MOTOR CONNECTIONS			
SPEED	SUPPLY LINES L1 L2 L3	OPEN	TOGETHER
SLOW	T1 T2 T3	NONE	T4 5 6
FAST	T6 T4 T5	T1 23	NONE

Connections for speed-indicating pilot lights. Can be added to any of the control schematics shown.

Control by an automatic "two-wire" device. A selector switch is used to determine speed.

Push button connections to allow starting in either speed and changing from one speed to another without first pressing the "Stop" button.

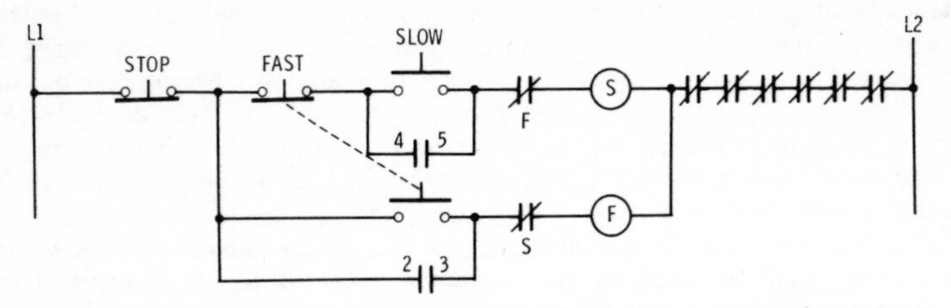

A standard connection used with a consequence pole, constant horsepower motor is shown above. The wiring diagram and line diagram illustrate connections for the following method of operation: Motor can be started in either "fast" or "slow" speed. The change from slow to fast can be made without first pressing the stop button. When changing from fast to slow, the stop button must be pressed between speeds. The pilot device diagrams shown illustrate other connections that can be made to obtain different sequences and methods of operation.

REDUCED VOLTAGE STARTERS— AUTOTRANSFORMER TYPE

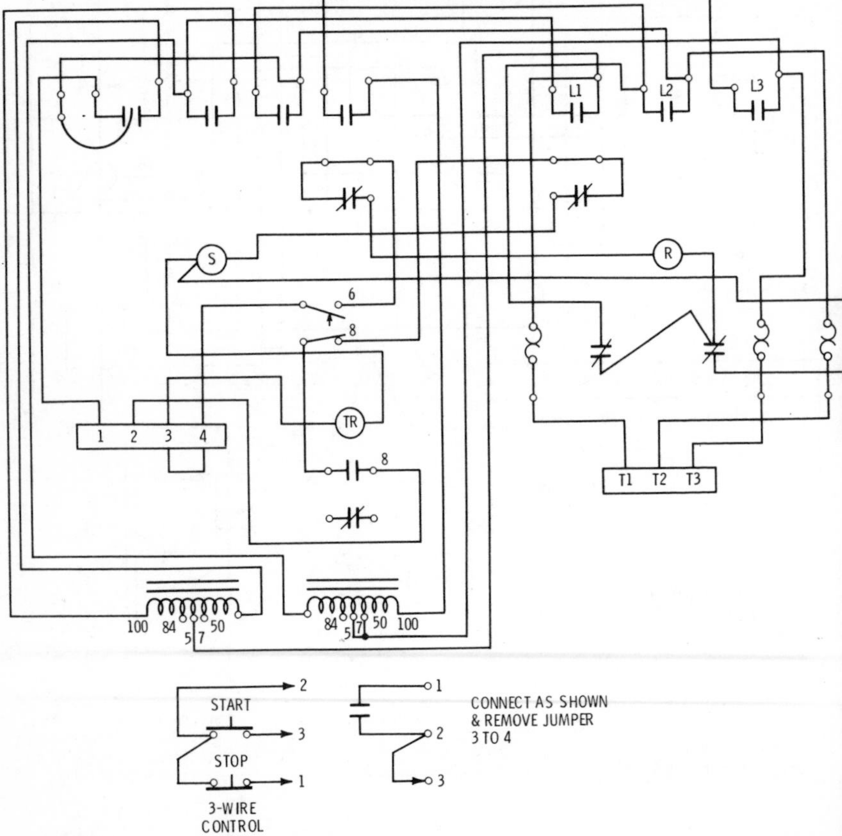

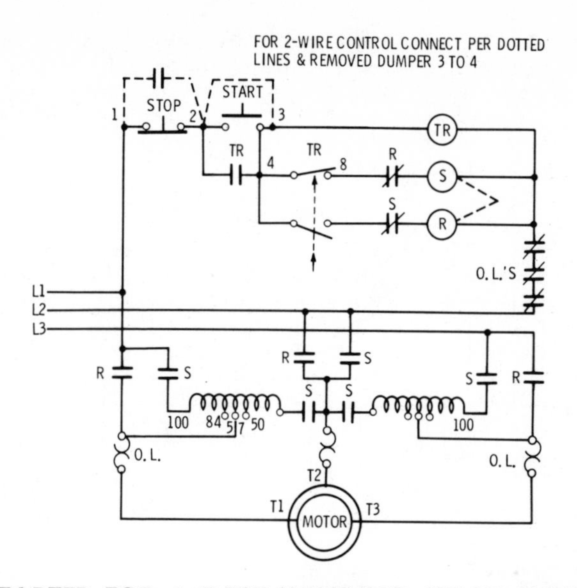

STARTER FOR A PART-WINDING START MOTOR

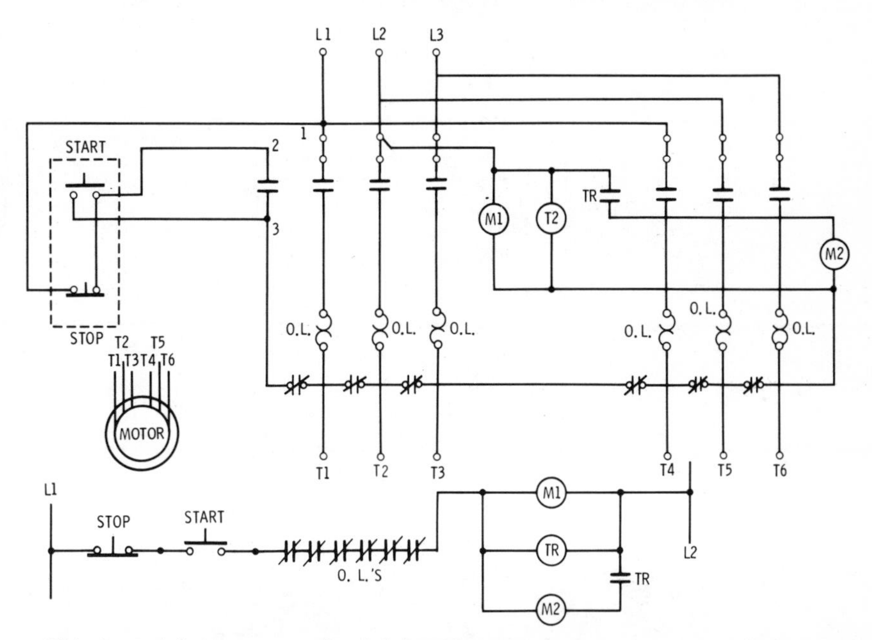

This is used for large centrifugal fans, refrigeration equipment, etc., which starts under a comparatively small load, but requires more H.P. when operating. The motor has two windings, T1, T2, and T3 and also another winding T4, T5, and T6. To reduce the starting current the motor is started on winding T1, T2, and T3. There is a time relay and contacts TR, which may be set for the time at which they cut into circuit. Thus the motor starts on one-half winding and the other winding cuts in later according to the setting of TR relay. If the O.L.'s in either winding trip the entire starter is deenergized and the motor stops.

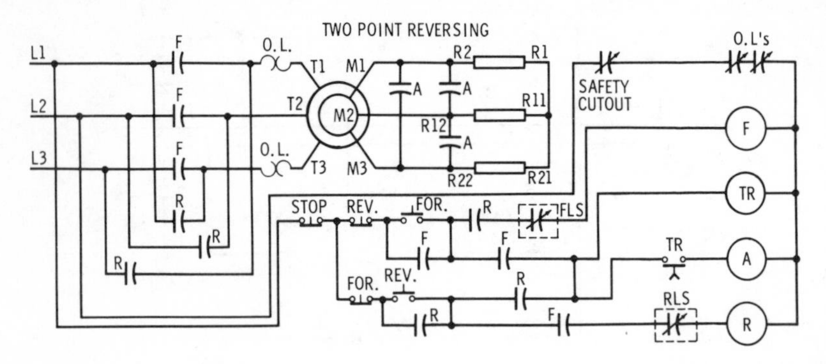

Starter for Wound-Rotor Induction Motor—The motor is started with the full amount of resistance in the rotor circuit. As the motor gains speed the rotor resistance is gradually cut out by decreasing the resistance.

MOTOR-CONTROL METHODS

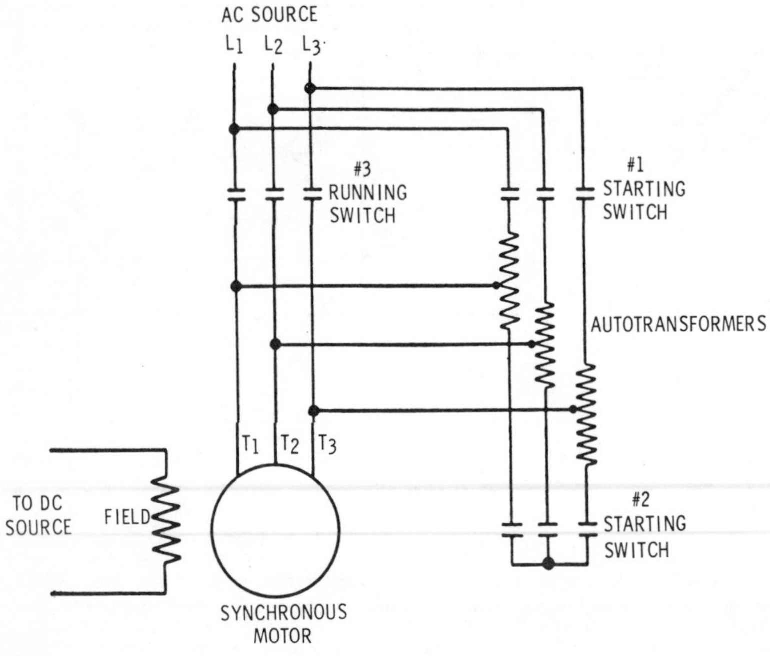

Method of Reduced Voltage Starting Using Three Switches or Contactors—In this starting method, switches #1 and #2 are first closed simultaneously connecting the motor to the line through the star-connected autotransformers. Power is then supplied to the motor at reduced voltage. Switch #1 is opened as soon as the speed becomes constant and switch #3 closes instantly, being electrically interlocked with switch #1. This connects the motor directly across the line, the terminal voltage having increased to normal without dropping to zero.

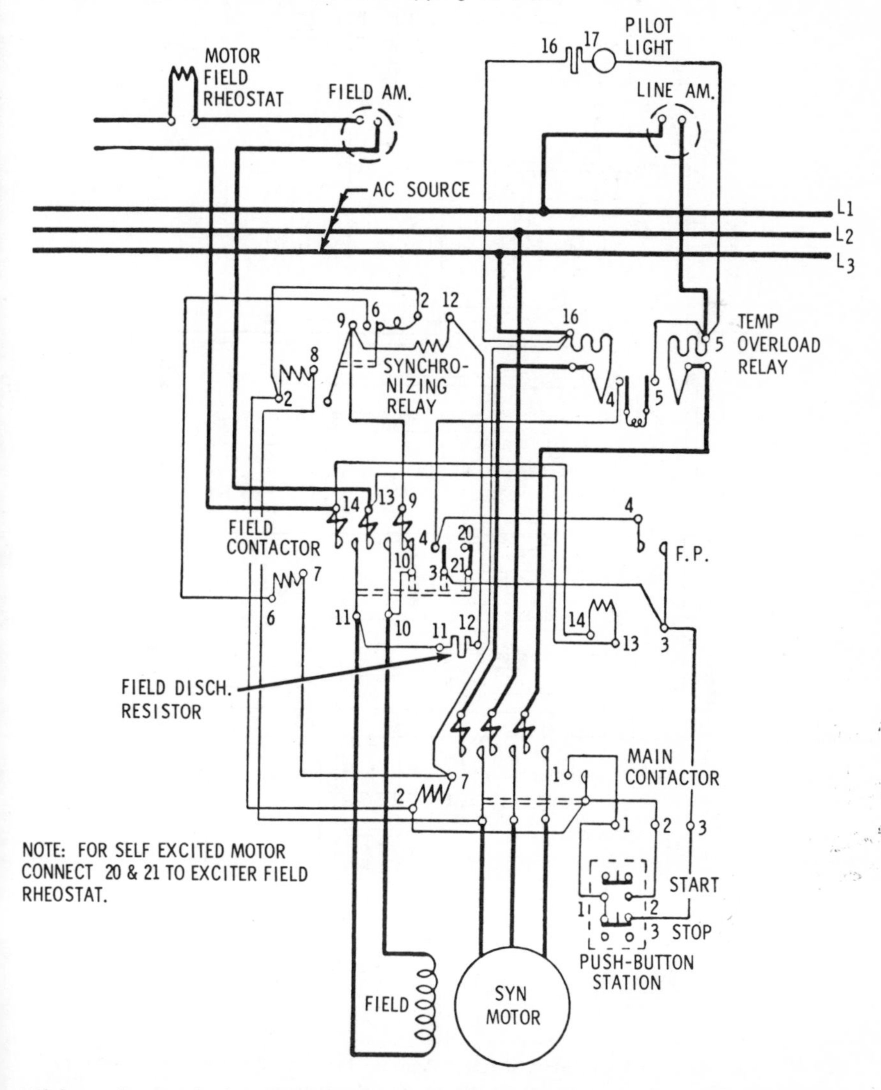

Wiring of a Typical Automatic, Across-the-Line, Synchronous Motor Starter.

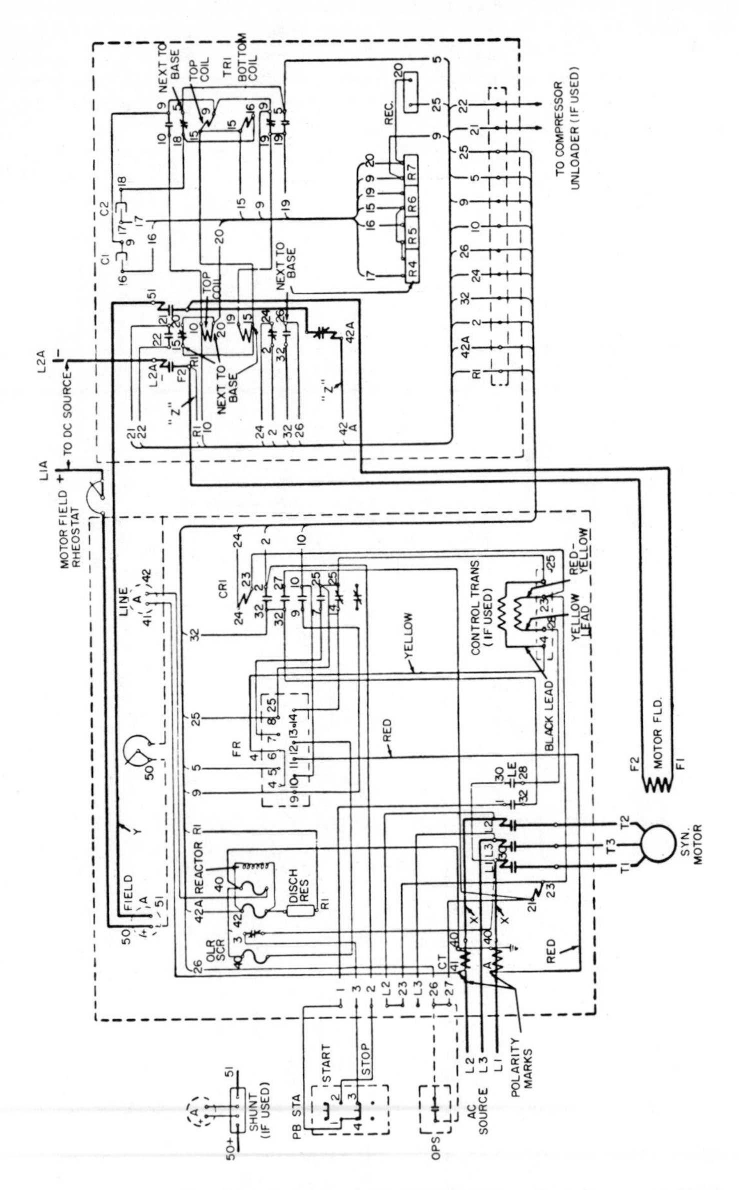

Diagram of General Electric Across-the-Line Synchronous Motor, Low-Voltage Magnetic Starter.

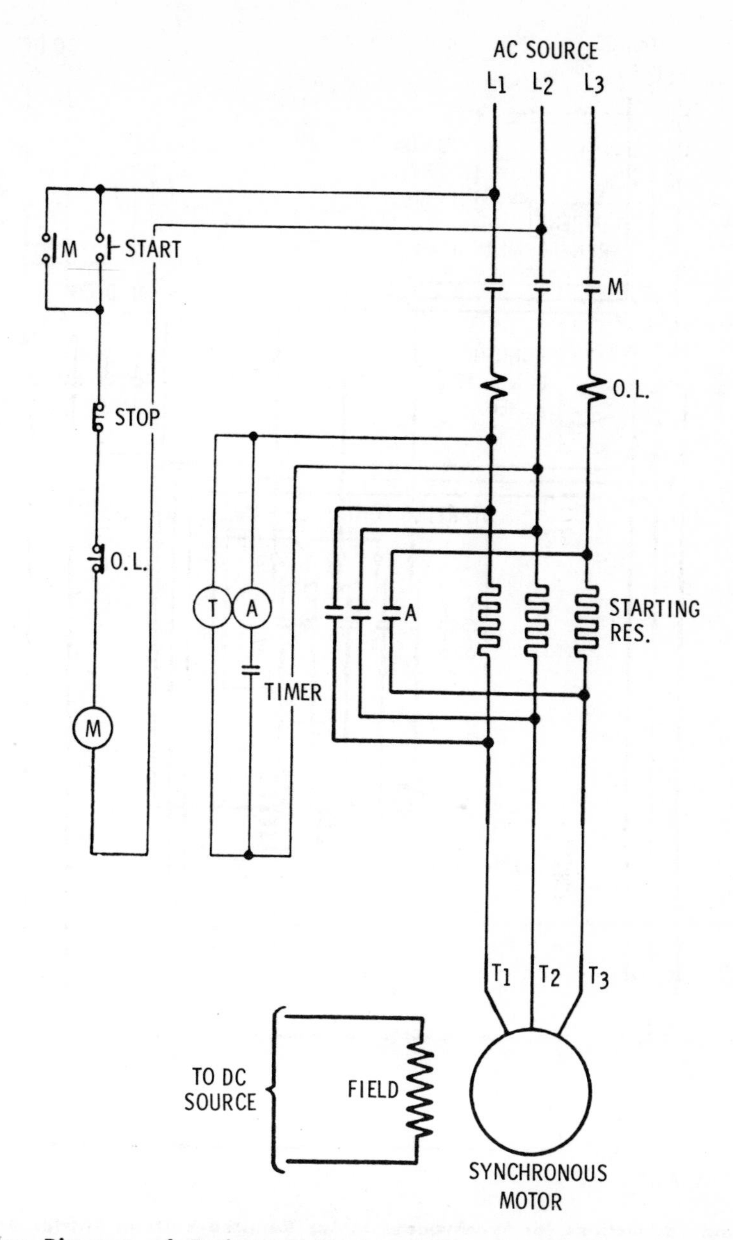

Wiring Diagram of Resistance Starter Equipped with Three-Wire Push Buttons Which Includes Low-Voltage Protection—In the diagram M represents main running contactor, OL, overload relays A, starting contactor and T timing relay.

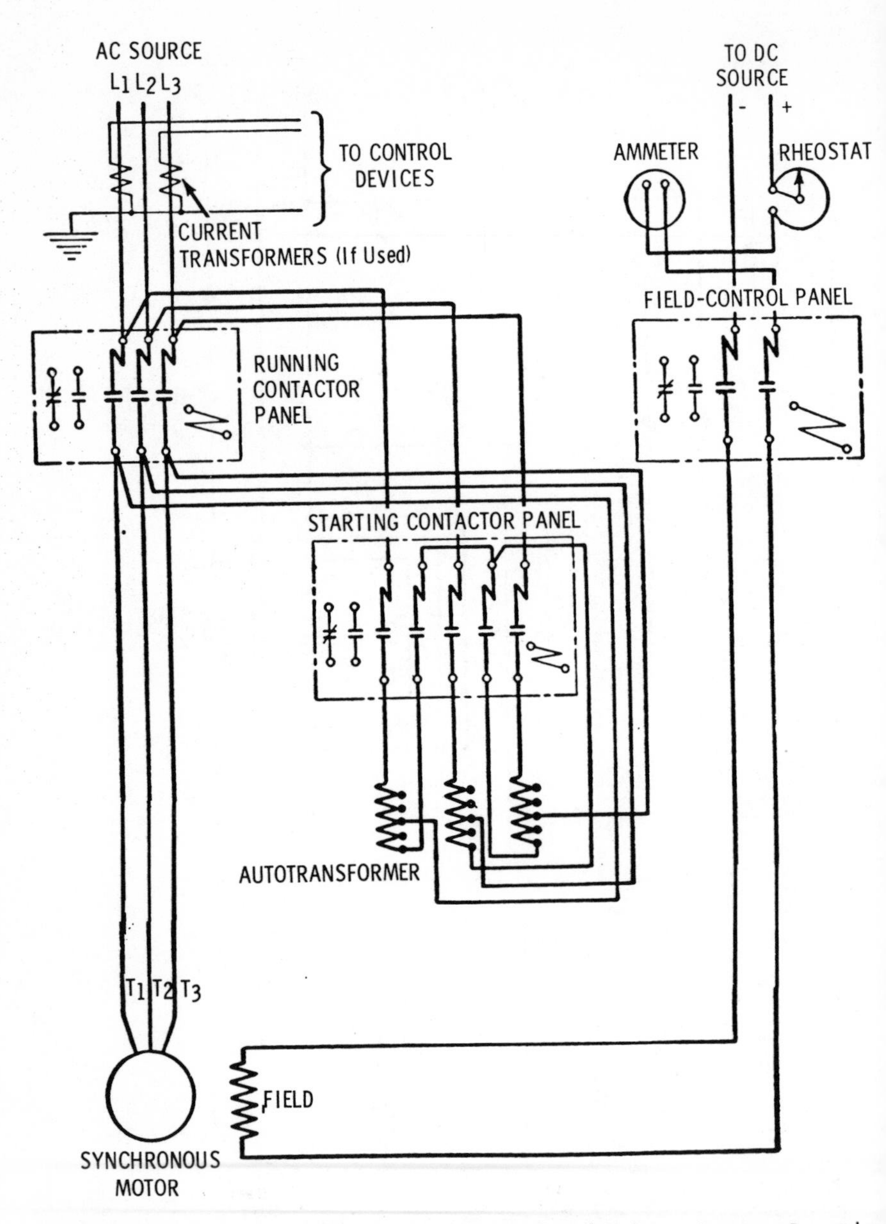

Main Connections for Synchronous Motor Reduced-Voltage Starter—Depending upon the size of the motor and the control scheme used, the complete control wiring may consist of three separate panels as illustrated.

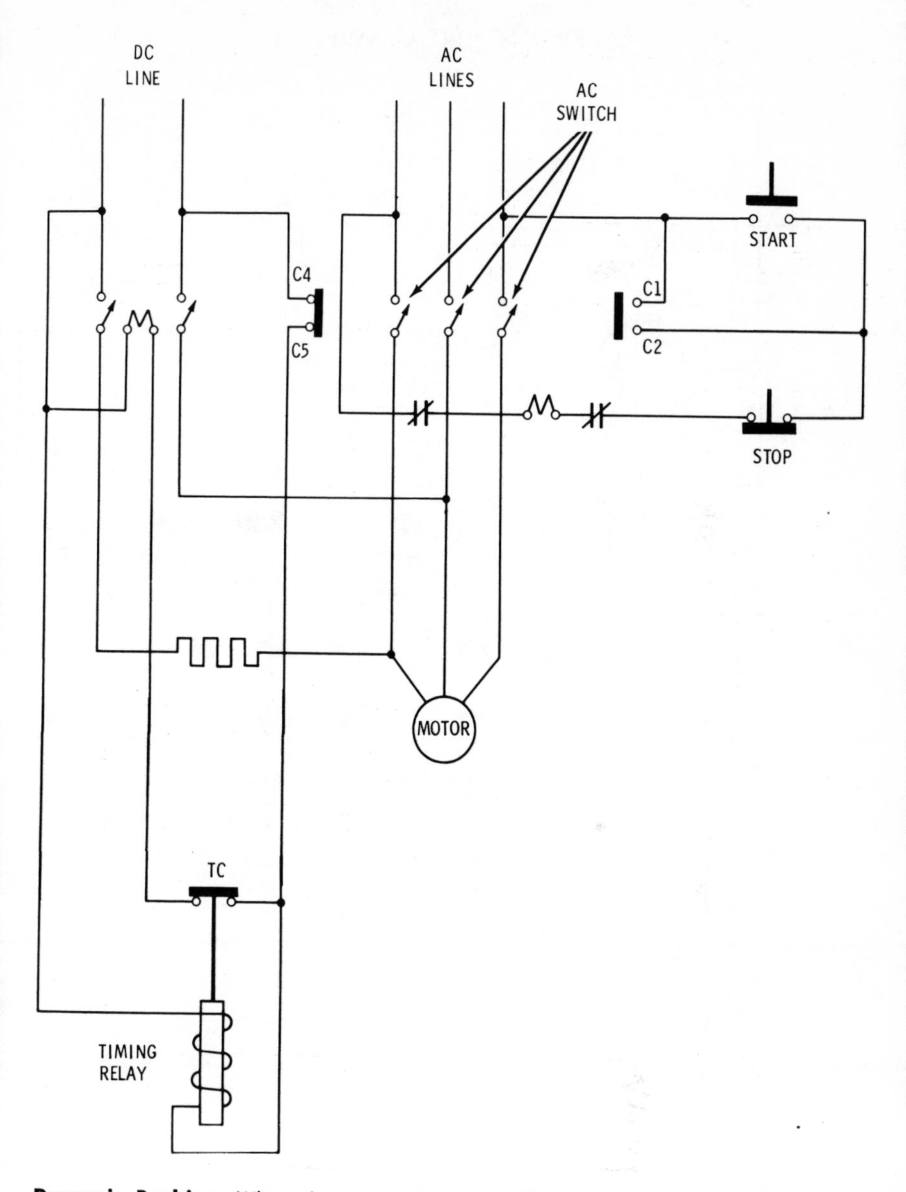

Dynamic Braking—When the stop button is pushed, contacts C4-C5 close, energizing the DC switch, and applying DC power to the stator winding of the motor. The stationary field of this winding induces a high current in the rotor bars. The rotor field reacts with the stator field to produce a retarding torque that stops the motor.

REVERSING DRUM SWITCH

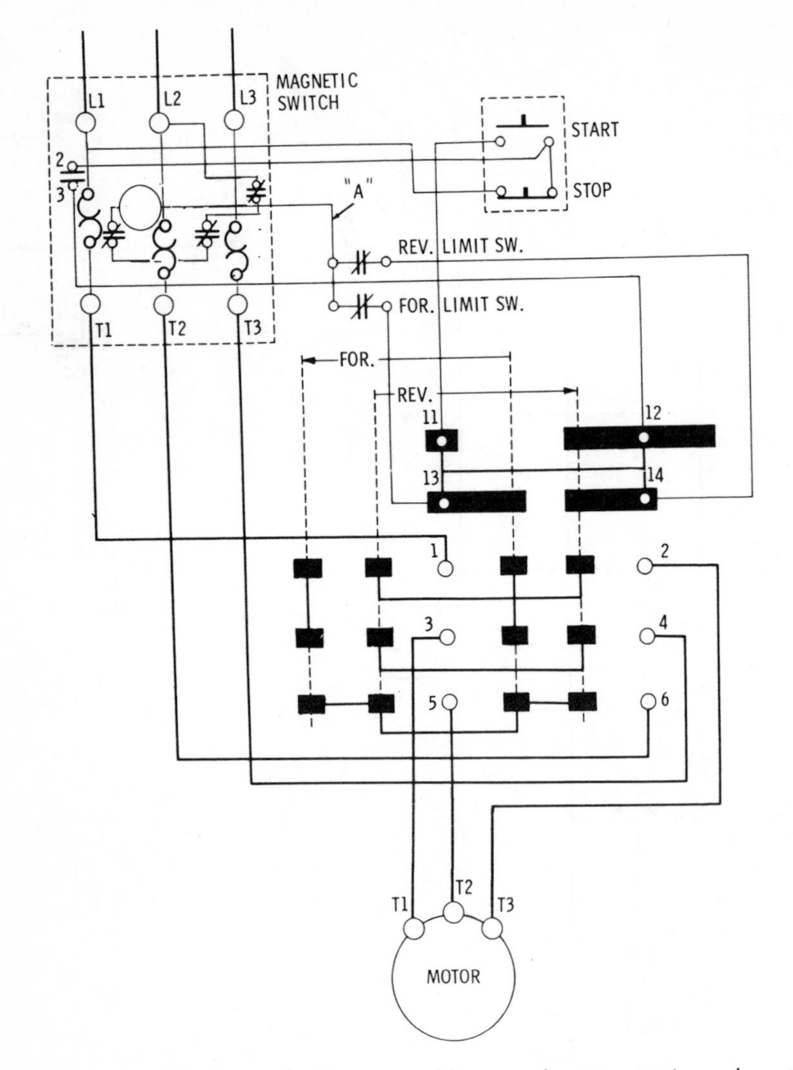

The reversing drum switch shown provides no-voltage protection when the drum switch is used with a full-voltage starter and a start-stop push button station. The starting contact of the interlock is open when the drum switch handle is in either the forward or reverse position and closed when the handle is in the off position. If a voltage failure occurs while the handle is in either position, the starter will not reclose when voltage returns because of the open interlock contact. The handle must be returned to the off position and the start button pressed.

MORE THAN ONE START-STOP STATION USED TO CONTROL A SINGLE STARTER

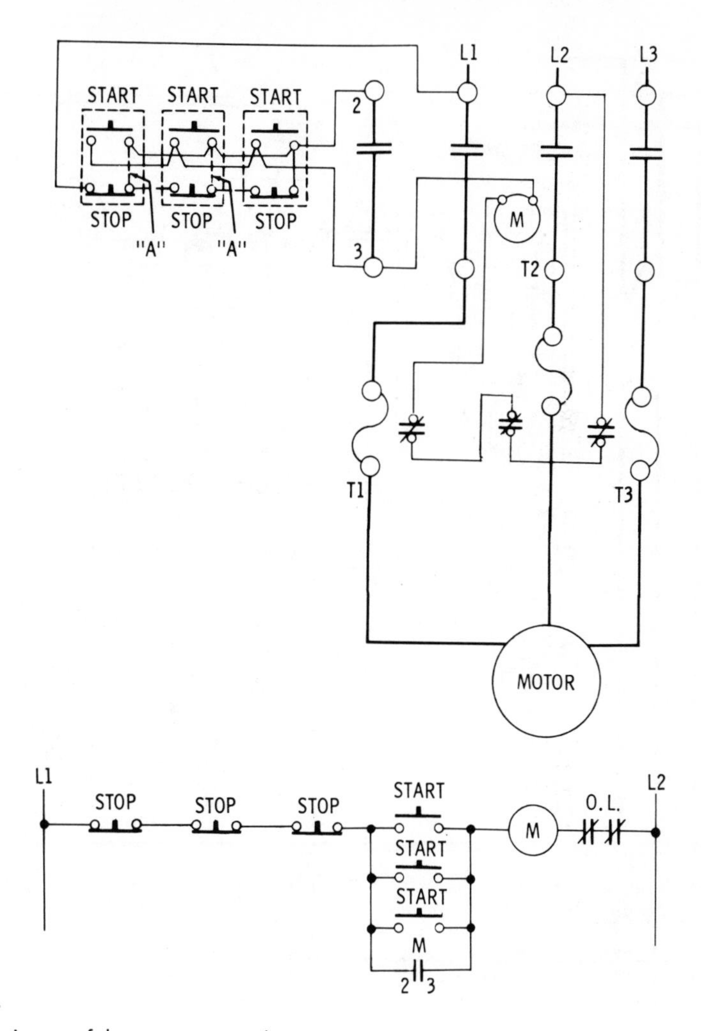

This is a useful arrangement when a motor must be started and stopped from any of several widely separated locations.

Notice that it would also be possible to use only one start-stop station and have several stop buttons at different locations to serve as emergency stops.

If start-stop stations are provided with the connections "A" shown in the diagram, this connection must be removed from all but one of the start-stop stations used.

WIRING DIAGRAM OF TYPICAL AC FEEDER PANEL

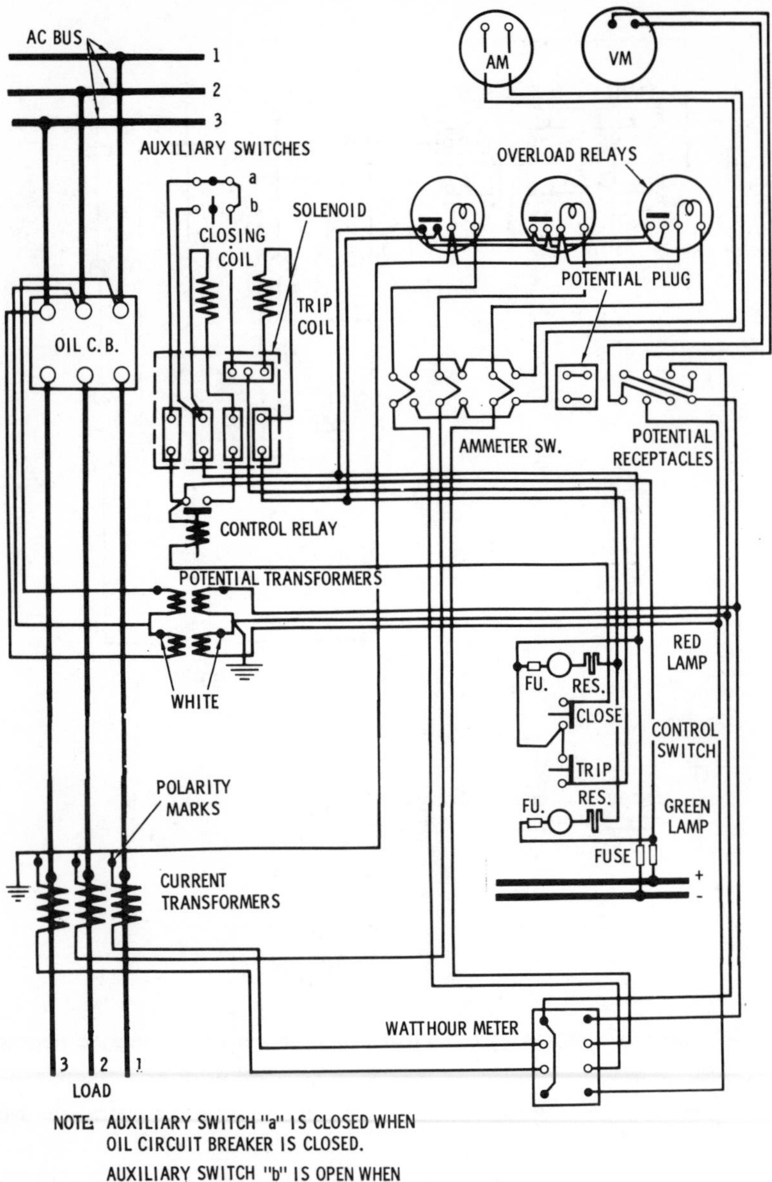

NOTE: AUXILIARY SWITCH "a" IS CLOSED WHEN OIL CIRCUIT BREAKER IS CLOSED.

AUXILIARY SWITCH "b" IS OPEN WHEN OIL CIRCUIT BREAKER IS CLOSED.

WIRING DIAGRAM FOR TYPICAL AC GENERATOR

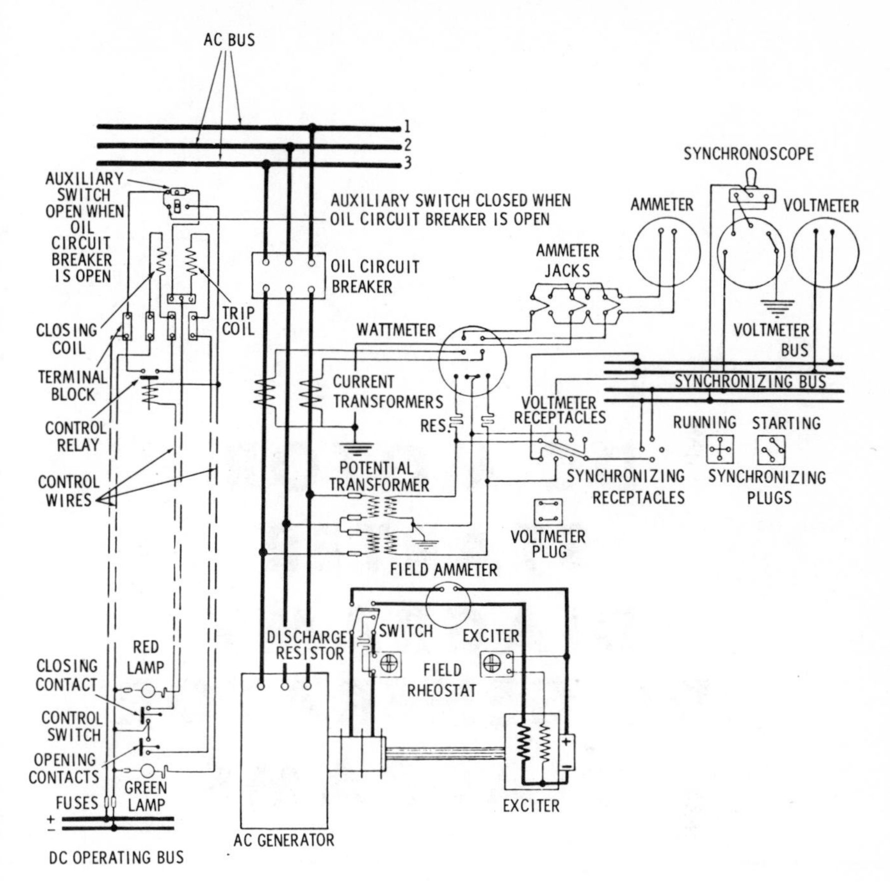

This diagram is typical only of the switching equipment and instruments usually found on an AC generator panel. The AC voltmeter and ammeter readings may be obtained by inserting plugs in their respective receptacles.

Oil circuit breaker control switches and indicating lamps are usually installed on the lower part of the panel, whereas meters and plug receptacles are located on the upper part.

The AC generator voltage may be controlled from two points:

(1) By the exciter field rheostat which controls the terminal voltage of the exciter or (2) By the AC generator field rheostat which varies the resistance of the AC generator field circuit.

DC MOTOR STARTER DIAGRAMS DC GENERATOR DIAGRAMS

The use of DC motors is coming back into use. They are very appropriate for elevators, as the speed control is much more accurate and useful for same. The DC is usually obtained for this purpose with an AC motor and a DC generator set. Since most if not practically all of our present power is alternating current, this makes a fine method of obtaining DC.

With the development of solid state rectifiers of large capacities, DC has become very popular where a number of motors operate together and their speed must be accurately controlled. This type control is used extensively in industry.

This portion will give the basic diagrams for DC generator connections, and basic DC motor diagrams, plus a few schematic diagrams of modern DC motor connections. The basic diagrams, will give an insight into the basic principles of DC motor connections.

DC GENERATOR WIRING DIAGRAM

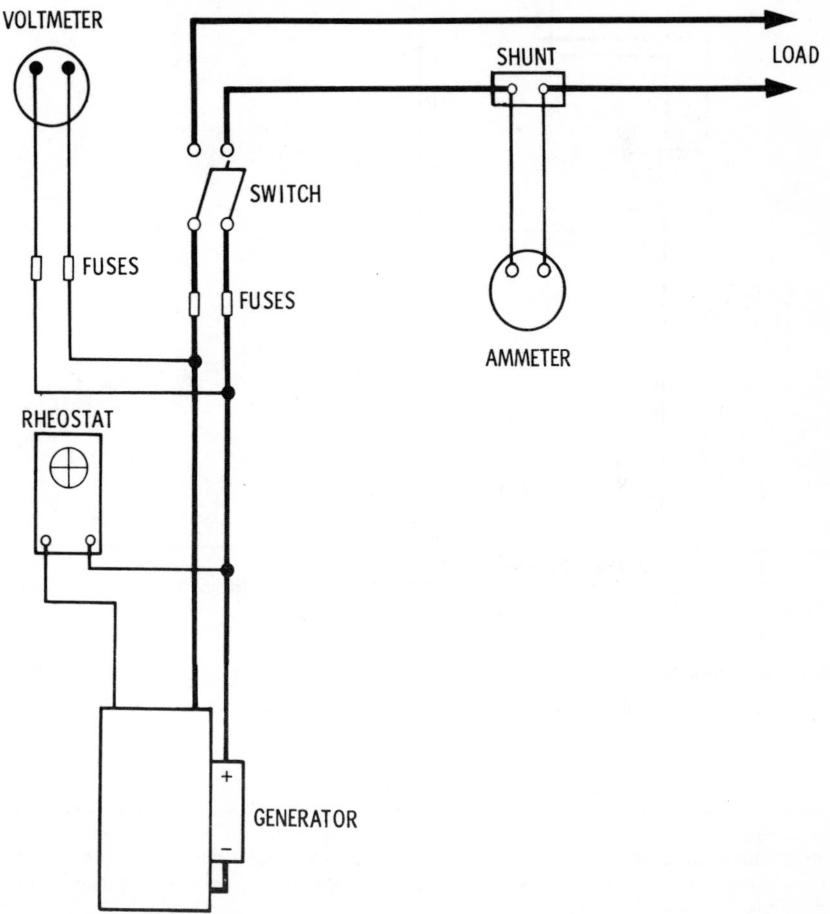

DIRECT-CURRENT GENERATOR CONNECTIONS

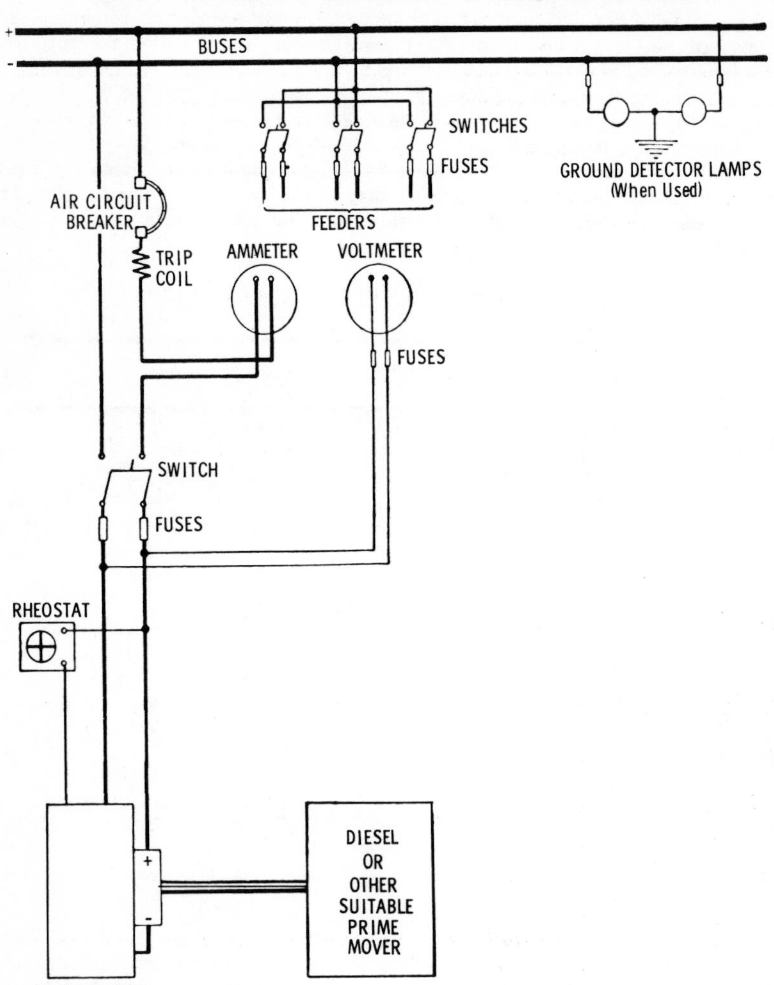

Connection of a Shunt-Wound Direct-Current Generator—The connections are largely self-explanatory, the voltmeter being connected across the main leads at the generator side of the double-pole knife switch. This will enable the operator to read the voltage of the machine at all times, regardless of the position of the main switches. The current indicator (ammeter) is connected in series with the positive lead connecting the machine to the load. The purpose of the overload coil on the circuit breaker is to prevent the current from reaching dangerous proportions, that is, when the current exceeds the calibrated settings of the coil, the breaker trips, disconnecting the generator from its load.

CONNECTION FOR PARALLEL OPERATION OF TWO SHUNT-WOUND DC GENERATORS

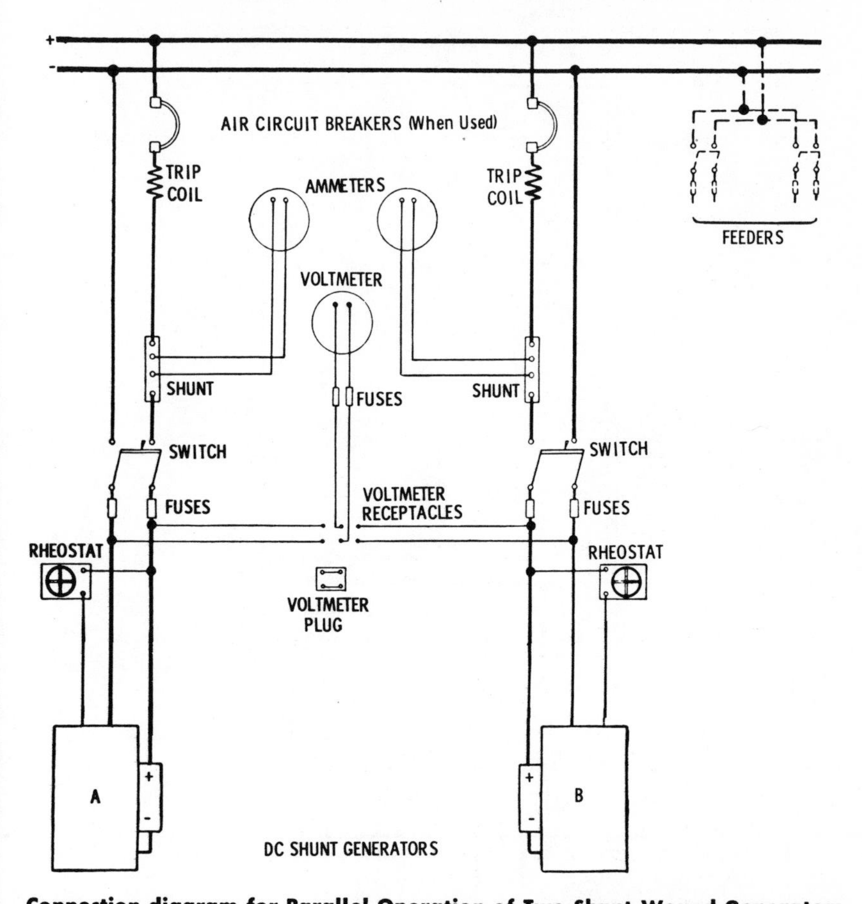

Connection diagram for Parallel Operation of Two Shunt-Wound Generators —It is customary to employ only one voltmeter with the addition of receptacles and a plug as shown. Sometimes a rotary switch arrangement is employed, in which case the receptacles and plug are omitted. In either case, the voltmeter may be connected at will, to read the voltage across the terminals of any one of two or more generators. Occasionally voltage readings across the bus-bars (load) may be included in the voltmeter-switch arrangement. The method for operating the two generators in parallel is as follows: Assume that generator **B,** by means of its prime mover has been brought up to normal speed and is already connected to the bus-bars. Then with the switch and circuit breaker of **A** open, start the prime mover of **A,** and bring it up to speed. Now adjust the field rheostat of **A,** and note the voltmeter reading on this machine. Finally close the circuit breaker and switch of generator **A.**

PARALLEL OPERATION OF 2 DC COMPOUND GENERATORS

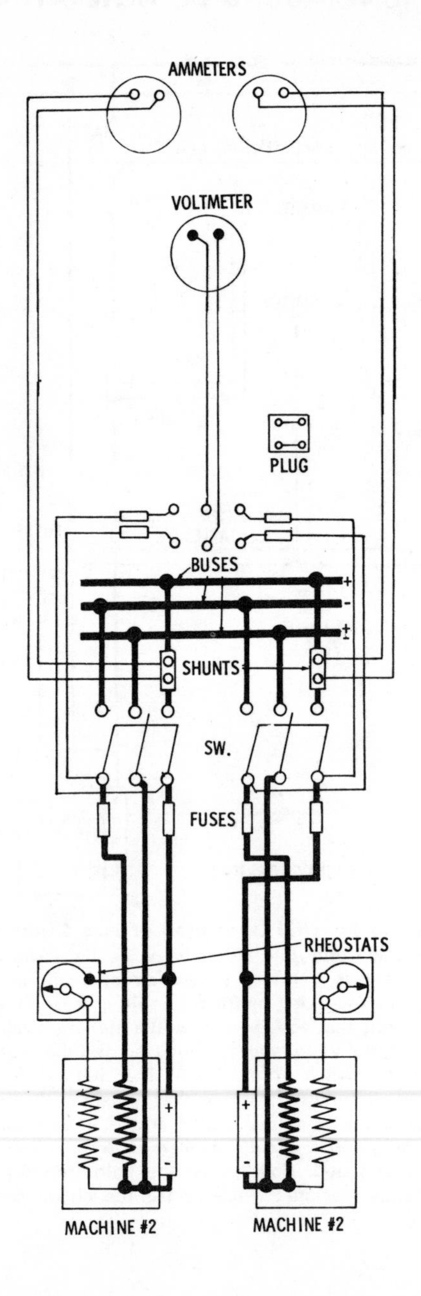

CONNECTION FOR PARALLEL OPERATION OF TWO COMPOUND-WOUND DC GENERATORS

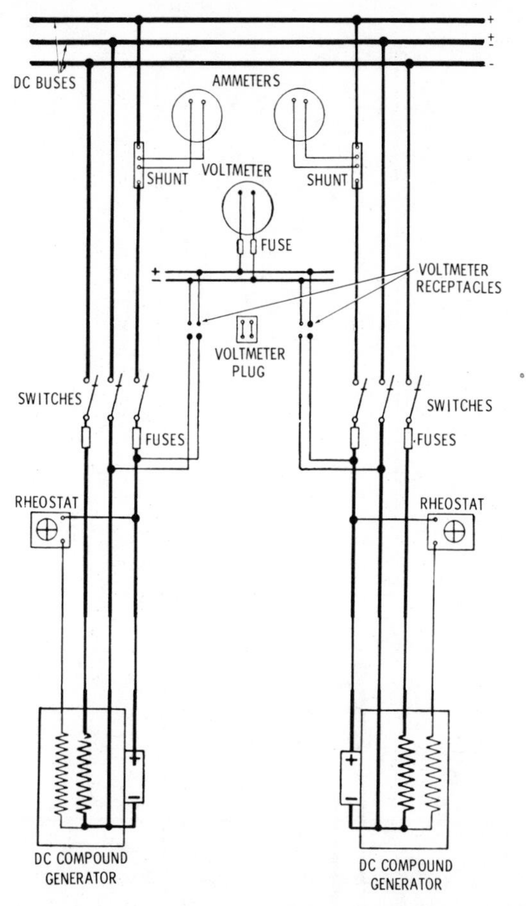

Detail of Connections for Two Compound Generators in Parallel—When two over-compounded generators are to be operated in parallel, it is necessary for a satisfactory division of loads, to parallel their respective series fields. This is accomplished by connecting their negatives together as indicated, and this common connector is usually referred to as the equalizer. The instruments and switches shown are connected in the usual manner, which is similar to that used for connection of shunt generators in parallel, the only addition being the equalizer and connections thereto. It should be noted, however, that the ammeter for each machine should be connected in the lead from the armature to the main bus, and not in the lead from the series field, because if the ammeter is placed in the latter it will read the series field current which may be quite different from the current supplied by the generator to the load connected to the buses.

3-WIRE DC GENERATOR DIAGRAM

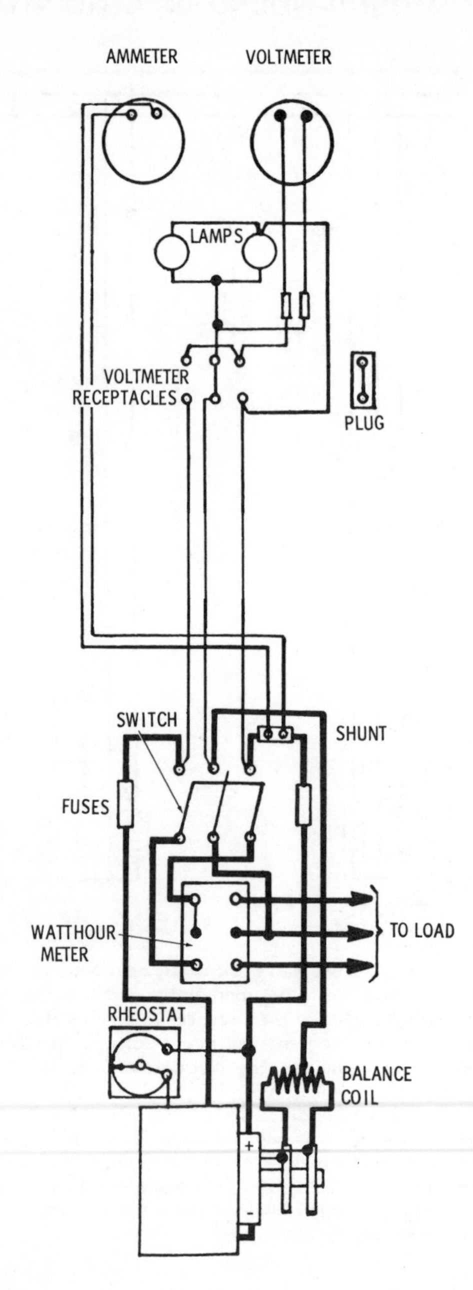

DIRECT-CURRENT, THREE-WIRE GENERATOR

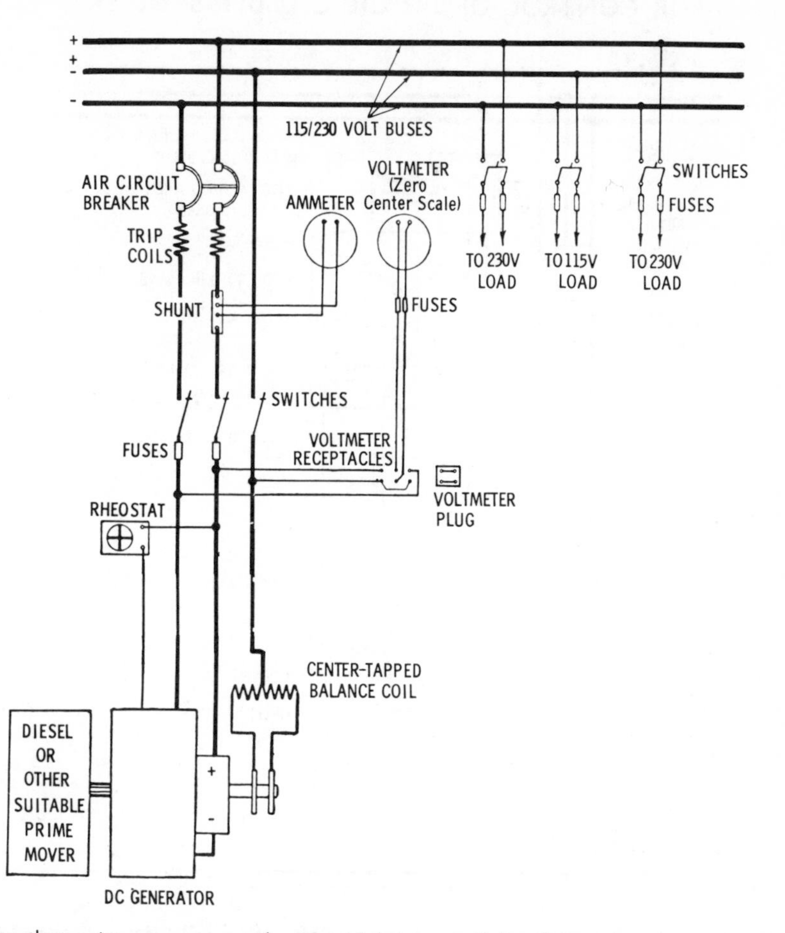

The three-wire generator with external balance coil is often resorted to when it is desired to obtain a three-wire system with a minimum of rotating machinery. The third wire (sometimes misleadingly called neutral) is obtained as follows: To an ordinary generator designed to give a terminal voltage equal to that between the two main wires, are added two slip rings as shown; from these slip rings two leads are brought out and connected to armature points located 180 electrical degrees apart (this connection is not shown in the diagram). Collectors from the slip rings are connected from the two ends of the balance coil wound on an iron core, and the middle point of this coil is finally connected to the third wire. It should be observed that in a system of this kind it is necessary to balance the load between the two main wires and the wire leading from the balance coil as closely as possible. The amount of unbalance allowed for a properly designed system (usually specified by the manufacturers) should not exceed approximately 10% of the total current.

CONNECTION FOR DC OPERATED SOLENOID USED FOR CONTROL OF DC OIL CIRCUIT BREAKER

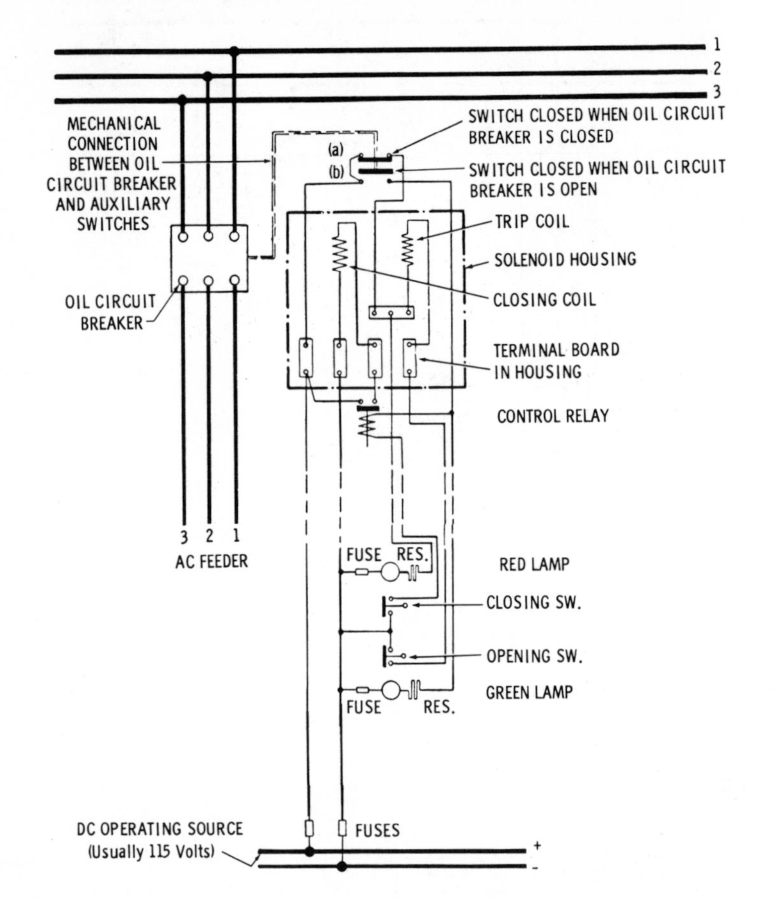

Typical Connection Diagram for a Remote Controlled Oil Circuit Breaker—In this method of operation, it is necessary, however, that an unfailing supply of direct current be available. The operation of the breaker is accomplished as follows: Assume the breaker is open and the condition for its closing has been established. When the main breaker is open, auxiliary switch marked (b) is closed, and the green lamp on the instrument board is lighted. When the closing switch is operated, the coil of the control relay whose contacts are normally open, becomes energized and closes its contact. This in turn actuates the closing coil (which is mechanically connected with the breaker contacts) closing the breaker. This closing of the breaker simultaneously reverses the position of the auxiliary switches, opening the previously closed switch marked (b) and closes switch marked (a), which in turn extinguishes the green lamp and lights the red. The breaker may be opened in a similar manner by operating the lower of the two switches on the control board.

ELECTROLYTIC GENERATOR WITH POLARITY-DIRECTIONAL PROTECTION

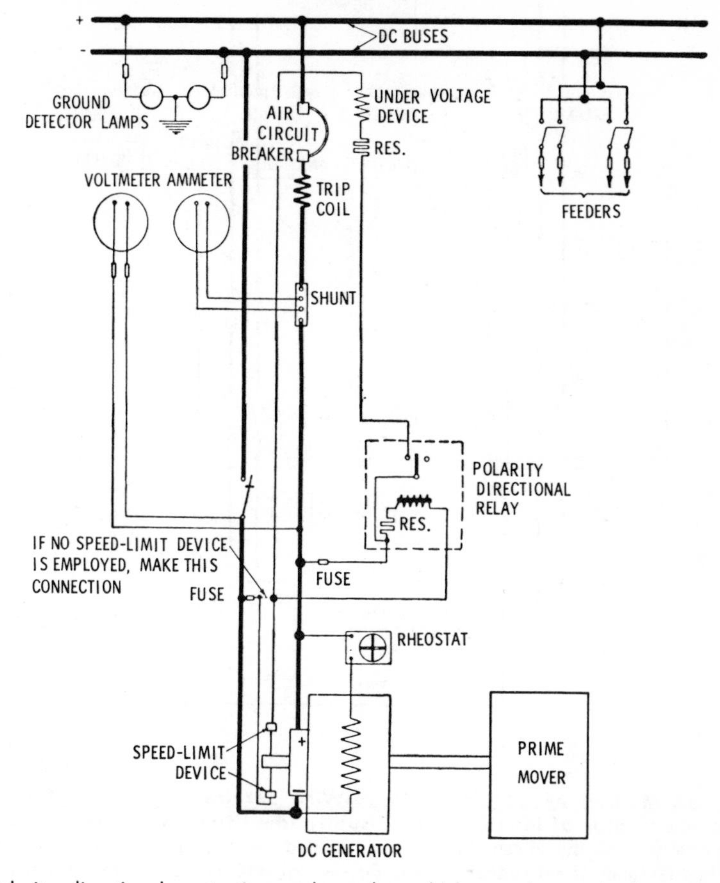

Polarity directional protection such as that which may be used where protection against sparks is of the utmost importance, as, for example, where hydrogen or other high explosive materials are manufactured. In such cases it is of the utmost importance that the polarity is not inverted, as the explosion resulting from such a condition might endanger both life and property. The polarity-directional relay consists essentially of a pair of stationary permanent magnets, a rotatable soft iron armature pivoted within a stationary coil and a double-throw set of contacts. The winding of the coil is of such direction that when potential is applied, connected with the proper polarity, the armature tends to rotate in a direction to keep the contacts closed to one side. A spring, in tension, tends to pull the armature back, open the closed contacts, and close the contacts on the other side. When an inversion of the polarity occurs, the spring overcomes the action of the magnet, which opens the circuit breaker.

DIRECT-CURRENT GENERATORS IN THREE-WIRE SERVICE

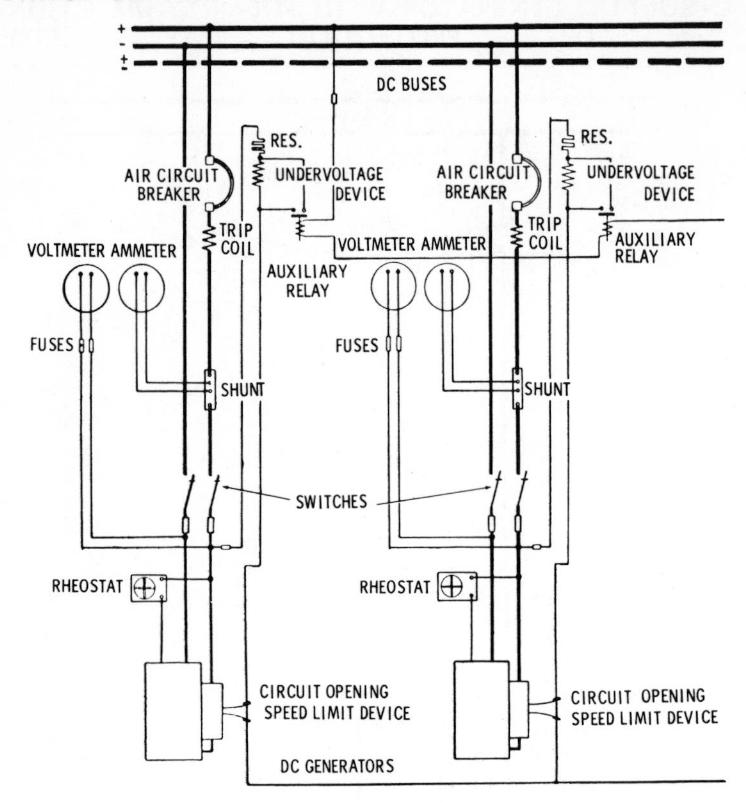

Common Method of Obtaining Three-Wire Service by Means of a Small Motor-Generator of Identical Size, Usually Identified as a Balancer Set—The additional wire or the so-called neutral is obtained and brought out from the common lead in the balancer set connecting the positive of one machine with the negative of the other. By the employment of a system of this kind, it is possible to establish better economy, in that the higher potential between the main generators positive and negative leads can be utilized for power service. The amount of this saving in copper may best be understood by the fact that the weight of the connectors (and therefore the cost) required to transmit a given amount of power at a given efficiency is inversely proportional to the square of the line voltage. When establishing such a system, however, it is necessary to employ some protective scheme to guard against the unbalance of voltage in case the balancer set should become disconnected. The voltage differential relay shown, will protect against unbalanced voltage, and as this relay is practically instantaneous in action, will protect against false operation caused by transitory disturbances. Definite time limit relays are utilized in the contact circuits.

METHOD OF UNBALANCED VOLTAGE PROTECTION

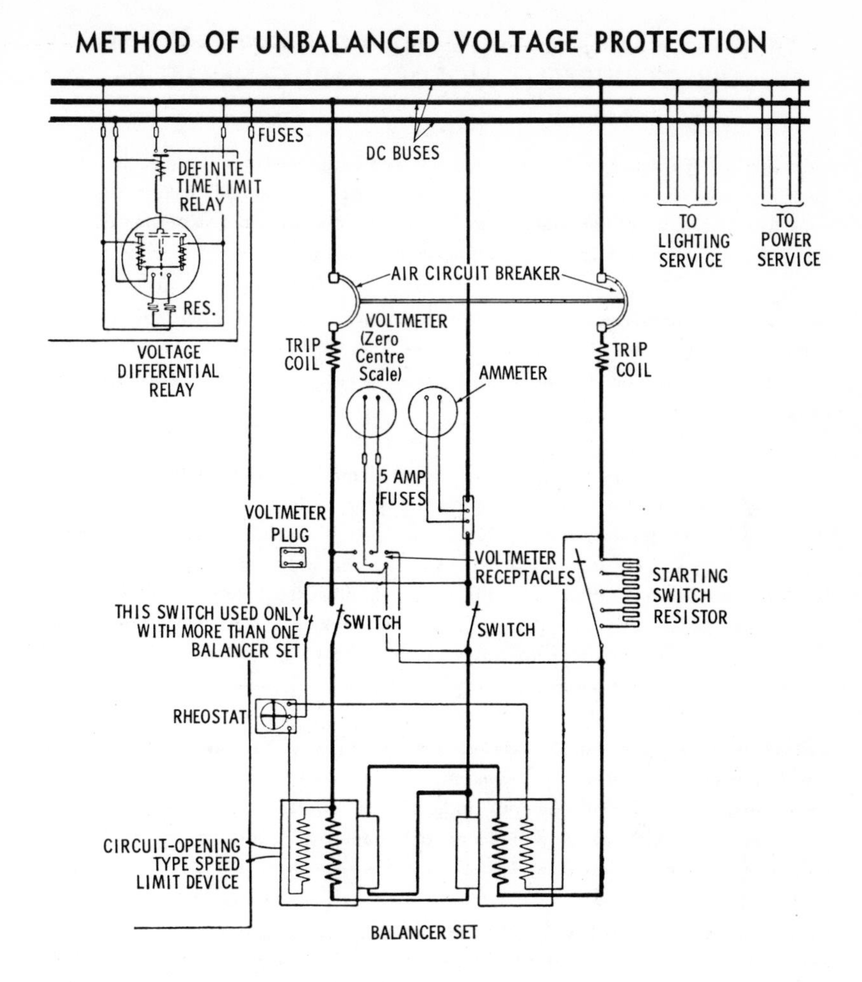

The voltage differential relay functions principally as follows: The relay consists essentially of a pair of solenoids of equal characteristics, and each with a plunger core connected to a balanced lever which actuates the contacts. One winding is connected across one circuit and the other winding across the other circuit of the two circuits to be differentially protected. As long as the voltages are equal, the balance lever is in equilibrium and the contacts remain open. When for any reason the voltage becomes unequal, the unequal pull of the two solenoids tends to close the contacts and when this difference in voltage reaches the value at which the relay is calibrated, the contacts close instantaneously energizing the definite time limit and auxiliary relays, which in turn shorts the coil of undervoltage device on the circuit breakers, tripping the breakers and disconnects the generators from the buses.

TERMINAL MARKINGS AND CONNECTIONS FOR DIRECT-CURRENT MOTORS AND GENERATORS

Marking of Terminals—The purpose of applying markings to the terminals of electric power apparatus according to a standard is to aid in making up connections to other parts of the electric power system, and to avoid improper connection that may result in unsatisfactory operation or damage.

Location of Markings—The markings are placed on or directly adjacent to terminals to which connections must be made from outside circuits, or from auxiliary devices which must be disconnected in order to facilitate shipments from the manufacturer.

Precautions—Although the system of terminal markings with letters and subscript numbers gives information, facilitating the connections of electrical machinery, there is the possibility of finding the terminals marked without system or according to some system other than standard (especially on old machinery or machinery of foreign manufacture). There is a further possibility that internal connections have been changed or that errors were made in markings. It is therefore advisable, before connecting apparatus to power supply systems, to make a check test for phase rotation, phase relation, polarity and equality of potential.

Subscript Numerals on Direct-Current Machinery Terminals—As applied to the terminals of direct-current windings of generators, motors and synchronous converters, the subscript numerals indicate the direction of current flow in the windings. Thus, with standard direction of rotation and polarity, the current in all windings will be flowing from 1 to 2, or from a lower to a higher subnumber.

Direction of Rotation for DC Motors—Connections shown on the following pages will give the standard **counter-clockwise** rotation facing the end opposite the drive. To obtain the opposite direction, or **clockwise** rotation, the **armature or main field leads must be reversed.**

Direction of Rotation for DC Generators—The standard direction of rotation for DC generators is **clockwise** when facing the end of machine opposite the drive, usually the commutator end of the machine. Direct-current generators with connections properly made up for standard rotation (clockwise) will not function if driven counter-clockwise, as any small current delivered by the armature tends to demagnetize the fields and thus prevents the armature from delivering current. If conditions call for reversed rotation, connections should be made up with either the armature leads transposed or the field leads transposed.

TERMINAL MARKINGS AND CONNECTIONS FOR DIRECT-CURRENT MOTORS AND GENERATORS —continued

Motor Generators—Any direct-current machine can be used either as a generator or as a motor. For desired direction of rotation, connection changes may be necessary and should be accomplished as previously described. The conventions for current flow in combination with the standardization of opposite directions of rotation for direct-current generators and direct-current motors works out so that any direct-current machine can be termed **generator** or **motor** without change in terminal markings. A direct-current motor or a direct-current generator, by direct coupling constitutes a motor generator. With such coupling, direction of rotation of motor and generator are necessarily reversed when each is from the **end opposite the drive.** The standard clockwise rotation for direct-current motors meets such coupling requirements without change in standard connections or rotation for either direct-current machine.

Coupling of AC and DC Motors and Generators—In the same manner as that already described for direct-current motors and generators, a direct-current motor may be coupled to an alternating-current generator without changing from the standard in either individual machine. When, however, the coupling of an alternating-current motor to a direct-current generator becomes necessary, this coupling cannot be made without rotation other than standard for one of the two machines. Since the rotation of the alternating-current machine is usually the more simply changed, it is general practice to operate a motor generator with clockwise rotation viewed from the generator end.

Example on How to Change Direction of Rotation in a DC Motor—When brushes are set for standard counter-clockwise rotation, it will be necessary to change assembly of brush holders for trailing operation. With reference to page 132 showing diagram and connections of a typical motor, proceed as follows: 1. Change connection as shown for clockwise rotation. 2. To change assembly of brush holders for trailing operation, first lock the armature in position of one brush on the commutator surface. Mark the brush holder stud of this brush X and studs of the opposite polarity Y. Raise all brushes in the holders, remove holders from the studs and reassemble them on the same studs in the reverse direction. Lower brush holders until the distance between the bodies and commutator surface is 3/32 in. then shift brush holder yoke until the nearest brush from either stud Y exactly fits over the space previously occupied by the brush X. The leads from the brush holder studs, one from the commutating fields and the other from the terminal board (lead A_1), should be interchanged in the studs. Erase paint mark on bearing housing and make a new mark to line up with the mark on the brush yoke.

TERMINAL MARKINGS AND CONNECTIONS FOR DC SHUNT-WOUND MOTOR

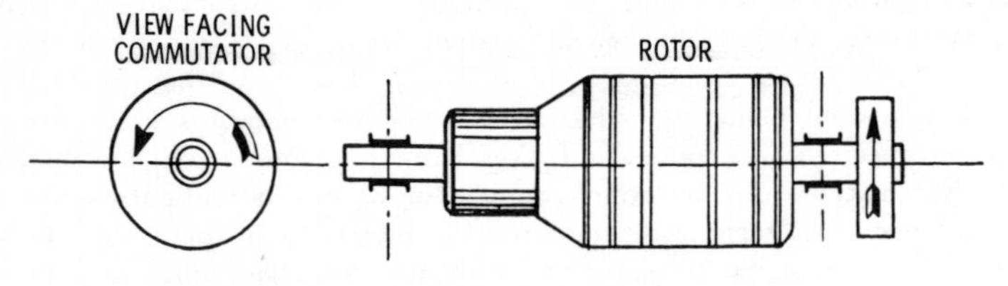

STANDARD DIRECTION OF ROTATION COUNTERCLOCKWISE WHEN FACING COMMUTATOR END OF MOTOR

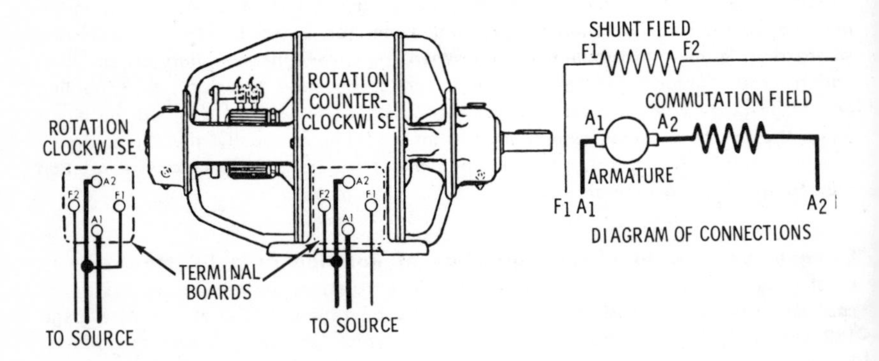

The above drawings represent a typical shunt-wound motor, with terminal connections for either the standard counter-clockwise rotation or clockwise rotation, which sometimes is utilized to facilitate the proper functioning of machinery to be operated.

All motor and control wiring should be carefully installed in accordance with the National Electrical Code and any local requirements, and should be of ample capacity based on a maximum line voltage drop of 2 per cent at full load current.

Before operation, make sure that voltage on motor and control nameplates corresponds with that of power supply.

TERMINAL MARKINGS AND CONNECTIONS FOR DC COMPOUND-WOUND MOTORS

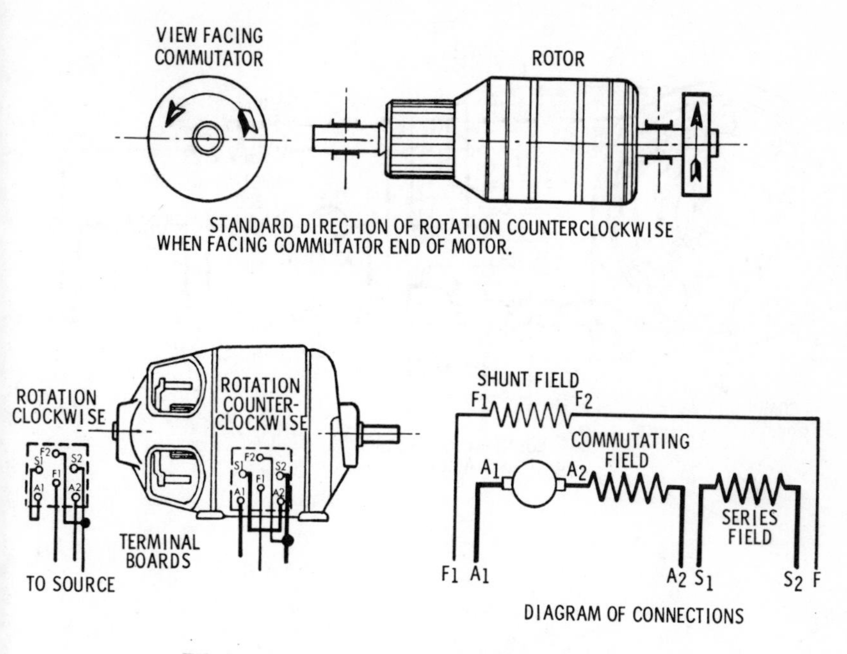

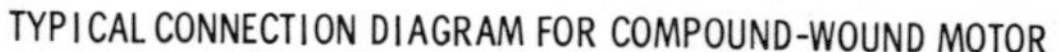
TYPICAL CONNECTION DIAGRAM FOR COMPOUND-WOUND MOTOR

Operation of Motors—Before placing the motor in service for the first time the following precautions should be observed:

Dry out all moisture. If the motor has been exposed to moist atmosphere for a long time while in transit or storage (or has been idle for a long period after installation in moist atmosphere) it should always be dried out thoroughly before being placed in service. If possible, place the motor in an oven and bake at a temperature not exceeding 85°C.

Fair results can be obtained by enclosing the motor with canvas or other covering, inserting some heating units or incandescent lamps to raise the temperature, and leaving a hole at the top of the enclosure to permit the escape of moisture. The motor may also be dried out by passing a current at low voltage (motor at rest) through the field windings to raise the temperature but not to exceed 85°C. The heat should be raised gradually until the whole winding is of this uniform temperature.

TERMINAL MARKINGS AND CONNECTIONS FOR STANDARD AND CLOCKWISE ROTATION

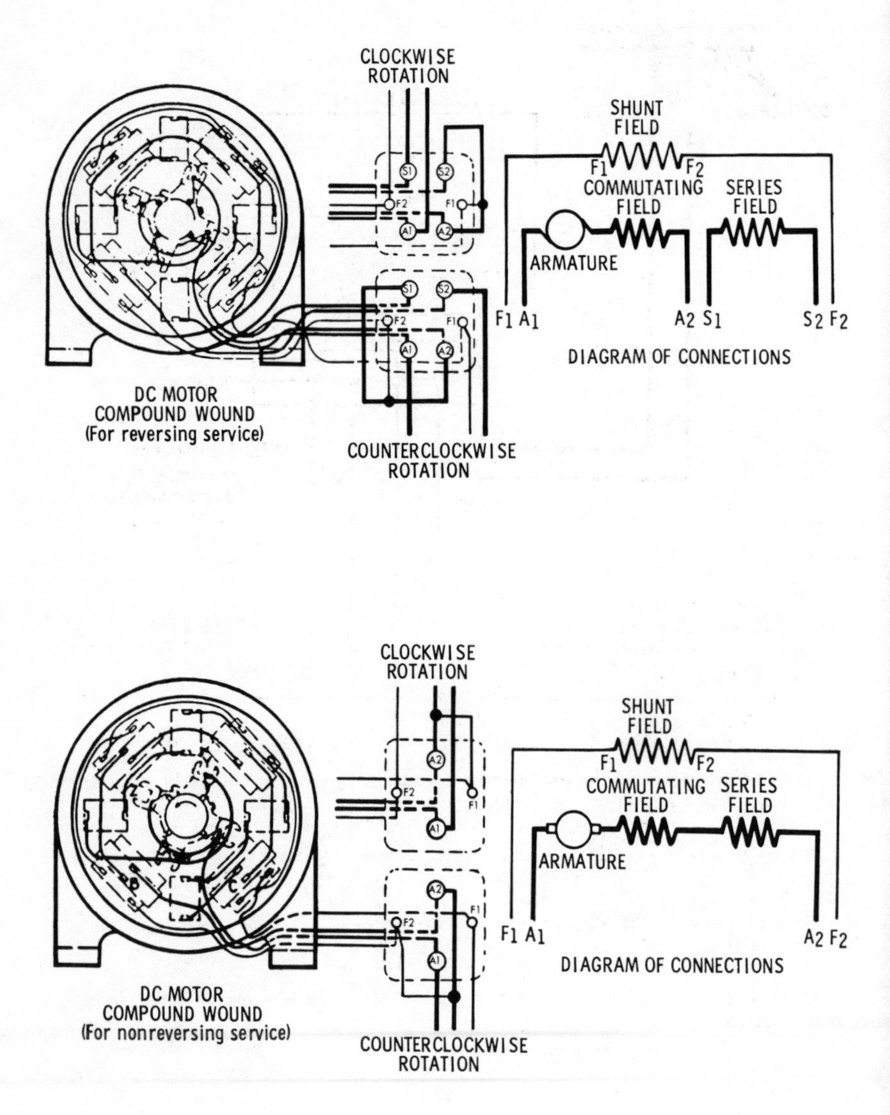

STARTING RHEOSTAT AND CONNECTION DIAGRAM FOR DC COMPOUND-WOUND MOTOR

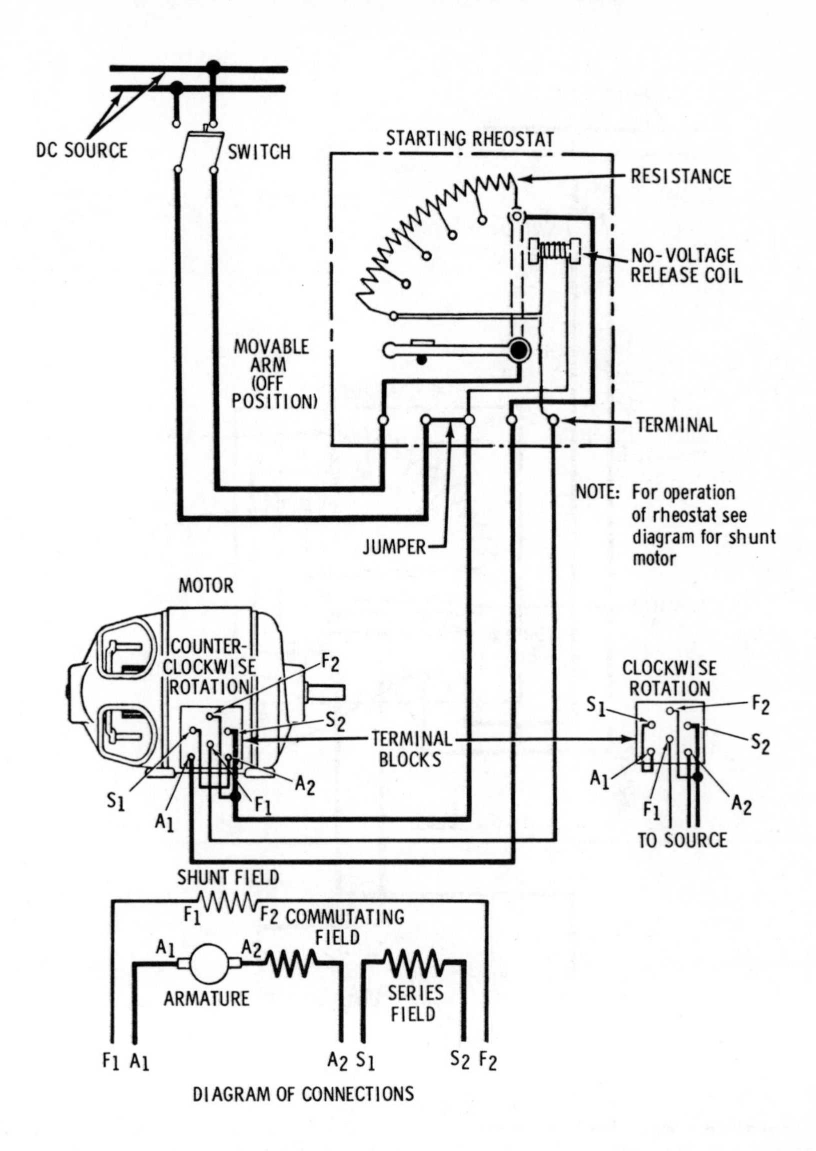

STARTING RHEOSTAT AND CONNECTION DIAGRAM FOR DC SERIES-WOUND MOTOR

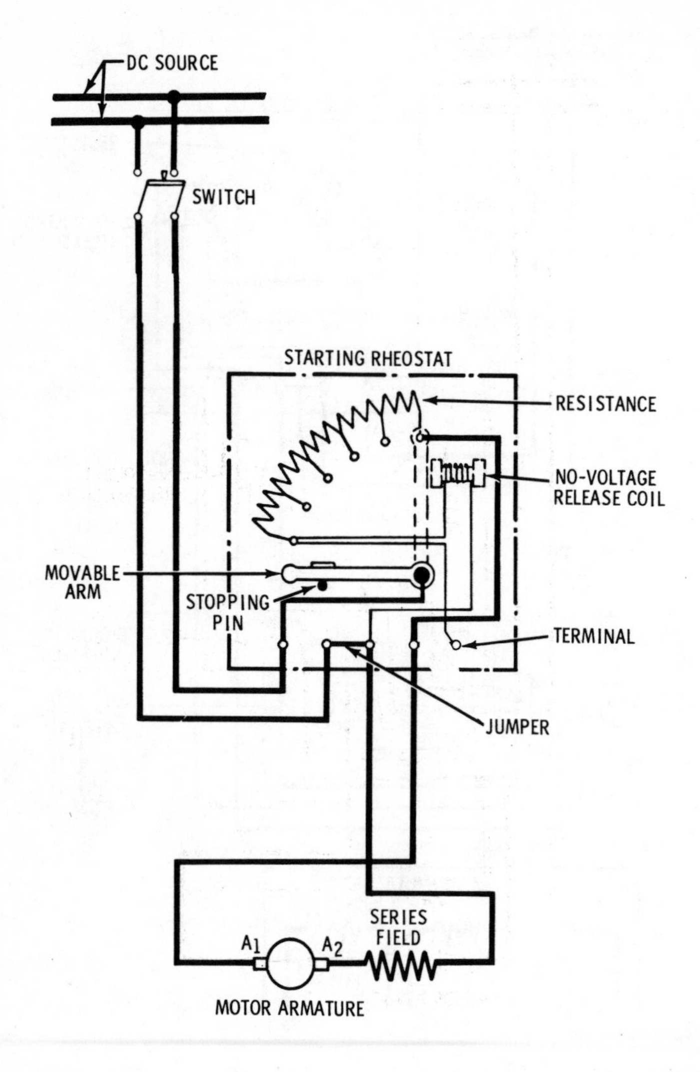

Operation—When field is interrupted due to an open connection or failure of source, the No Voltage release coil will automatically release the moveable arm, which by action of its spring (not shown in diagram) returns to its "Off" or starting position.

STARTING RHEOSTAT AND CONNECTION DIAGRAM FOR DC SHUNT-WOUND MOTOR

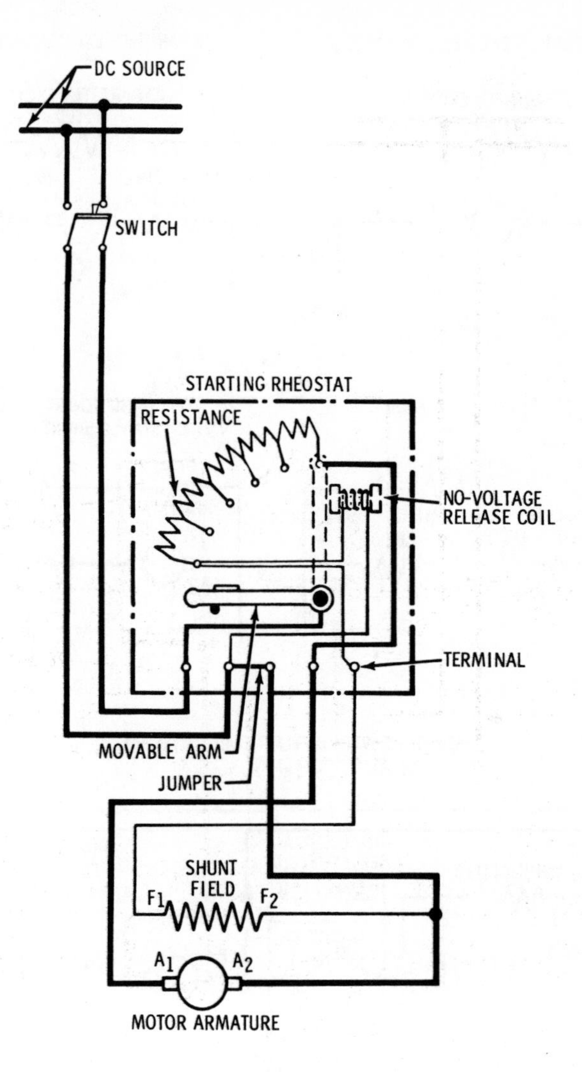

Operation—When voltage fails or shunt field is interrupted the No Voltage release coil will automatically release the moveable arm, which is returned to its starting position by action of holding spring (not shown in diagram). This method of starting will prevent accidental application of a heavy current through the motor armature, causing fuses to blow, or serious damage to motor.

CONNECTION DIAGRAMS FOR DC GENERATORS (TWO-WIRE)

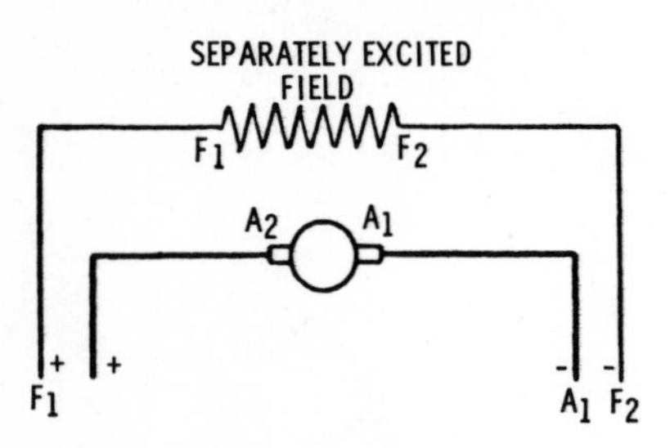

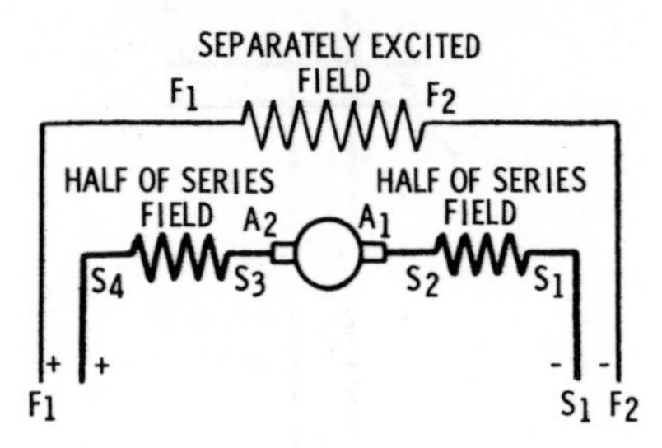

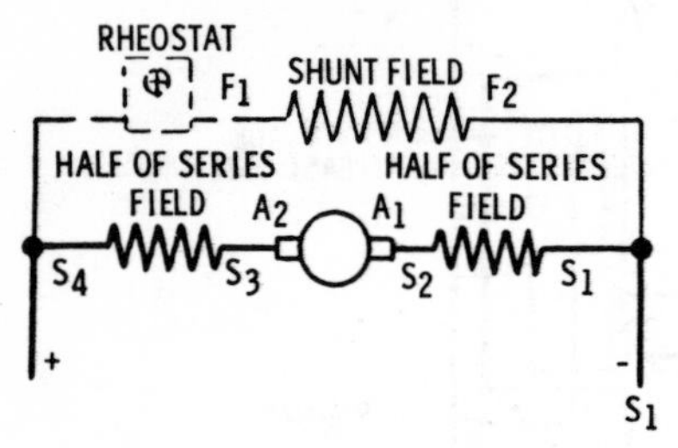

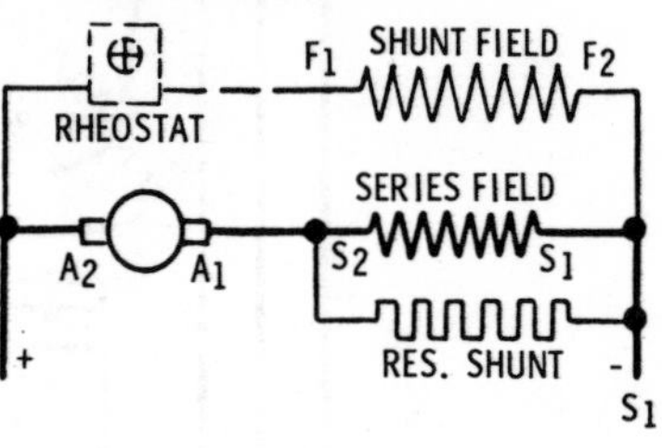

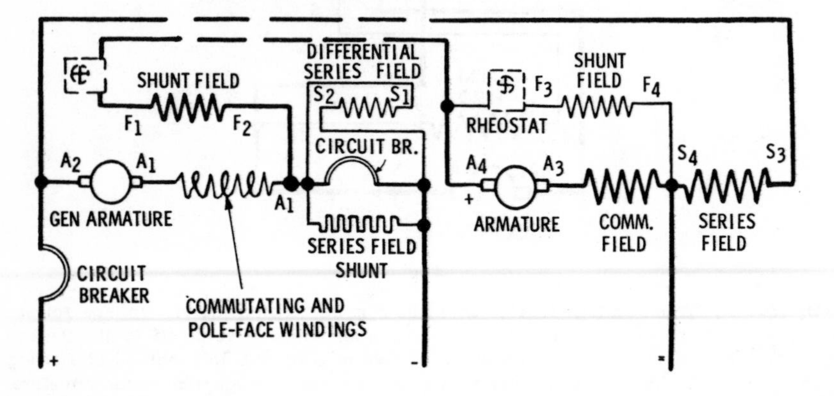

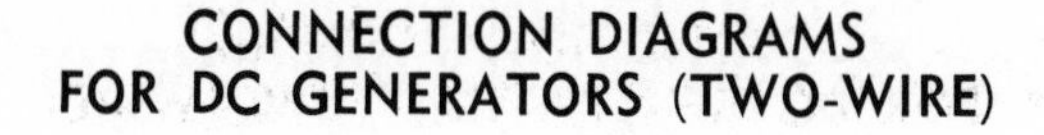

CONNECTION DIAGRAMS FOR DC GENERATORS (TWO-WIRE)

SERIES GENERATOR WITHOUT COMMUTATING POLES

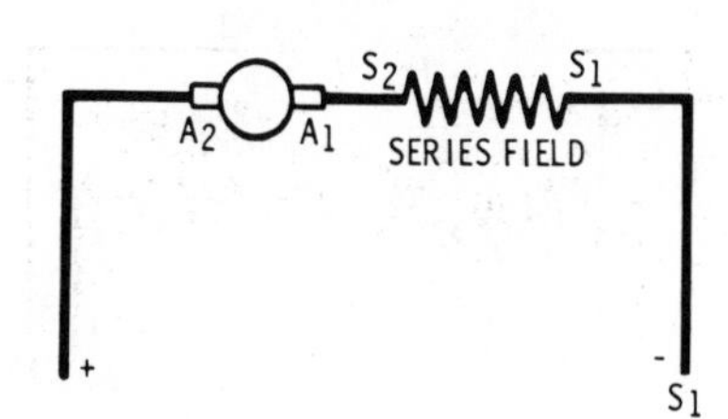

SHUNT GENERATOR WITHOUT COMMUTATING POLES

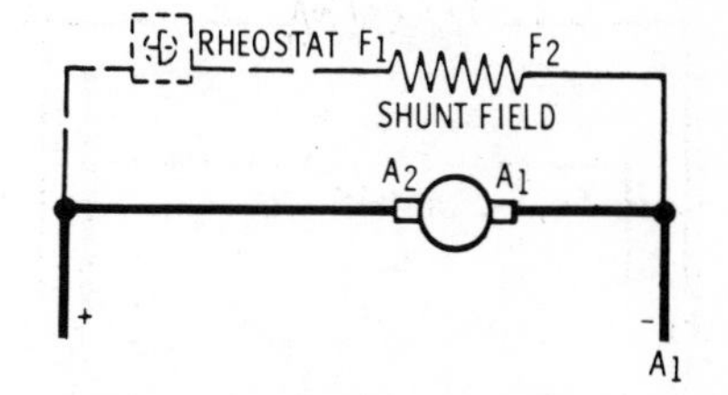

SHUNT GENERATOR WITH COMMUTATING POLES

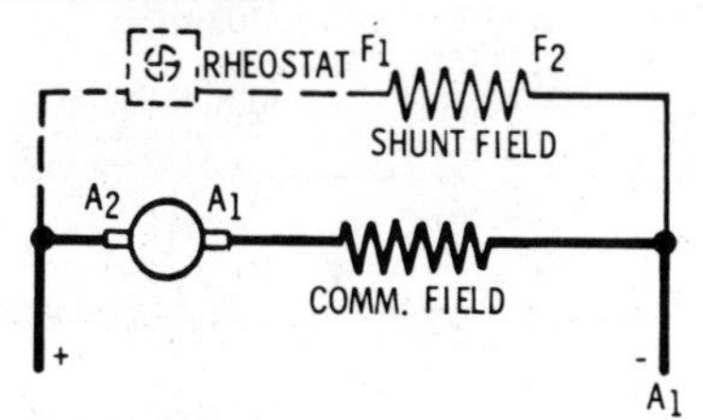

SHUNT GENERATOR WITH COMMUTATING AND COMPENSATING FIELDS

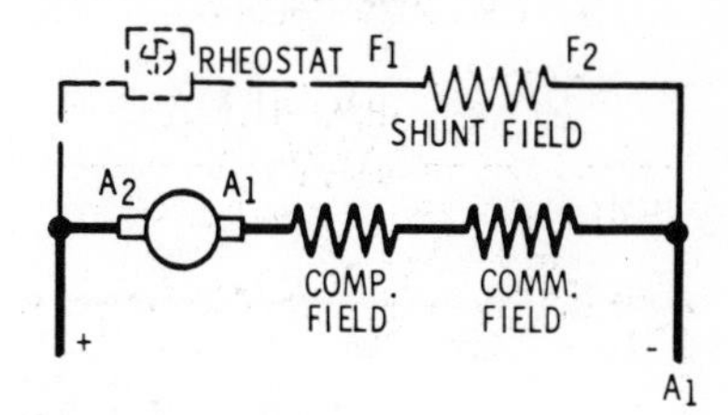

COMPOUND GENERATOR WITHOUT COMMUTATING POLES

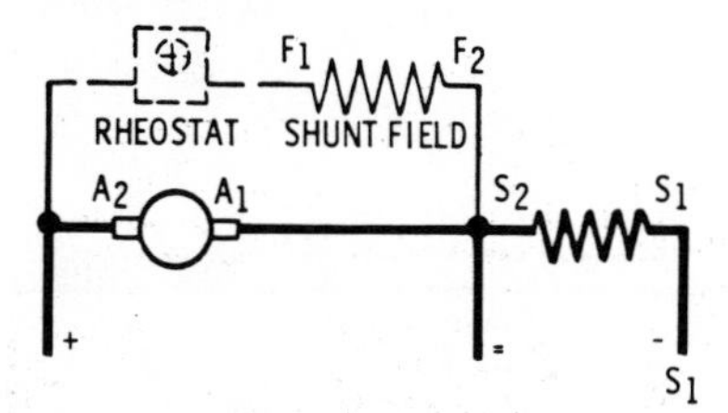

COMPOUND GENERATOR WITH COMMUTATING POLES

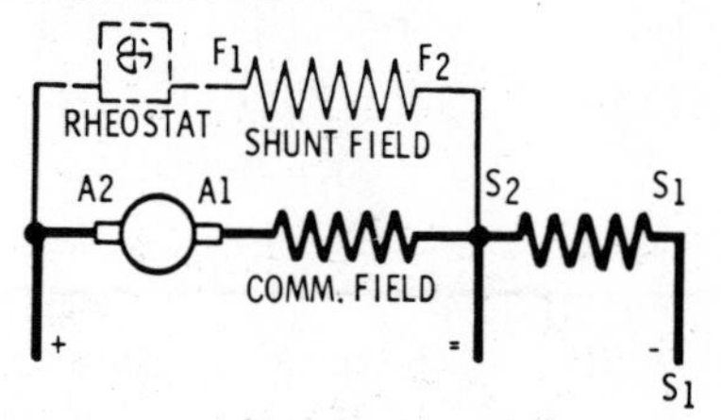

COMPOUND GENERATOR WITH COMMUTATING POLES AND COMPENSATING FIELD

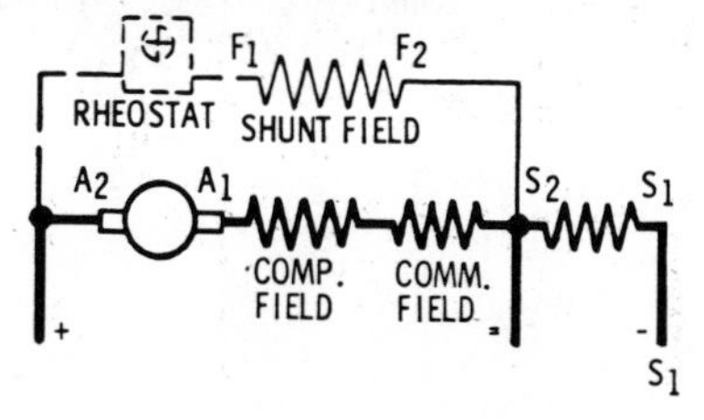

SEPARATELY EXCITED GENERATOR WITH COMMUTATING POLES

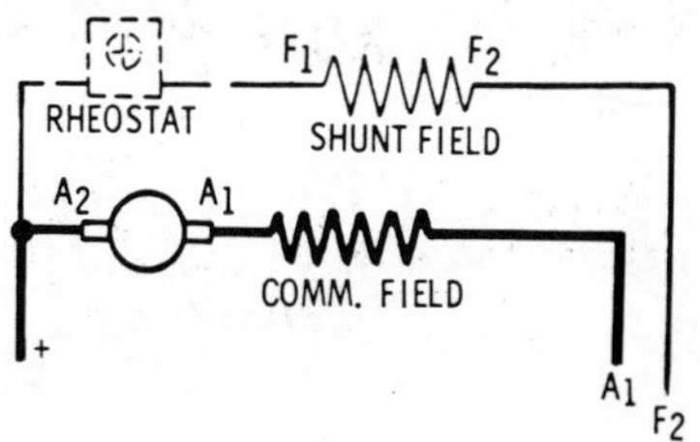

CONNECTION DIAGRAMS FOR DC COMPOUND-WOUND MOTORS

NONREVERSING COMMUTATING POLE TYPE

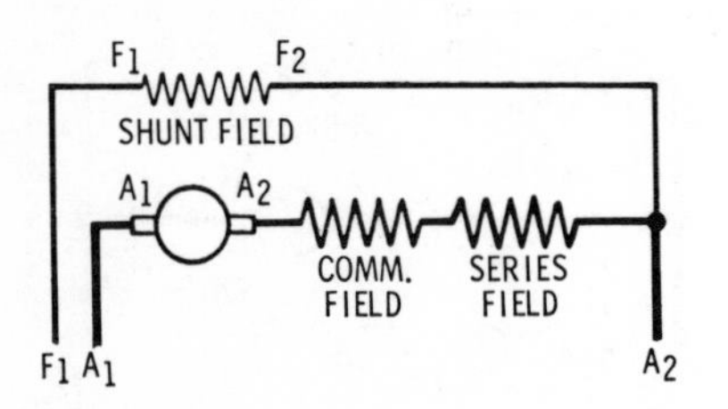

NONREVERSING NONCOMMUTATING POLE TYPE

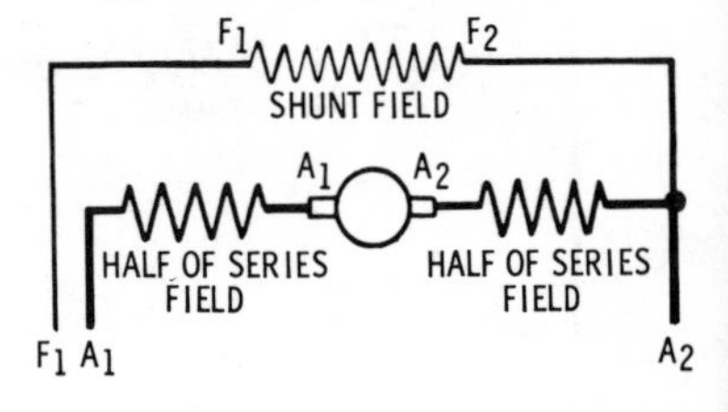

REVERSING COMMUTATING POLE TYPE

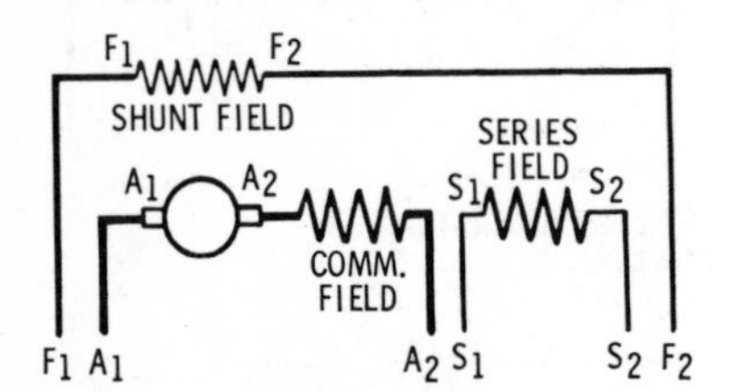

NONREVERSING NONCOMMUTATING POLE TYPE

F1 F2 SHUNT FIELD A1 A2 HALF OF SERIES FIELD HALF OF SERIES FIELD A1 A2

REVERSING NONCOMMUTATING POLE TYPE

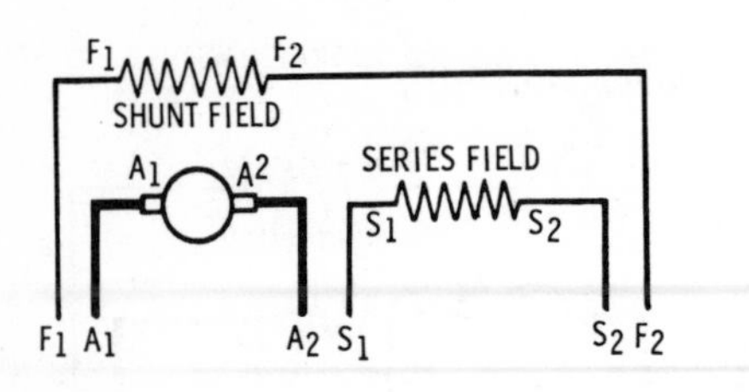

NONREVERSING NONCOMMUTATING POLE TYPE

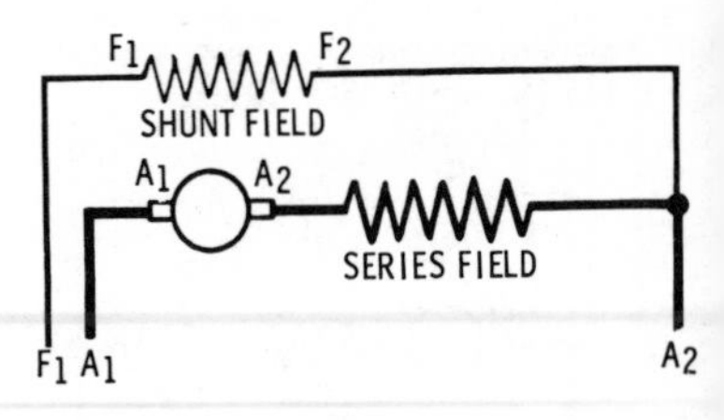

CONNECTION DIAGRAMS FOR DC SHUNT-WOUND MOTORS

NONREVERSING COMMUTATING POLE TYPE

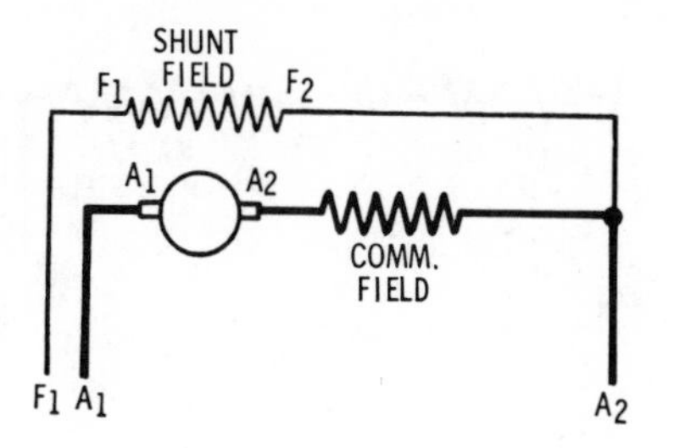

REVERSIBLE WITH COMMUTATING AND COMPENSATING FIELDS

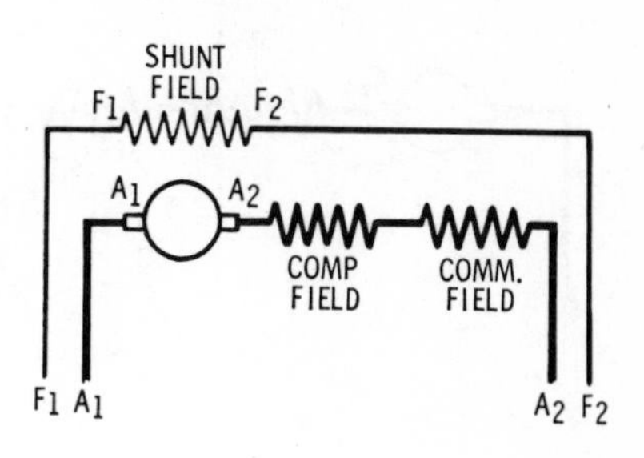

REVERSING COMMUTATING POLE TYPE

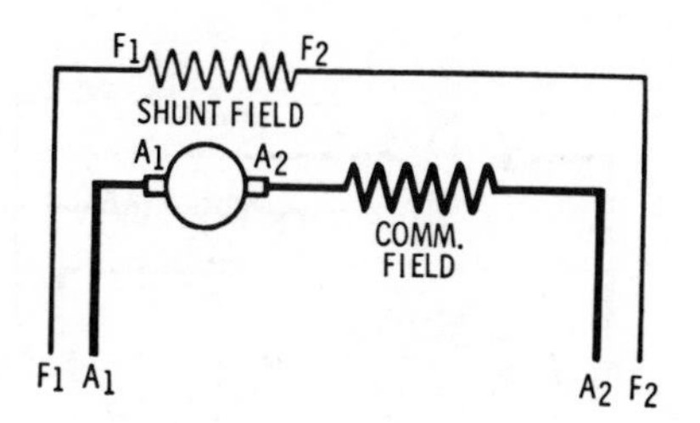

NONREVERSING NONCOMMUTATING POLE TYPE

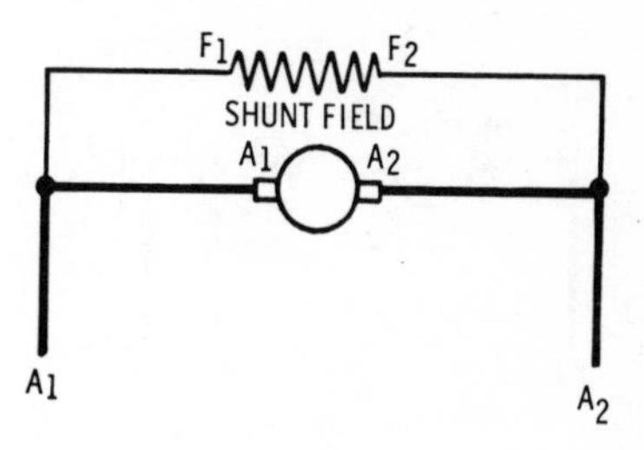

NONREVERSING NONCOMMUTATING POLE TYPE

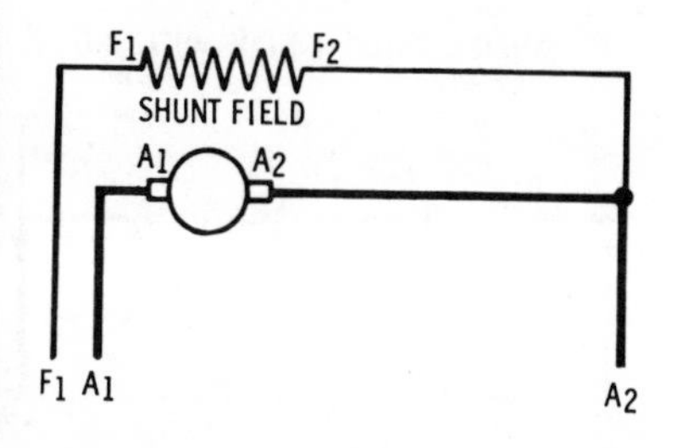

REVERSING NONCOMMUTATING POLE TYPE

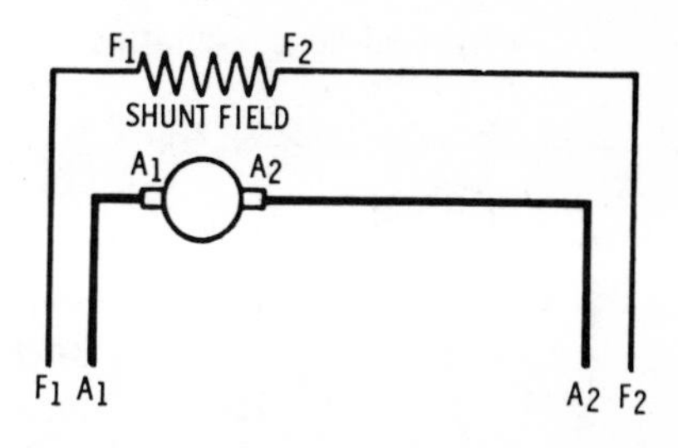

CONNECTION DIAGRAMS FOR DC SERIES-WOUND MOTORS

NONREVERSING COMMUTATING POLE TYPE

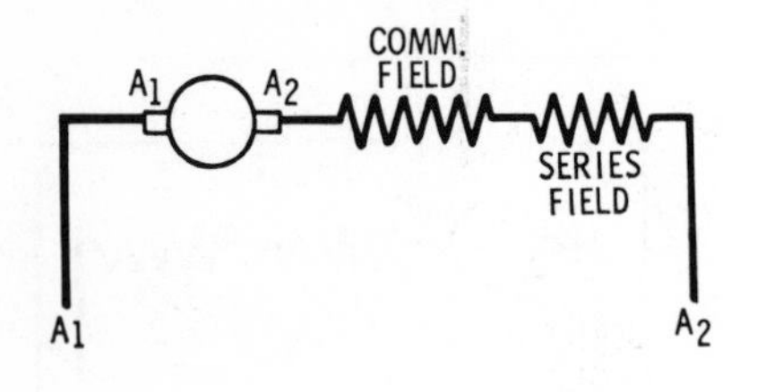

NONREVERSING NONCOMMUTATING POLE TYPE

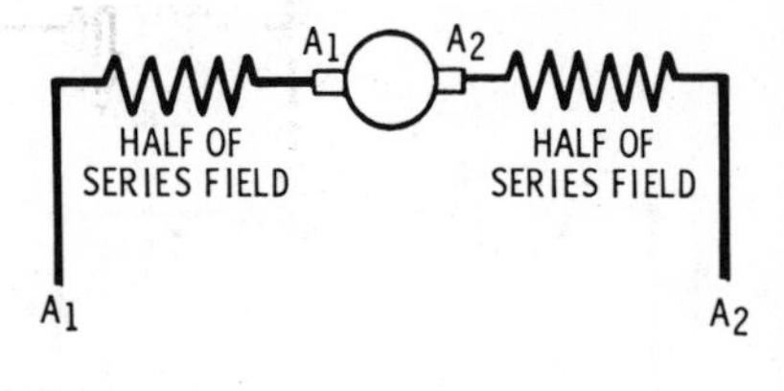

REVERSING COMMUTATING POLE TYPE

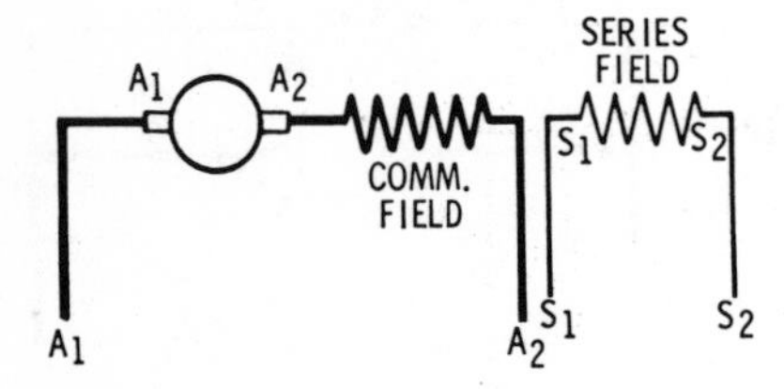

REVERSING NONCOMMUTATING POLE TYPE

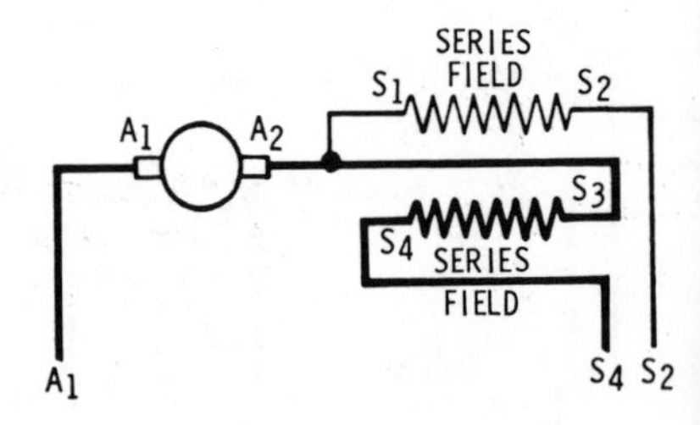

REVERSING NONCOMMUTATING POLE TYPE

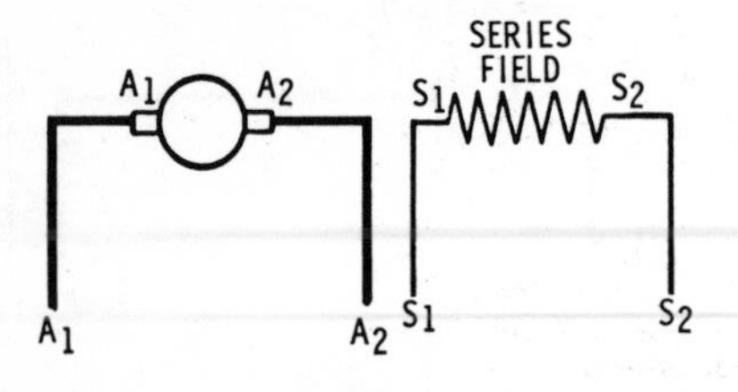

NONREVERSING NONCOMMUTATING POLE TYPE

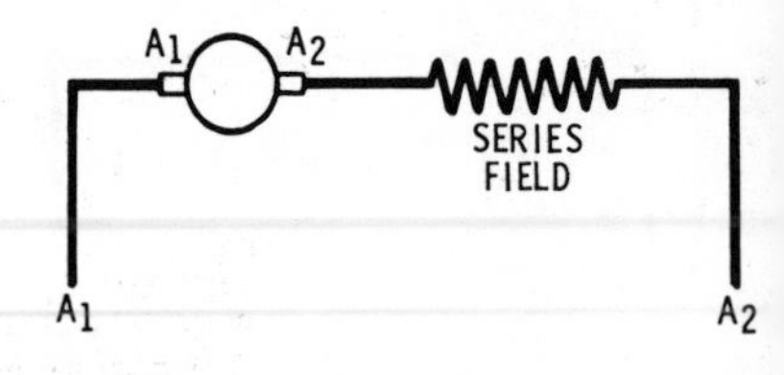

CONNECTION DIAGRAMS FOR UNIVERSAL-TYPE MOTORS (AC OR DC)

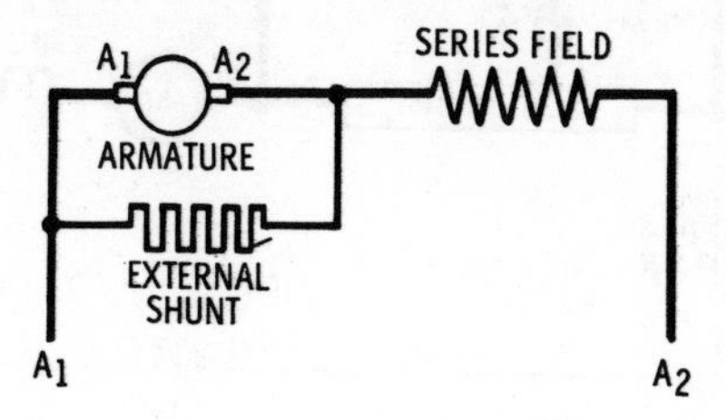

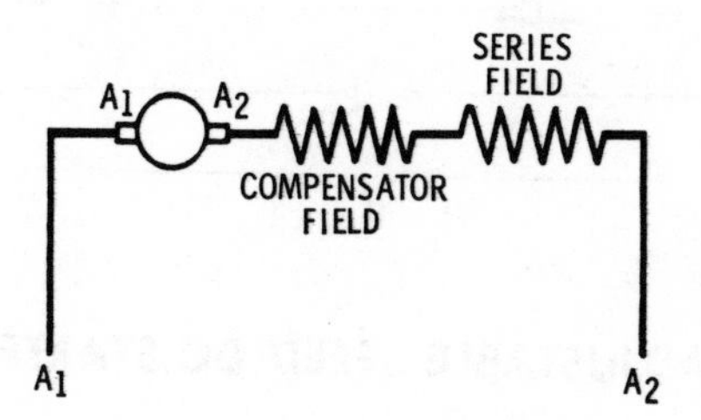

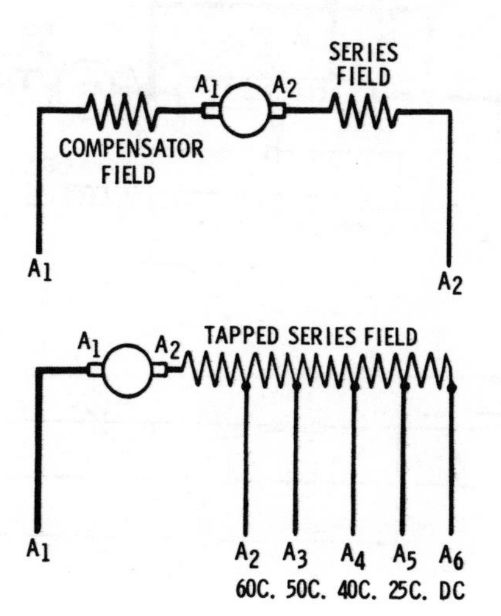

NOTE:
TAPS MAY BE OMITTED OR OTHER TAPS BROUGHT OUT WHEN FREQUENCIES ARE DIFFERENT FROM THOSE SHOWN.

CONSTANT SPEED DC STARTER

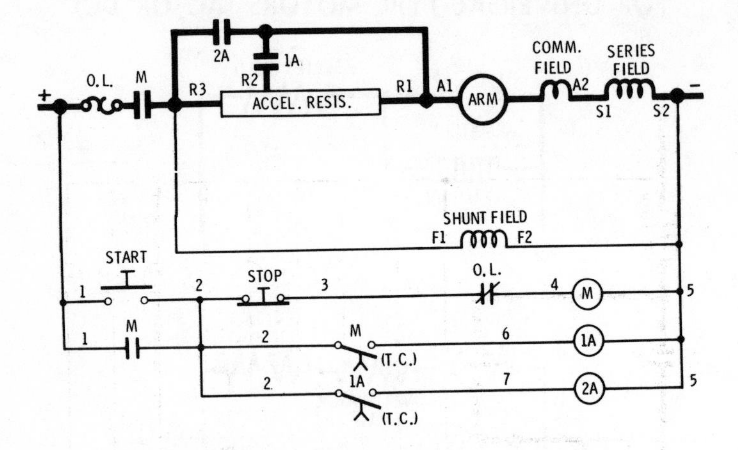

ADJUSTABLE SPEED DC STARTER

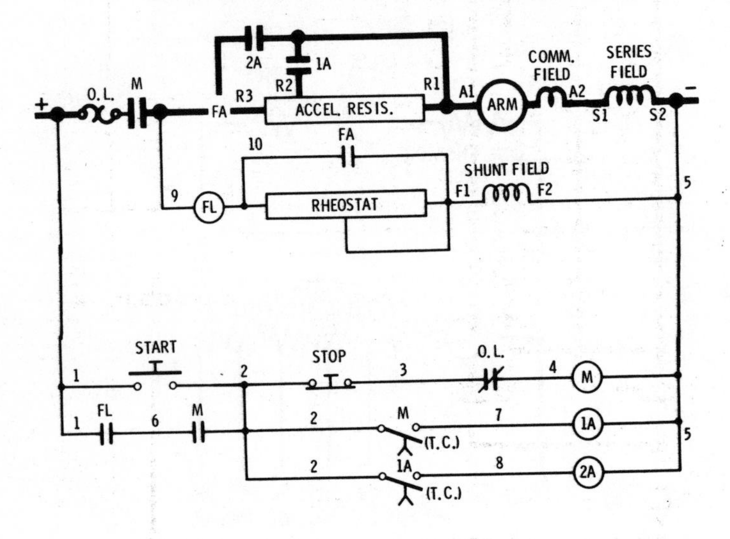

REVERSING CONSTANT SPEED DC STARTER

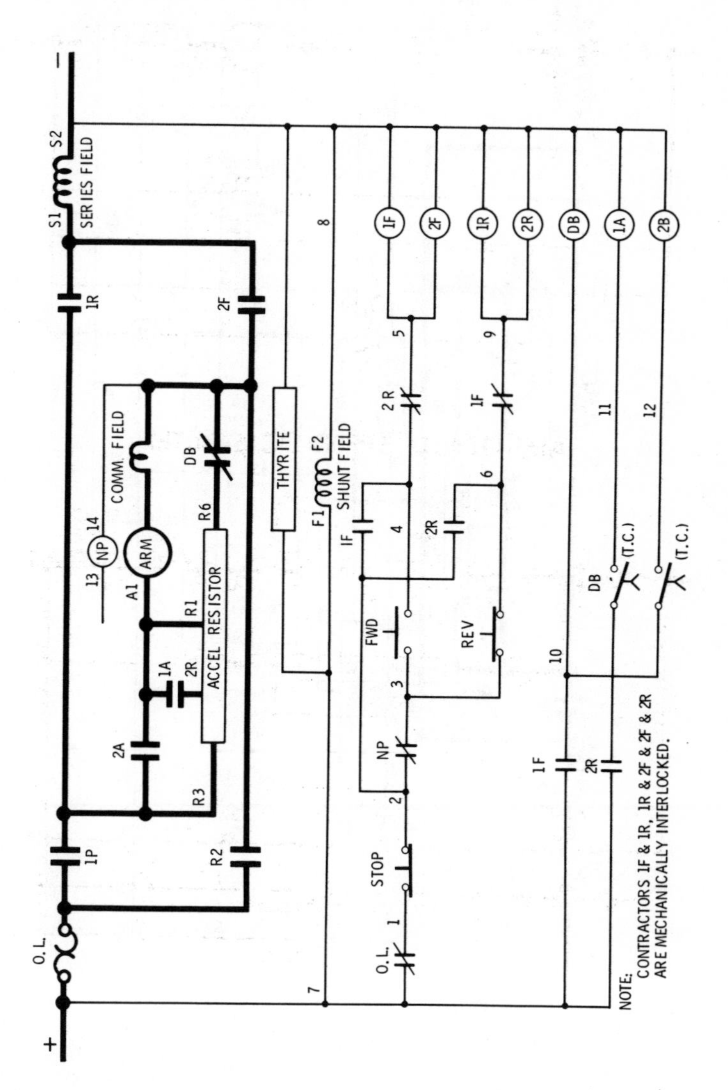

REVERSING ADJUSTABLE SPEED DC STARTER

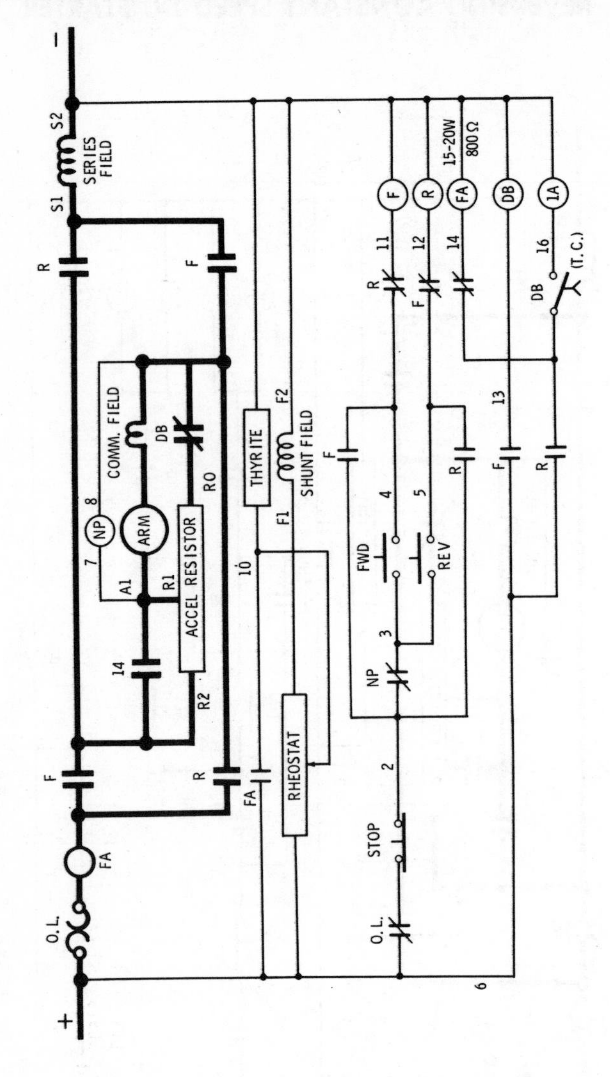

DC COMPOUND-WOUND MOTOR CONTROL FOR STEEL MILL, MAIN ROLL DRIVE

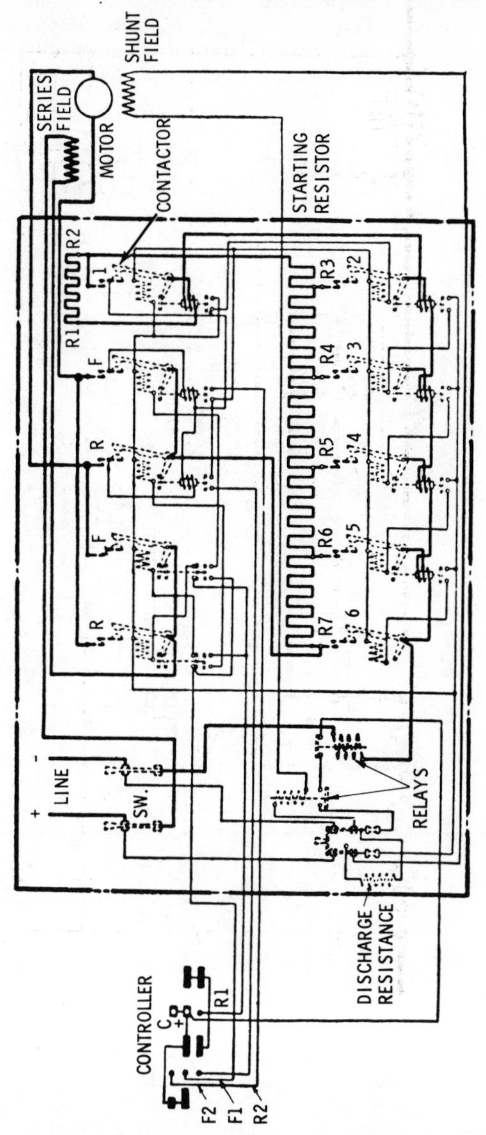

Operation—Direct-current motors for motor-driven reversing mills, are used only on low-voltage circuits usually not exceeding 250 volts, consequently remote control is easily secured and contactors can be used with entire satisfaction for closing, opening and reversing the armature circuit and for cutting out resistance. In addition to the disconnecting switches shown, a combined shunt-field and control switch is supplied which renders it impossible to open the shunt field without at the same time disconnecting the armature from the line. An overload relay and shunt-field relay afford protection respectively in case of overloads or accidental opening of the shunt field. Current-limit relays insure uniform acceleration, while no-voltage protection is secured automatically since the contactors are actuated by line potential.

WIRING DIAGRAM AND FRONT VIEW OF TYPICAL FACE-PLATE CONTROLLER

NOTE:
WHEN PERMANENT SLIP RESISTANCE IS FURNISHED, CONNECT AS PER DOTTED LINES AT A, B & C.

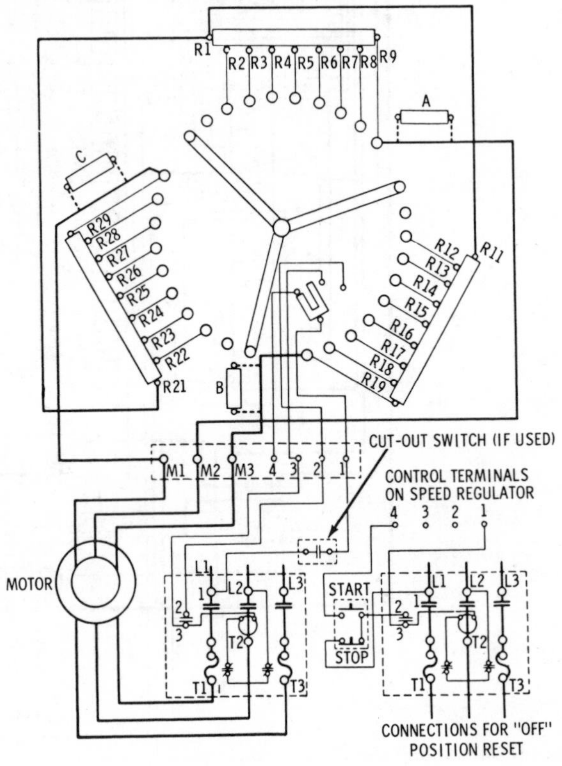

Operation—As shown in the illustrations, the segments are installed in a circle on the face of the controller. Connections from these segments to the collector plates (designated A and A_1) are established by means of two arms which by a movement in either direction change the resistance in the circuit to be operated, as indicated in the wiring diagram. Thus when the handle is moved in a forward direction sections of the starting resistance are cut out in steps, causing the armature to increase its speed.

On the other hand when the arm is moved in a reversed direction, the current becomes reversed in the armature (only) which causes the motor to rotate in the opposite direction. Finally, when the arm is in the off position, the arm rests on two insulating plates, in which case the operating handle is the vertical position.

WIRING DIAGRAM AND INSIDE VIEW OF TYPICAL DRUM CONTROLLER

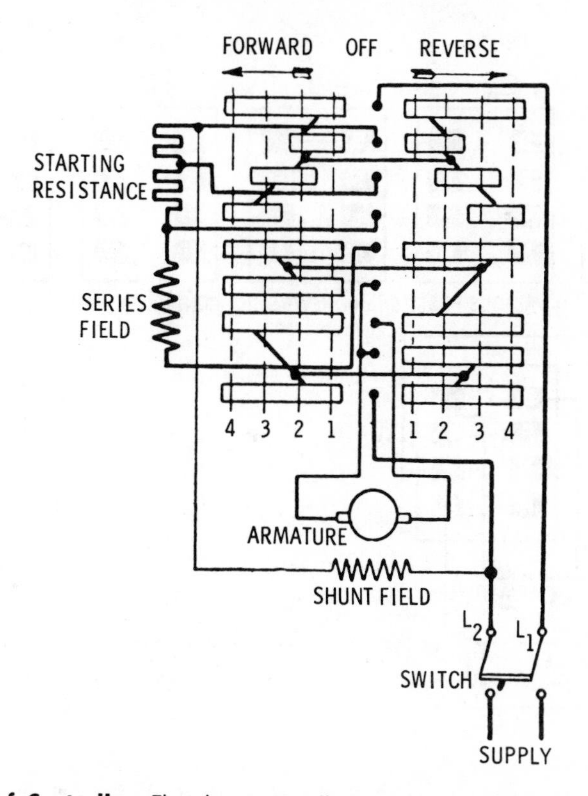

Operation of Controller—The drum controller consists generally of a drum cylinder insulated from its central shaft to which the operating handle is attached. To facilitate the operation, copper segments are attached to the drum. These segments are connected to and/or insulated from one another as shown in the diagram of connections. A series of stationary fingers are arranged to contact with the segments. These fingers are insulated from one another but interconnected to the starting resistance and the motor circuit. The drum assembly has a notched wheel keyed to the central shaft, the function of which is to indicate to the operator when complete contacts are made. With reference to the diagram, when the controller is moved forward one notch, the fingers are in position 1. The current then flows from L_1, through all the series resistance, to L_2, and the motor starts rotating. When the handle is moved further the resistance is gradually cut out of the armature circuit and inserted in the field circuit. Finally, when the handle is turned to notch 4, all the resistance has been transferred from the armature to the field circuit, and the motor is rcnning at full speed.

REVERSING CONTROLLER FOR DC MOTOR

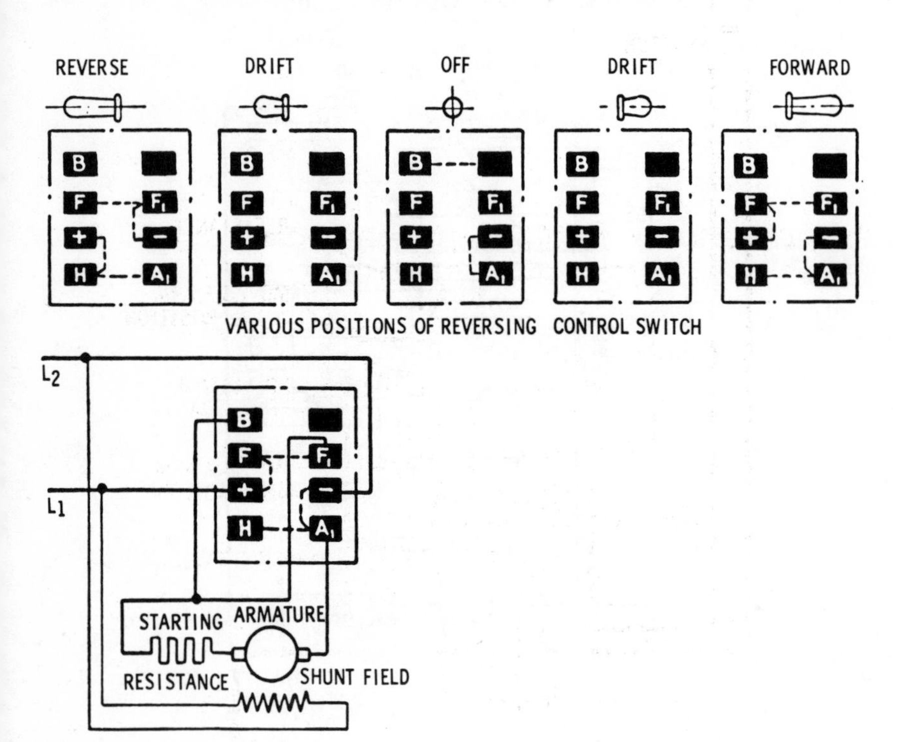

Operation—In some industrial processes a combination of dynamic braking and reversing is required. A common type of switch for this requirement is shown, together with a sequence in which the contacts are made. With reference to diagrams, L_1 and L_2 are connected to the positive and negative terminals. When the handle is in the reverse position, H and A_1 are positive, while F and F_1 are negative. If the handle is turned to forward position F, F_1 and H, A_1 exchange polarities. Suppose that the motor armature is connected between A_1 and F_1; the current through it and hence the direction of rotation will be reversed as the handle is turned from reverse to forward. Therefore this switch is fundamentally a reversing switch and may be utilized in connection with any starter.

STARTING RHEOSTAT AND CONNECTION DIAGRAM FOR DC COMPOUND-WOUND MOTOR

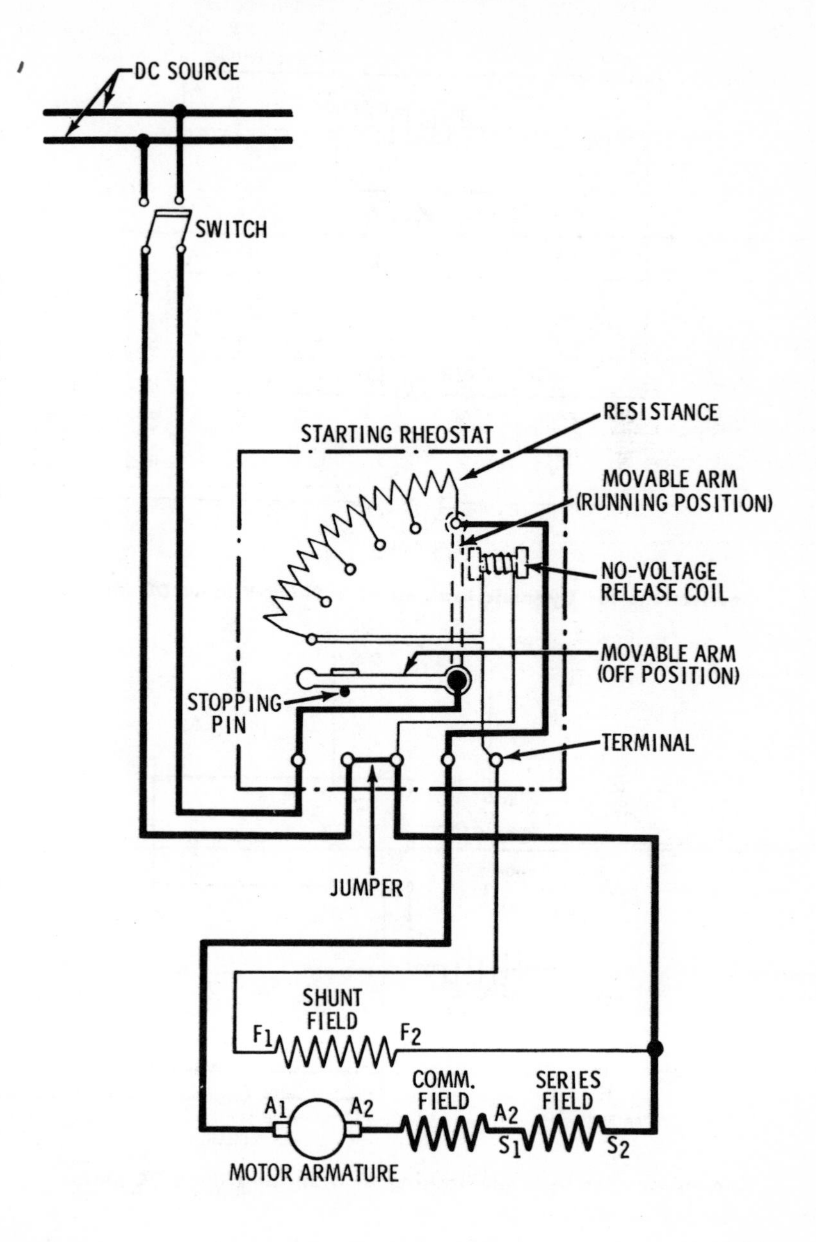

DYNAMIC BRAKING

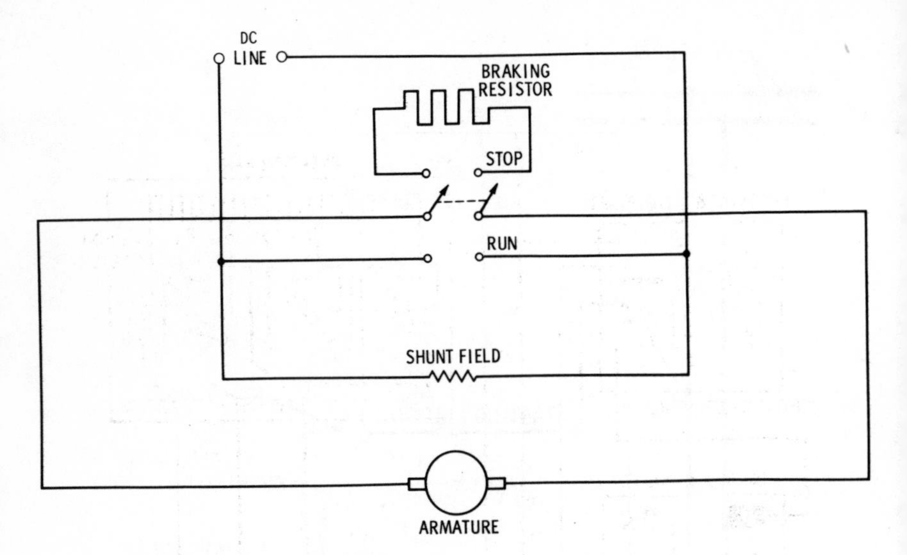

Connections for Dynamic Braking of a Shunt-Wound DC Motor.

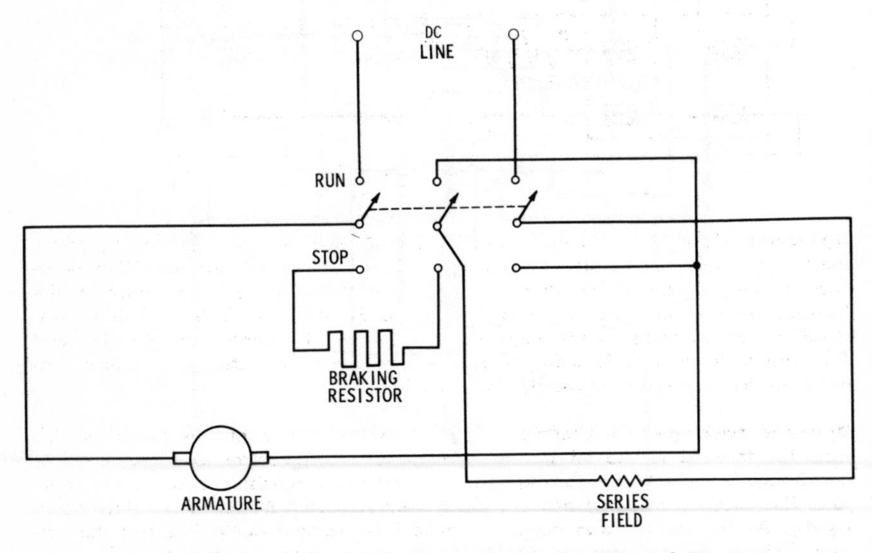

Connections for Dynamic Braking of a Series-Wound DC Motor.

WIRING DIAGRAM OF REVERSING AND DYNAMIC BRAKING CONTROLLER FOR DC MOTOR

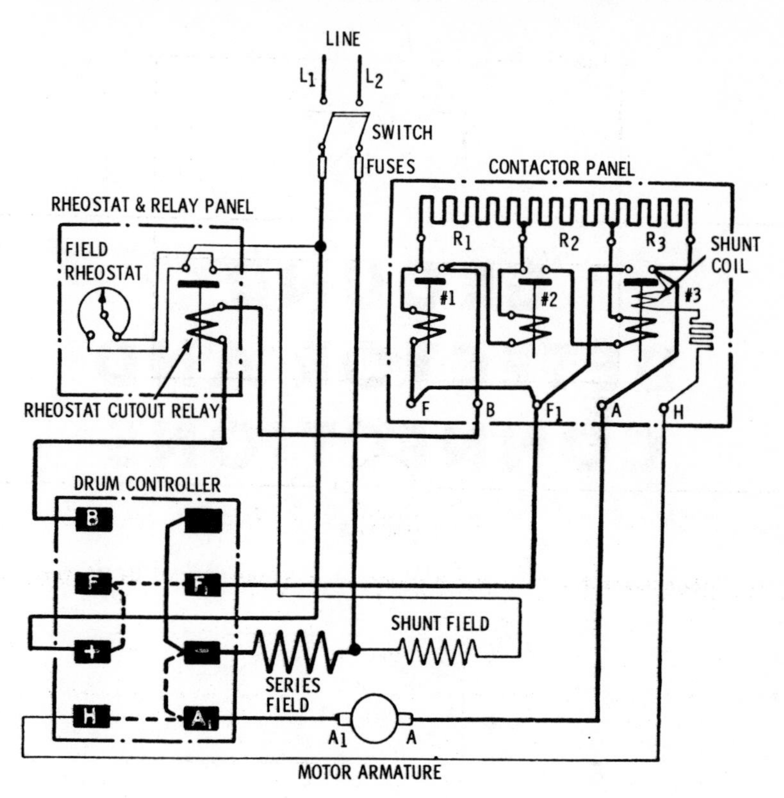

Operation—As indicated in the diagram the shunt field is connected directly across the line through the rheostat when the switch is in a closed position. With reference to the diagram of the drum controller wiring, with the main switch in the forward position the line circuit is from L_1 to F_1, through all the starting resistance, motor armature to the negative terminal and L_2. Contractors #1, #2 and #3 close their contacts in sequence as the armature current decreases, which connects the motor directly across its supply source.

Dynamic Braking—This is effected by turning the handle of the control switch from the forward to the off position. Segments in the control switch connect A_1 and B together, placing sections R_2 and R_3 of the starting resistance across the armature. This circuit provides dynamic braking action, causing the motor to slow down rapidly. As the motor slows down the generation voltage diminishes and the current through the resistance decreases as a consequence, tending to reduce the braking torque.

GROUND DETECTOR AND CONNECTIONS

In portions of the National Electrical Code, grounded detectors are recommended on underground systems, both AC and DC, for the detection of leakages to ground of the ungrounded conductors. On grounded systems, if a ground occurs, a fuse will blow or a circuit breaker will trip. This is not the case on ungrounded systems, as two phases will have to go to ground before the circuit is interrupted.

For low-voltage two-wire systems, the simplest method is to connect two lamps of the system voltage in series across the two lines, with a connection between the two lamps to ground.

For operating rooms of hospitals, no ground detector shall be used except one approved for the purpose. What we are discussing here is detection for regular systems.

A ground on one side will obviously short circuit the line to ground and the lamp on that side will be darkened. This gives the maintenance people an opportunity to correct the ground before serious trouble occurs. Lamps of course burn out eventually due to use and sometimes will give a false indication. Voltmeters of the right voltage ratings are more appropriate.

Industry often times uses a Wye system with a high-resistance neutral ground. This is basically not a neutral and is not run with circuit conductors, but is for the purpose of ground detection.

With this type system, if a ground occurs, the plant is not shut down, and shut downs may be arranged to coincide with the least stoppage of production. There is usually a ground relay on the switchboard, which will dispatch a signal to some point on the plant site, indicating the ground and on what building.

GROUND-DETECTOR CONNECTION

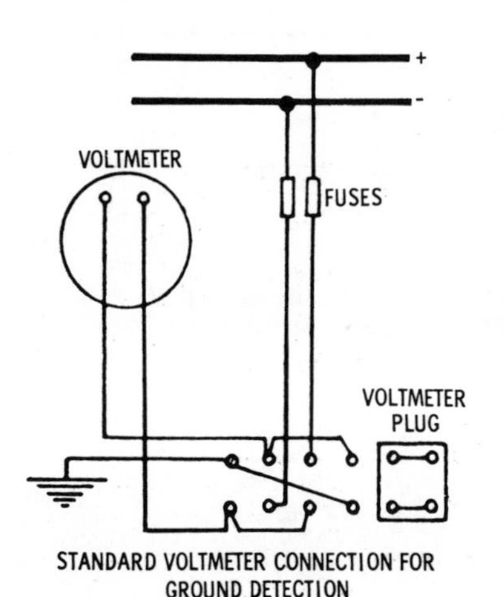

STANDARD VOLTMETER CONNECTION FOR GROUND DETECTION

A A_1 B B_1

2-PHASE, 4-WIRE

(BACK VIEW)

LOAD

ELECTROSTATIC GROUND DETECTORS

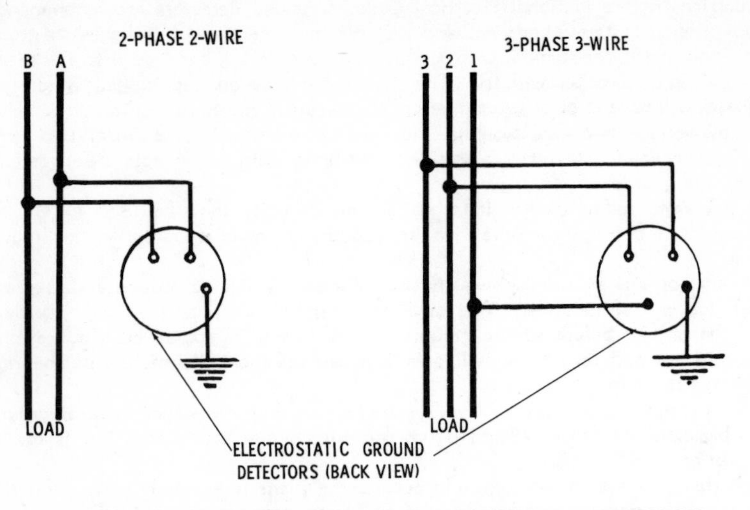

CONNECTION FOR GROUND DETECTION ON AC CIRCUITS, FOR VOLTAGES UP TO 3300.

GROUND-DETECTOR CONNECTION

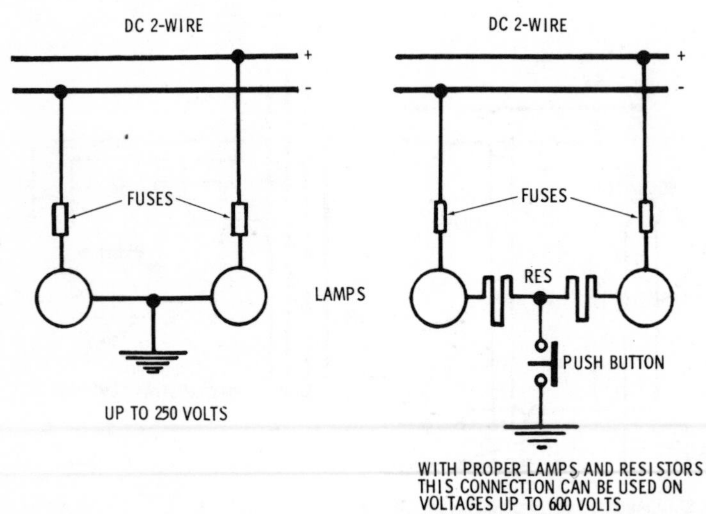

AC 3-PHASE 3-WIRE

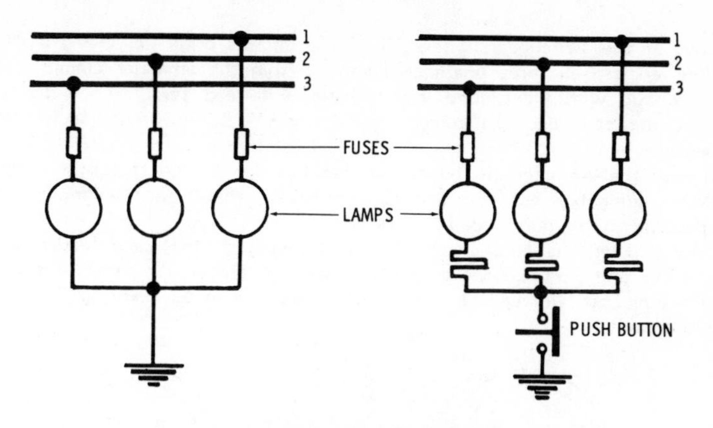

WITH PROPER LAMPS AND RESISTORS
THIS CONNECTION CAN BE USED ON
VOLTAGE UP TO 650.

CONNECTION OF GROUND DETECTOR LAMPS ON AC AND DC CIRCUITS.

WYE SYSTEM WITH HIGH RESISTANCE GROUNDING

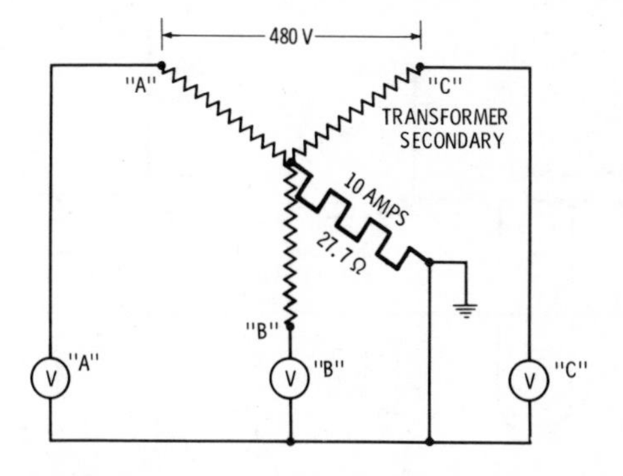

If no grounds are present on any of the phase conductors, A, B or C, the voltmeters will all read 277 volts. If a ground occurs on the system somewhere from "C" on down the line, current will flow from that ground back to the high-resistance ground point, causing a voltage-drop in the resistor and voltmeter "C" will read lower than 277 volts, depending upon the resistance to ground where the fault occurs.

Voltmeters A and B will indicate higher than 277 volts, thus we know that the grounded fault occurs in phase C.

LOCATING GROUND ON A WYE SYSTEM WITH A HIGH-RESISTANCE GROUND POINT

You start at the switchgear, where all of the circuits originate. Using a clamp-on-ammeter, enclose all three phase conductors within the ammeter clamp. If that is not the circuit with the ground, the ammeter will read zero. This is the same principle as the zero-sequence current transformers used in ground-fault circuit-interrupters.

If you have the ammeter on the circuit that has the ground fault, the ammeter will show a reading between zero and 10 amperes, as this is the maximum current that can possibly go to ground, due to the resistor.

Then going down the line, check at various points on the same feeder or any branch circuits, and by the same principle, when you are on the down side of the ground, the ammeter will read zero. This aids materially in pin-pointing the locality of the ground fault.

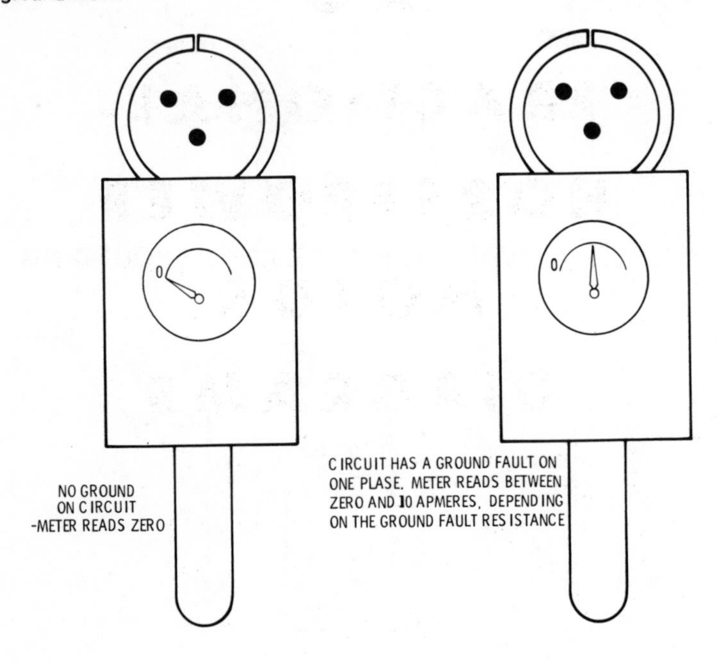

AC FRACTIONAL-HORSEPOWER MOTOR DIAGRAMS

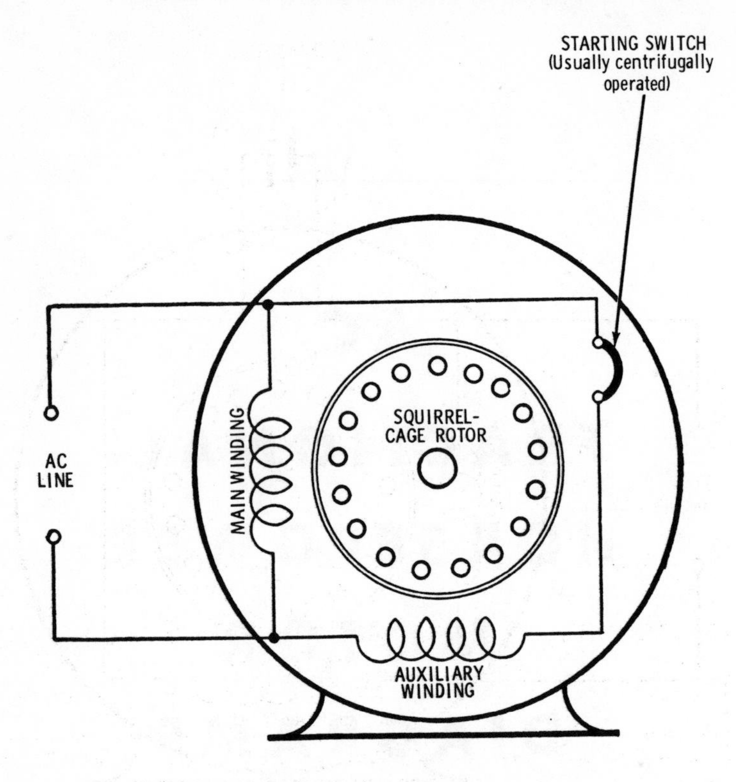

The Split-Phase Induction Motor—This motor type is commonly manufactured in fractional horsepower sizes. It is equipped with a squirrel-cage rotor for constant speed operation and has a starting winding of high resistance (commonly termed auxiliary winding), which is physically displaced in the stator from the main winding. This displacement produced by the relative electrical resistance values in the two windings, creates starting ability similar to that of a polyphase motor.

In series with the auxiliary winding is a starting switch (usually centrifugally operated) which opens the circuit when the motor has attained approximately 75 to 80 per cent of synchronous speed.

The function of the starting switch is to prevent the motor from drawing excessive current from the line and also to protect the starting winding from damage due to heating. The motor may be started in either direction by reversing either the main or auxiliary winding.

Single-phase, split-phase motors are suitable for oil burners, blowers, business machines, buffing machines, grinders, etc.

FRACTIONAL-HORSEPOWER MOTORS

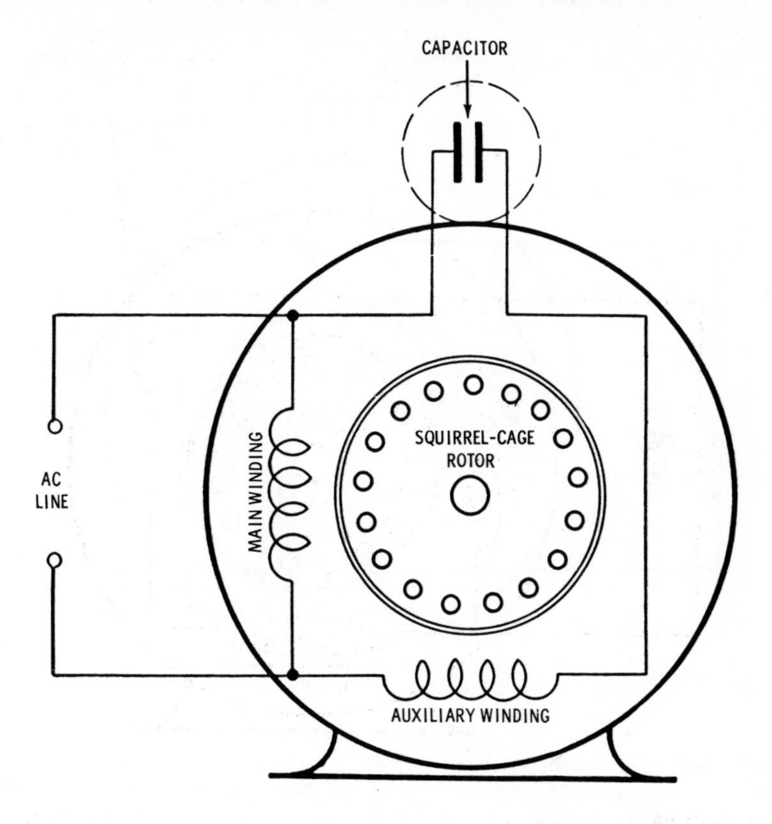

The Split-Phase, Permanently Connected Capacitor Motor—This type of split-phase motor, is commonly manufactured in fractional horsepower sizes.

In common with other types of split-phase motors, it is equipped with a squirrel-cage rotor and a main and auxiliary winding. A capacitor is permanently connected in series with the auxiliary winding, thus a motor of this type starts and runs with a fixed value of capacitance in series with the auxiliary winding.

The motor obtains its starting torque from a rotating magnetic field produced by the two stator windings physically displaced. The main winding is connected directly across the line, while the auxiliary or starting winding is connected to the line through the capacitor, giving an electrical phase displacement.

A motor of the permanent split-phase capacitor type is suitable for direct connected drives requiring low starting torque, such as fans, blowers, certain types of centrifugal pumps, etc.

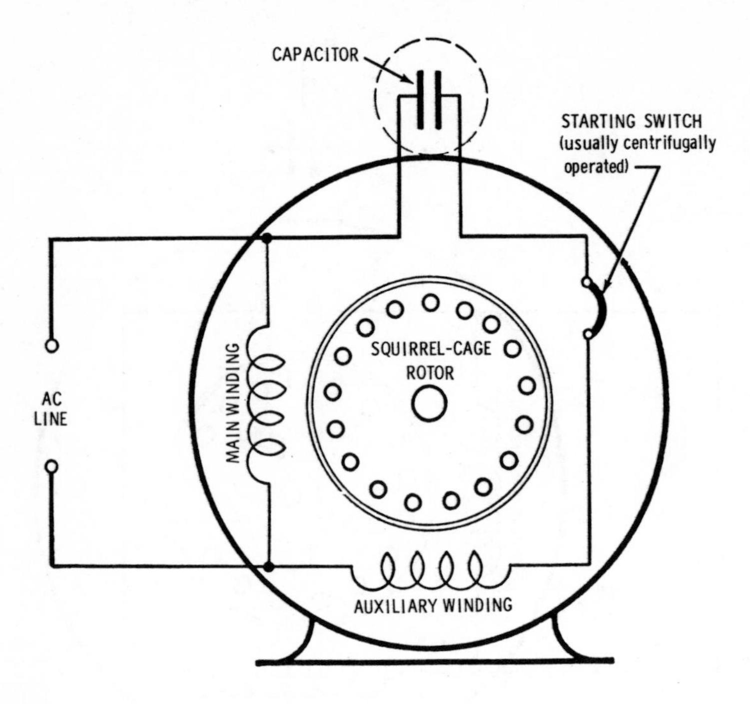

The Split-Phase, Capacitor-Start Motor—This fractional horsepower motor may be defined as a form of split-phase motor having a capacitor connected in series with the auxiliary winding.

The auxiliary circuit is opened by means of a centrifugally operated switch when the motor has attained a predetermined speed (usually approximately 70 to 80 per cent of synchronous speed).

A motor of this type is sometimes termed a capacitor-start, induction-run motor in contrast to the straight capacitor run type which is termed a capacitor-start, capacitor-run motor.

The rotor is of the squirrel cage type as in other split-phase motors. The main winding is connected directly across the line, while the auxiliary or starting winding is connected through a capacitor which may be connected into the circuit through a transformer with suitable designed windings and capacitor of such values that the two windings will be approximately 90 degrees apart.

This type of motor is particularly suited for such applications as air conditioning, domestic and commercial refrigeration, belt driven fans, etc.

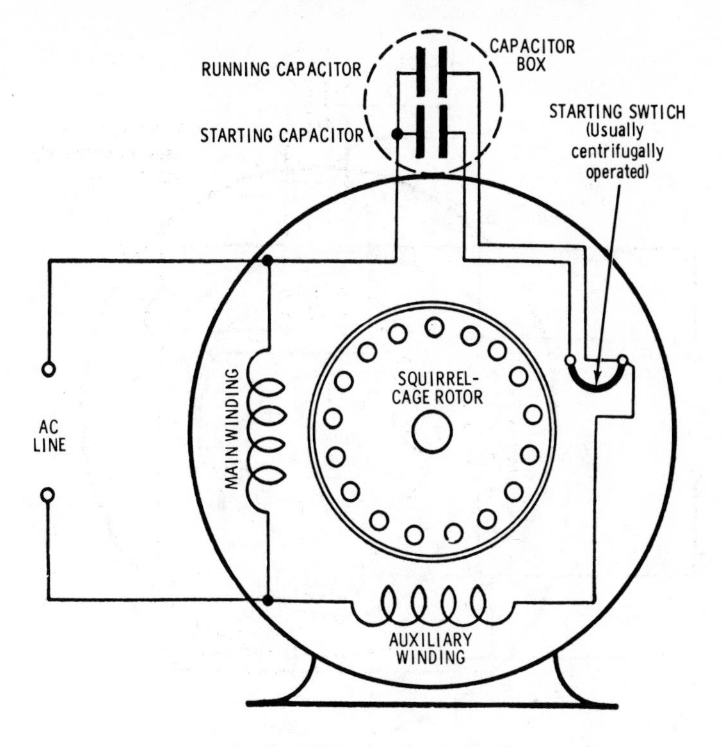

The Split-Phase, Capacitor-Run Motor—This type of motor, also termed two-value capacitor-motor has a running capacitor permanently connected in series with the auxiliary winding, the starting capacitor being in parallel with the running capacitor only during the starting period.

In operation the motor starts with the starting switch closed. After the motor has attained a speed of approximately 70 to 80 per cent of synchronous, the starting switch opens, thus disconnecting the starting capacitor.

The running capacitor is usually of the paper-spaced oil filled type, normally rated at 330 volt AC for continuous operation. They usually range from 3 to 16 microfarads, depending upon the size of the motor.

The starting capacitor is generally of the electrolytic type and may range in sizes of from 80 to 300 microfarads approximately, for 110 volt, 60 cycle motors.

This type of motor is designed for applications requiring high starting torque, such as compressors, loaded conveyors, reciprocating pumps, refrigeration compressors, stokers, etc.

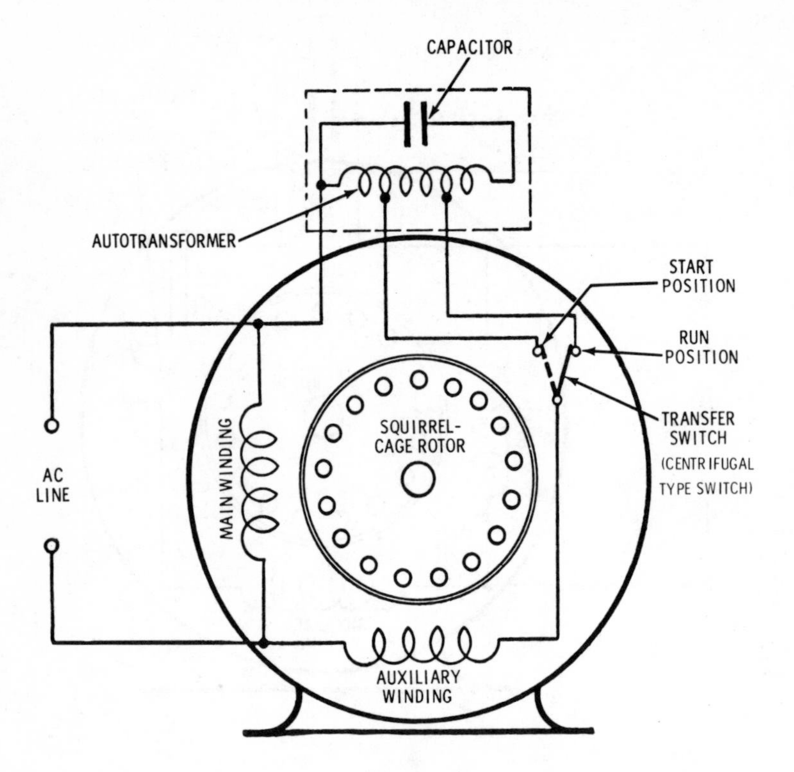

The Split-Phase, Capacitor-Run Motor—Another type of two-value capacitor motor uses a capacitor transformer unit. The motor is of the split-phase squirrel cage type with the main and auxiliary winding physically displaced in the stator.

This type of motor employs a **transfer switch** which is equivalent to a single-pole, double-throw switch, by means of which a high voltage is impressed across the capacitor during the starting period.

After the motor has attained a speed of 70 to 80 per cent of synchronous ,the transfer switch operates to change the voltage taps on the transformer. The voltage impressed upon the capacitor by means of the transformer will be of a value of from 600 to 800 volts during the starting period and approximately 350 volts for continuous operation.

This type of motor is designed for applications requiring high starting torque such as compressors, loaded conveyors, reciprocating pumps, refrigeration compressors, stokers, etc.

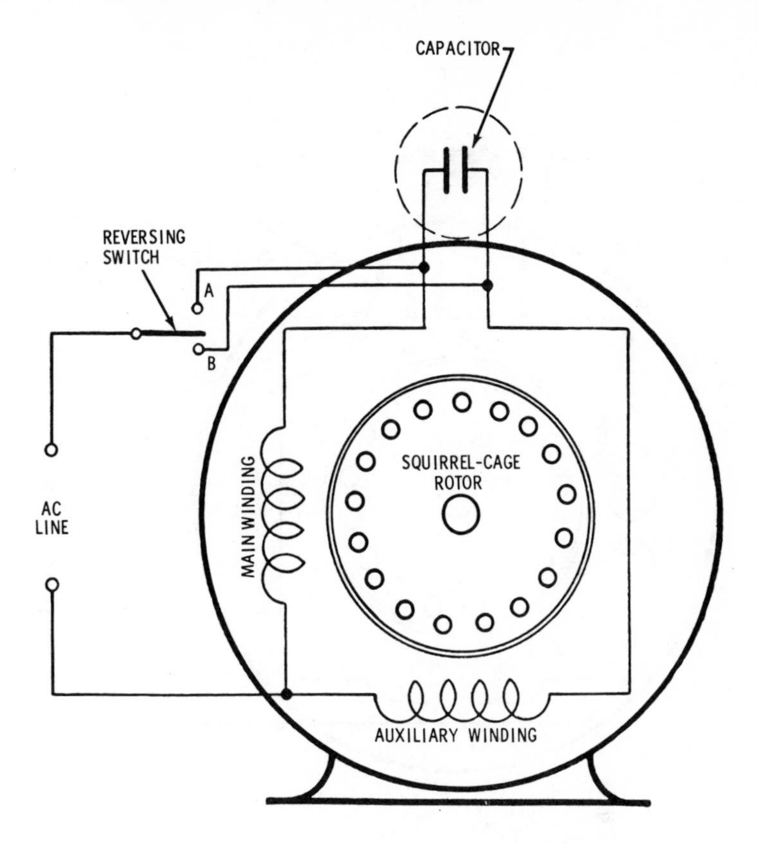

The Split-Phase, Capacitor-Run, Induction Motor (Reversible Type)—In applications where it is necessary to employ a reversible high-torque intermittently rated capacitor type, a motor connected as shown has found employment.

When the reversing switch is in the **B** position, the auxiliary winding becomes the main winding, and the main winding becomes the auxiliary winding. With the switch in the **A** position, both windings function as shown in the diagram.

Since the direction of rotation in split-phase motors is always from the auxiliary winding toward the main winding, it follows that an interchange of the windings will also reverse the direction of rotation.

From the foregoing it follows that with motor connections arranged in this manner, the main and auxiliary windings in the motor must be identical, both as to size of wire and number of effective turns.

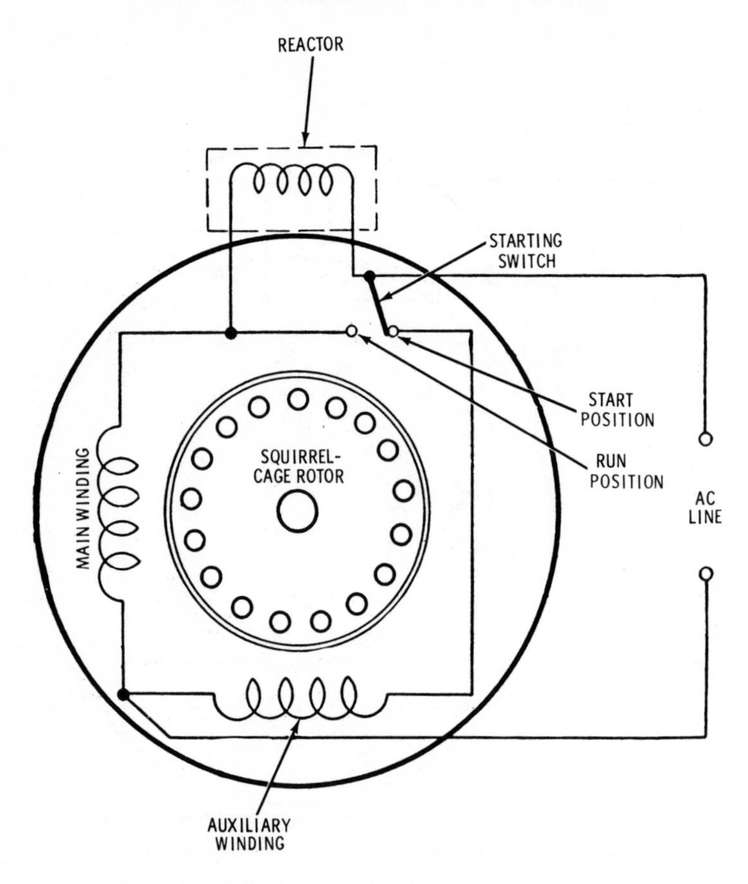

The Reactor-Start, Split-Phase, Induction Motor—This type, in common with other types of split-phase motors, is equipped with an auxiliary winding, displaced in magnetic position from, and connected in parallel with the main winding.

The function of the reactor is to reduce the starting current and increase the current lag in the main winding. At approximately 75 per cent of synchronous speed the starting switch operates to shunt out the reactor, disconnecting the auxiliary winding from the circuit.

The starting switch must be of such construction so as to be equal to a single-pole, double-throw switch.

This is a constant speed motor and lends itself best for such applications as light running machines such as fans, small blowers, business machines, grinders, etc.

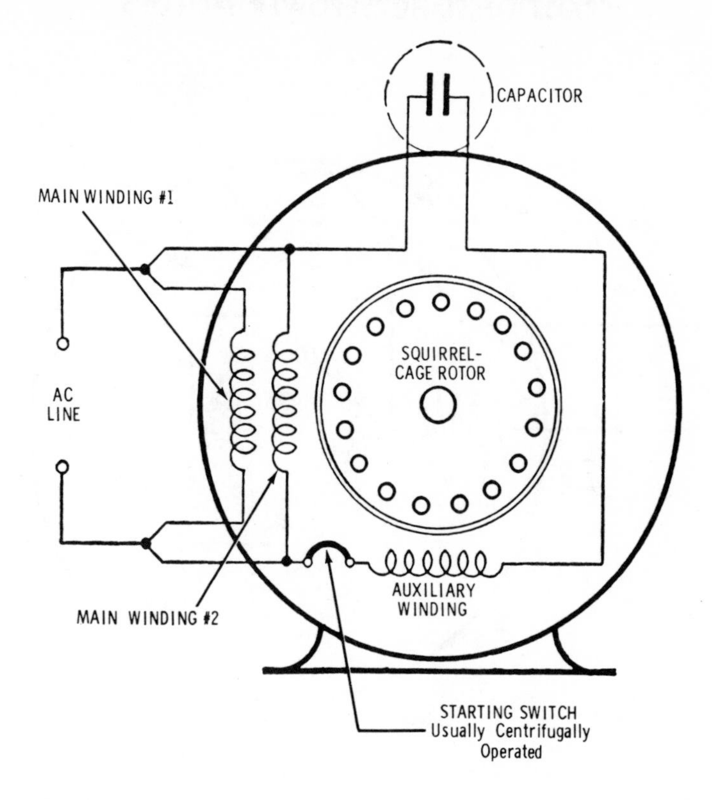

The Split-Phase, Single-Value Capacitor Motor (Dual-Voltage Type)—A motor of this type differs from the conventional type of capacitor motor in that it has two identical main windings arranged for either series or parallel connections.

With the main windings connected in parallel the line voltage is usually 115 volts, whereas when the main windings are connected in series a line voltage of 230 volts is usually employed.

The starting switch and operating characteristics of the motor do not differ from single-value capacitor-start motors previously described.

In common with other types of split-phase motors, the auxiliary winding is a separate winding displaced in space from the main winding 90 degrees. Also in series with the auxiliary winding is the usual centrifugal switch and starting capacitor.

A winding arrangement of this type gives only half as much starting torque on 115 volts as on a 230 volt winding.

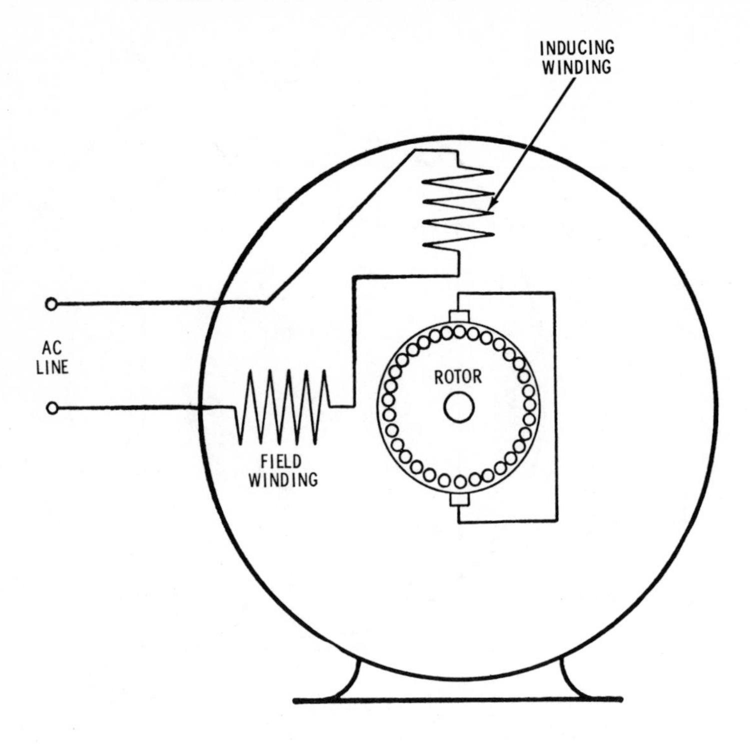

The Repulsion Motor—By definition, a repulsion motor is a single-phase motor which has a stator winding arranged for connection to the source of power and a rotor winding connected to a commutator. Brushes and commutators are short-circuited and are so placed that the magnetic axis of the rotor winding is inclined to the magnetic axis of the stator winding.

It has a varying speed characteristic, a high starting torque and moderate starting current.

Due to its low power factor except at high speeds, it is often modified into the **compensated repulsion motor,** which has another set of brushes placed midway between the short-circuited set. This added set is connected in series with the stator winding.

The Repulsion-Start Induction Motor—By definition, a repulsion-start induction motor is a single-phase motor having the same windings as a repulsion motor, but at a predetermined speed the rotor winding is short-circuited or otherwise connected to give the equivalent of a squirrel-cage winding. This type of motor starts as a repulsion motor, but operates as an induction motor with constant-speed characteristics.

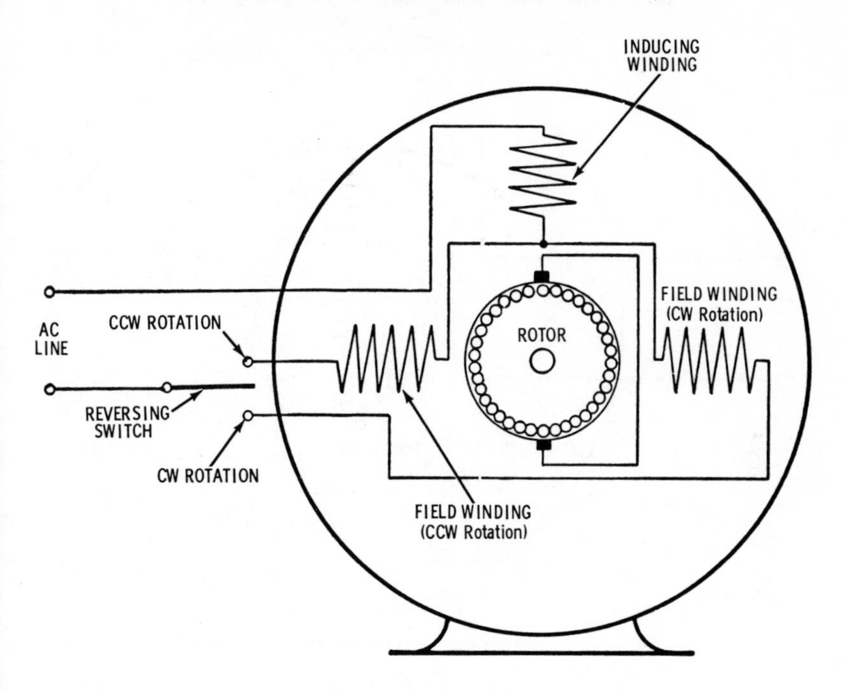

The Repulsion-Start Induction Motor (Reversible Type)—In certain applications, it is necessary to reverse the direction of rotation. A motor of this type has two stator windings displaced as indicated. Reversal of the motor can be accomplished by interchanging the field winding connections.

Thus, for example, with the switch in the upper position, the motor will rotate in a counter-clockwise direction, whereas if the switch is in the lower position the motor will run in the opposite direction, or clock-wise.

The current induced in the armature is carried by the brushes and commutator, resulting in high starting torque. When nearly synchronous speed is attained the commutator is short-circuited so that the armature is then similar in its functions to a squirrel-cage armature.

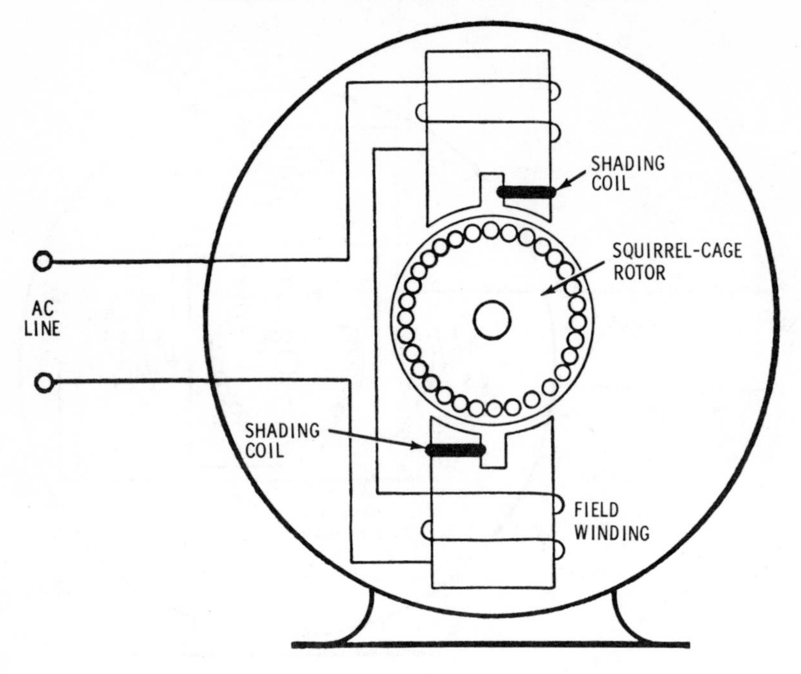

The Shaded-Pole Motor—By definition a shaded-pole motor is a single-phase induction motor provided with an auxiliary short-circuited winding or windings displaced in magnetic position from the main winding.

Although there are a number of different construction methods employed, principally the motor operates as follows: The shading coil (from which the motor has derived its name) consists of low resistance copper links embedded in one side of each stator pole, and are used to provide the necessary starting torque. When the current increases in the main coils a current is induced in the shading coils that opposes the magnetic field building up in part of the pole pieces they surround.

When the main coil current decreases, that in the shading coil also decreases, until the pole pieces are uniformly magnetized. As the main coil current and the pole piece magnetic flux continue to decrease, current in the shading coils reverses and tends to maintain the flux in part of the pole pieces.

When the main coil current drops to zero, current still flows in the shading coils to give the magnetic effect which causes the coils to produce a rotating magnetic field which makes the motor self starting.

This motor is used largely where the power requirements are small, such as in electric clocks, instruments, toys, hair dryers, small fans, etc. It is simple in construction and low in cost and is in addition very rugged and reliable.

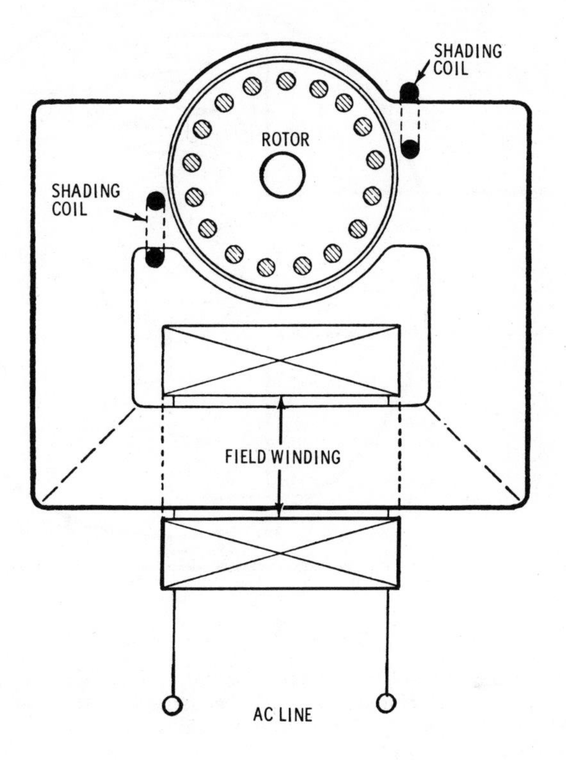

The Shaded-Pole Motor (Skeleton Type)—This type of motor is built for applications where the power requirements are very small. The field circuit with its winding, is built around the conventional squirrel-cage rotor and consists of punchings that are stacked alternately to form overlapping joints, in the same manner that small transformer cores are assembled.

Motors of this class will operate only on alternating current; they are simple in construction, low in cost and extremely rugged and reliable. Their principal limitations are, however, low efficiency and a low starting and running torque.

A shaded-pole motor is not reversible, unless shading coils are provided on each side of the pole, and means for opening one and closing the other coil are provided.

The inherently high slip of a shaded-pole motor makes it convenient to obtain speed variation on a fan load, for example, by reducing the impressed voltage.

FRACTIONAL-HORSEPOWER MOTORS

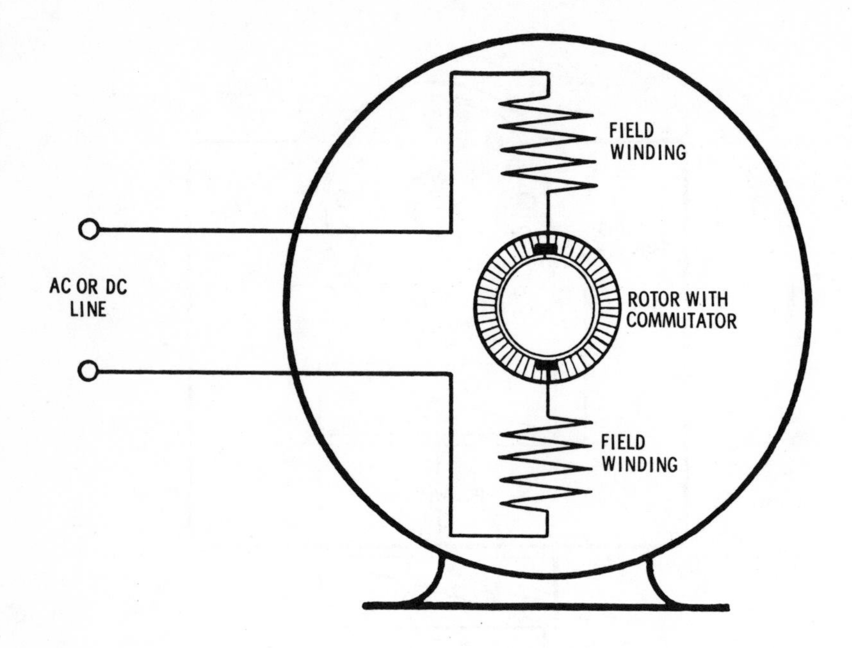

The Universal Type Motor—The universal motor is designed for operation on either alternating or direct current. It is of the series-wound type, that is, it is provided with a field winding on the stator which is connected in series with a commutating winding on the rotor.

Universal motors are commonly manufactured in fractional horsepower sizes and are preferred because of their use on either AC or DC currents, particularly in areas where power companies supply both types of current.

Full-load speeds generally range from 5,000 to 10,000 r.p.m. with no-load speeds from 12,000 to 18,000 r.p.m. Typical applications are portable tools, office appliances, electric cleaners, kitchen appliances, sewing machines, etc.

The speed of universal motors can be adjusted by connecting a resistance of proper value in series with the motor. Advantages of this speed control method is obvious, in such applications as motor operated sewing machines, where it is necessary to operate the motor over a wide range of speed.

In such applications adjustable resistances are used and the speed varied at will of the operator. Universal motors may be either compensated or uncompensated, the latter type being used for the higher speeds and lower ratings only.

SYNCHRONISM INDICATOR AND WIRING DIAGRAMS

GENERAL METHOD OF SYNCHRONIZING

Parallel Operation of Synchronous Generators—Before generator #2 can be connected in parallel with generator #1, the following conditions must be obtained.

(1) Both machines must have the same frequency and waveform.
(2) Their terminal voltages must be equal.
(3) Their sequence of maximum potential values must be the same.

When synchronizing proceed as follows:

I. Lamp Synchronizing—Machine #1 is running and supplying the load. Its oil circuit breaker is closed and the running plug is inserted. Bring machine #2 up to speed by slowly increasing the speed of its prime mover. As the speed of machine #2 increases, insert the starting plug; when the machines are running at nearly the same speed, the synchronizing lamps light up then go out, light up again, etc. If the machines are in step with lamps out or lamps in, (depending on whether light or dark lamp connections are used) wait until they go out for a few seconds then close the oil circuit breaker on machine #2 and the machines are now in parallel. Voltages must be the same on both generators.

II. Indicator Synchronizing—Proceed same as before: The rotary motion of the pointer on the indicator indicates whether the generator to be synchronized is running too slow or too fast. When the pointer remains stationary in the vertical position, the two machines are in synchronism and the oil circuit breaker can be closed.

After paralleling the two machines, adjust the mechanical power input and the generated emf until each machine supplies its share of the total load, and the power factor of each machine is the same and equal to that of the total load.

When preparing to add another generator, the present generators must be phased out with the new generator, so that phase A will connect to phase A and phase B connect to phase B and phase C to phase C. This is done by potential transformers and voltmeters.

METHOD OF SYNCHRONIZING 2 AC GENERATORS

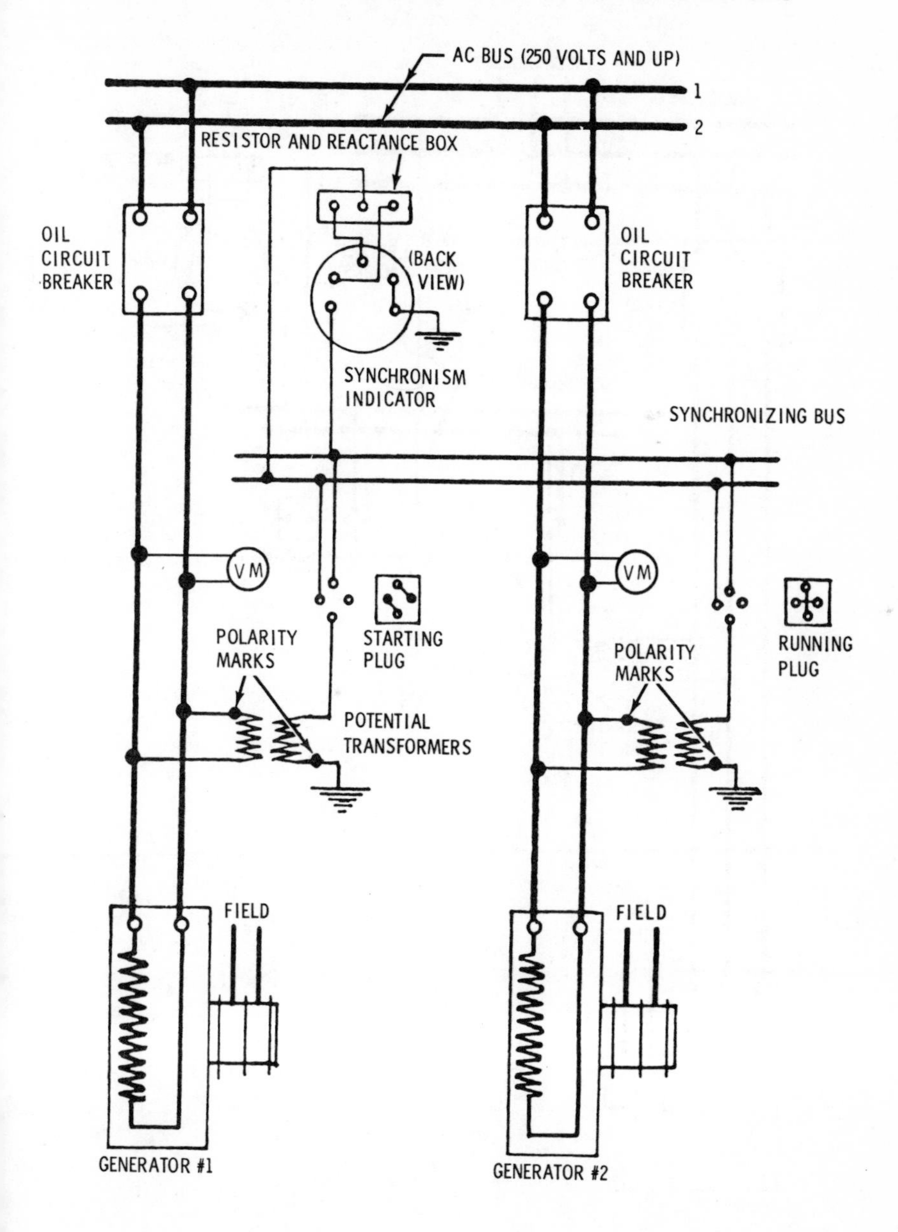

METHOR OF SYNCHRONIZING BUS AND MACHINE

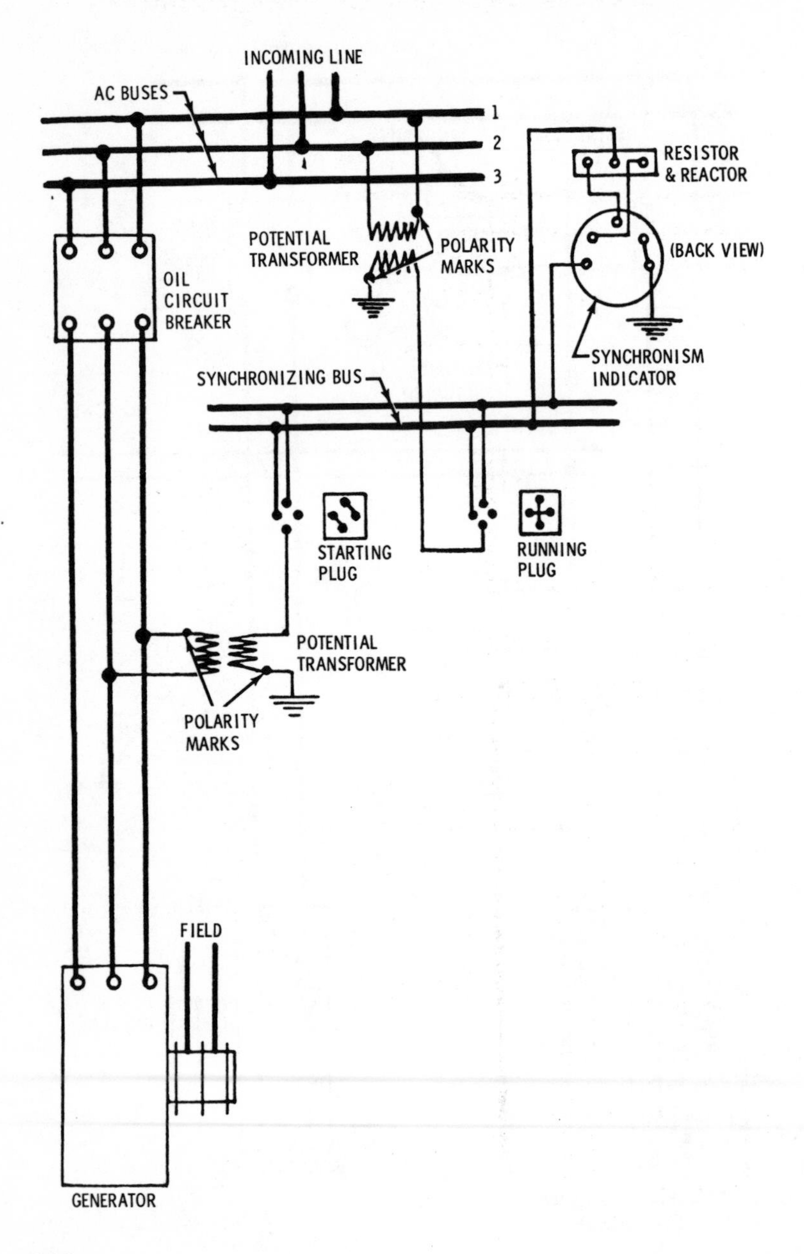

SYNCHRONIZING CONNECTION ACROSS A DELTA DIAMETRICAL BANK OF TRANSFORMERS

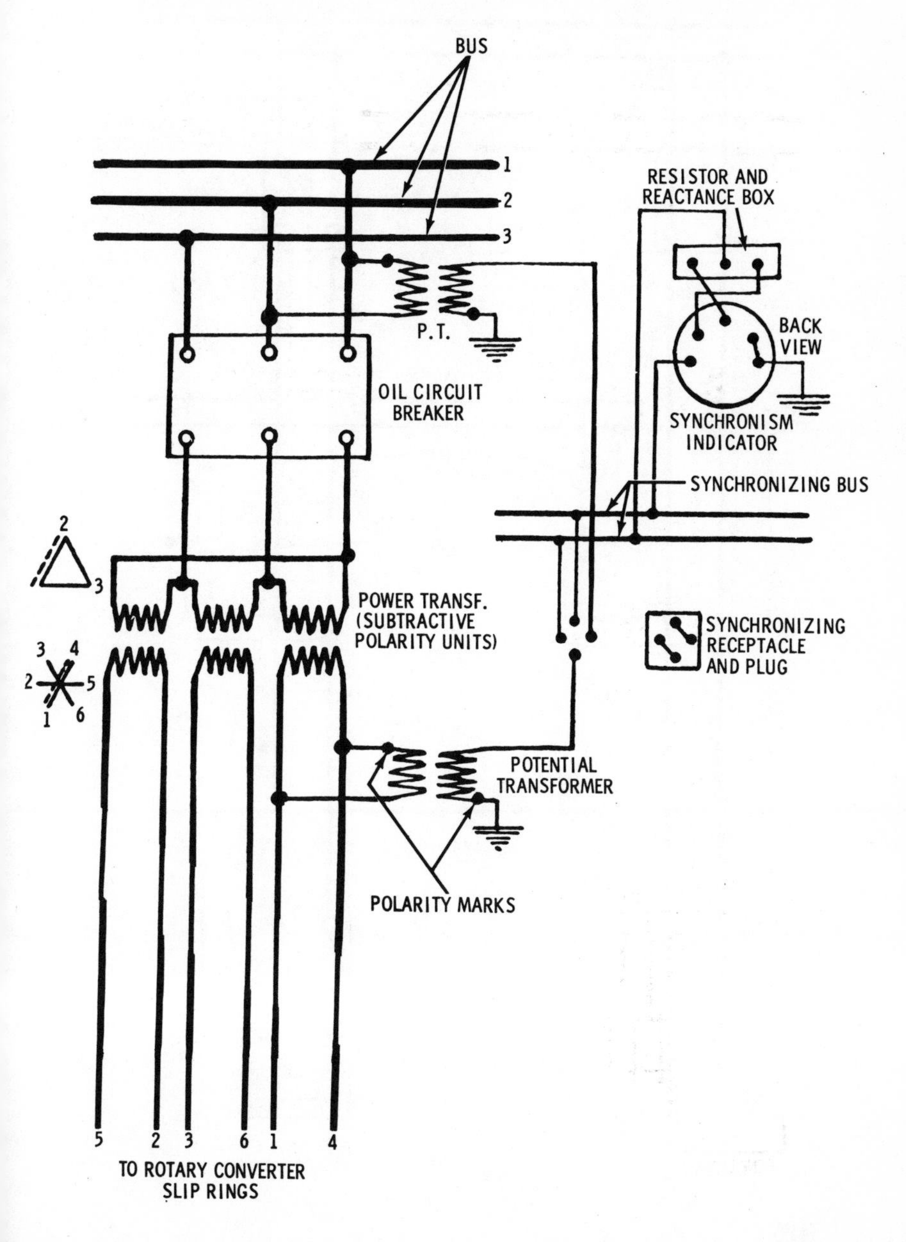

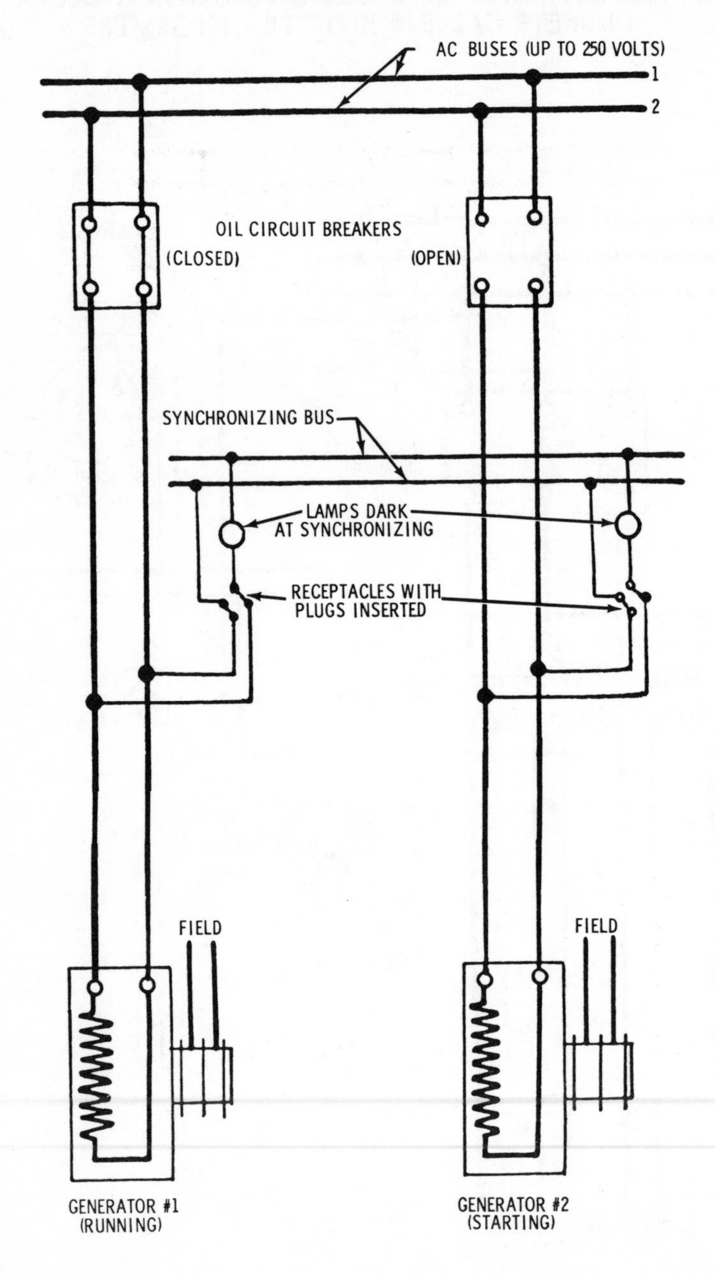
METHOD OF SYNCHRONIZING (LAMPS DARK)
AC BUSES (UP TO 250 VOLTS)
1
2
OIL CIRCUIT BREAKERS
(CLOSED)
(OPEN)
SYNCHRONIZING BUS
LAMPS DARK
AT SYNCHRONIZING
RECEPTACLES WITH
PLUGS INSERTED
FIELD
FIELD
GENERATOR #1
(RUNNING)
GENERATOR #2
(STARTING)

METHOD OF SYNCHRONIZING (LAMPS BRIGHT)

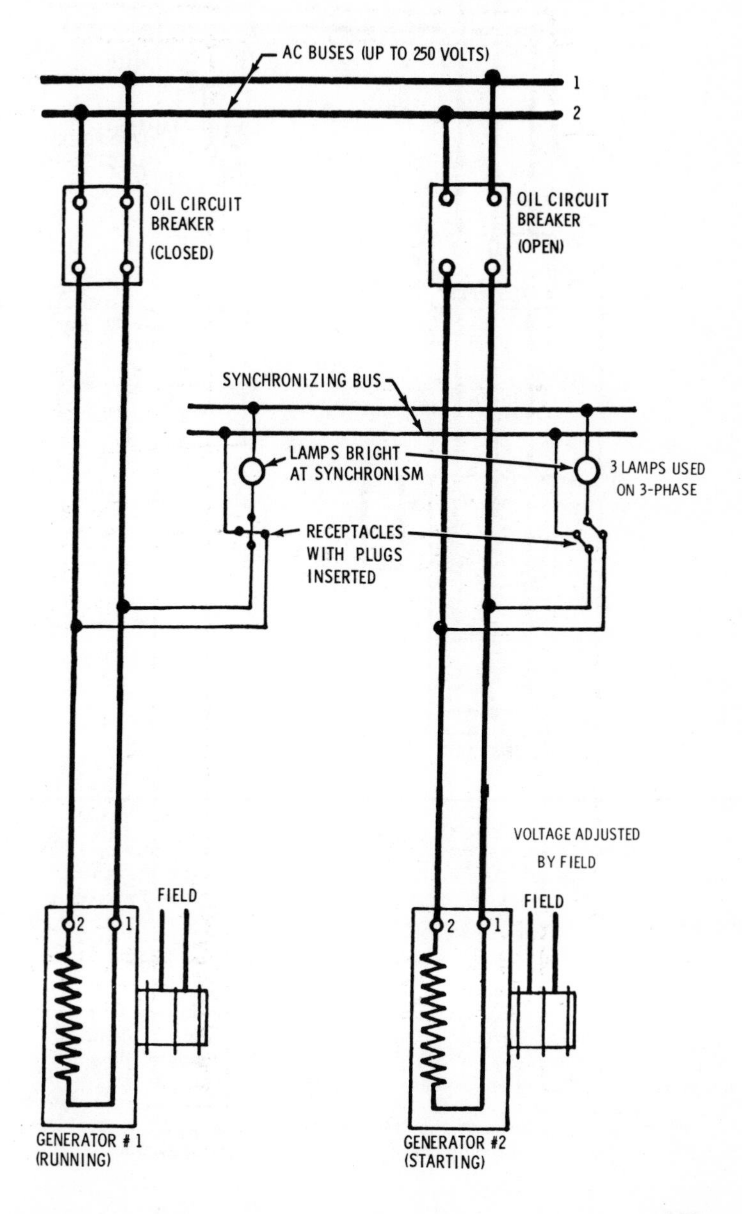

PHASE-ROTATION TEST

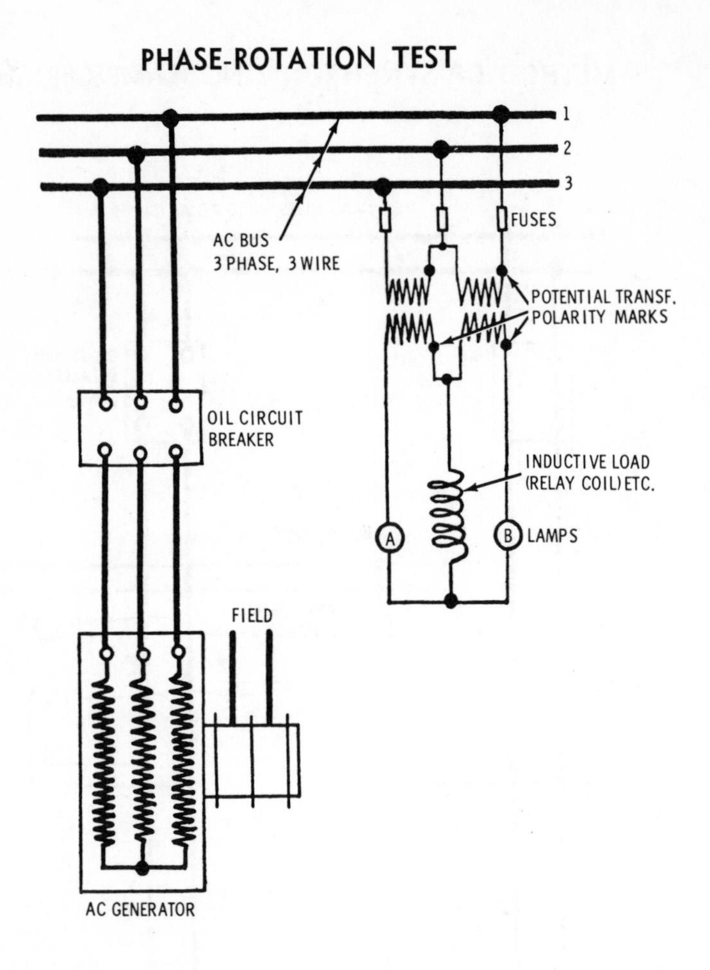

Before two AC generators can be synchronized, it is necessary, in addition to having one phase of the machine in synchronism with the phase of the other machine, that the sequence of maximum potential values in various phases shall be the same.

Adjustments of the phases to obtain this sequence is known as "phase rotation test." To obtain the phase rotation of the machine, proceed as follows: Connect two lamps "A" and "B" and an inductive load as shown. One lamp will glow brighter than the other; and if that lamp should be the "B" lamp, the phase relation is clockwise or 1-2-3. If on the other hand "A" should glow the brighter, the phase rotation is counterclockwise or 3-2-1.

After a check, the leads of the machine may be transposed to conform with phase rotation of the other machine, or the bus, as the case may be.

TYPICAL SYNCHRONISM INDICATOR

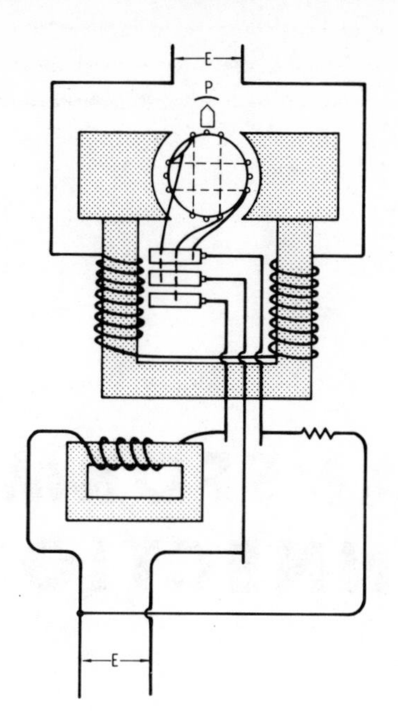

TRANSFORMER CONNECTIONS

A transformer may be defined as a stationary means of transforming power by electro-magnetic induction. This is the most efficient way to transform power, running close to the 100% efficiency mark.

There are the two wing type, which is the one we ordinarily think of, in which the two wingings are insulated from each other and there is the autotransformer, in which the windings are connected.

A very common error is calling the high-voltage winding the primary and the low voltage winding the secondary. This is often the case but in reality the primary is the winding connected to the power source and the secondary the winding connected to the load.

Transformers transform voltage from one voltage to the other in direct ratio to the number of turns on each winding and the current is inverse to the voltage. The high-voltage side is the lower current and the low-voltage side is the higher current, but the currents are inversely proportional to the voltage. The three-phase transformer consists of three primary and three seconday windings (see A below) usually connected in star or delta respectively.

Single-phase transformers (see B below) connected in star or delta are often preferable to three-phase transformers because single-phase reverse units are less expensive also because damage to one single-phase transformer may be repaired while another identical spare transformer is interconnected in the three-phase unit without loss of service.

When two sets of transformers are connected in parallel to the primary and secondary circuits of a three-phase system, any combination of delta and star may be used in each set except that, with one set of transformers connected in delta-star or star-delta, the other set may not be connected delta-delta or star-star.

For examples of transformer connections, see the following pages.

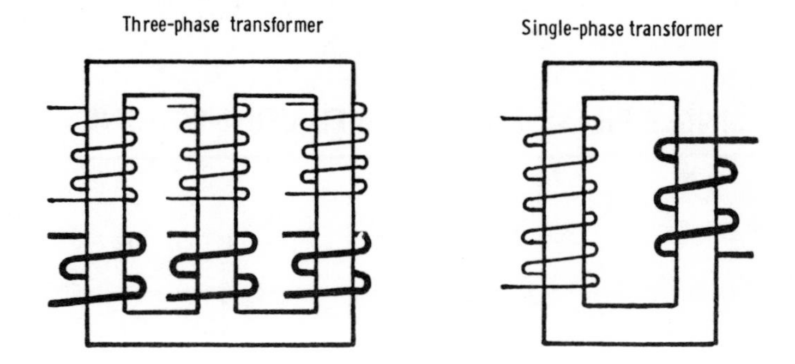

Transformers are usually marked whether they are "additive or subtractive polarity." The polarity make no difference, except if transformers are being paralleled, or if single-phase transformers are being connected for two-phase, three-phase or six-phase connections, and then the polarity must be our concern.

The high-side has the leads usually marked H_0, H_1, H_2, H_3, etc. and the low side marked V_0, X_1, X_2, X_3, etc.

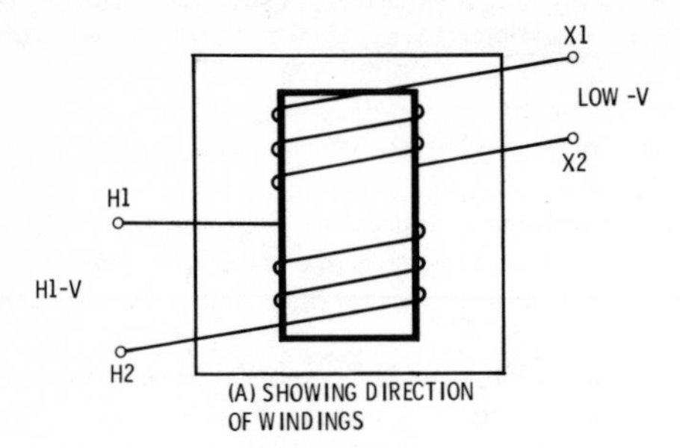

(A) SHOWING DIRECTION OF WINDINGS

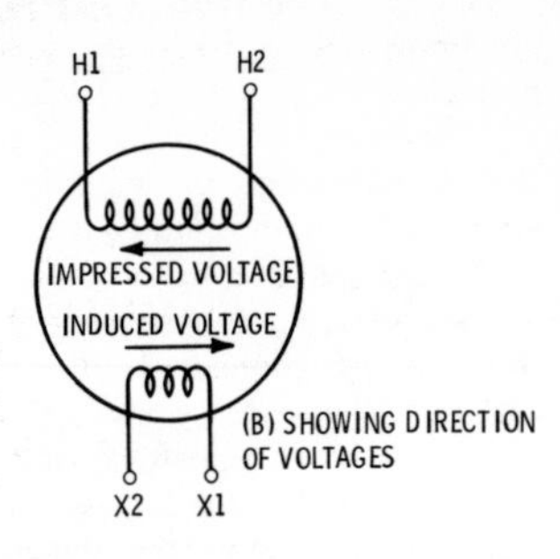

(B) SHOWING DIRECTION OF VOLTAGES

ADDITIVE POLARITY TRANSFORMERS

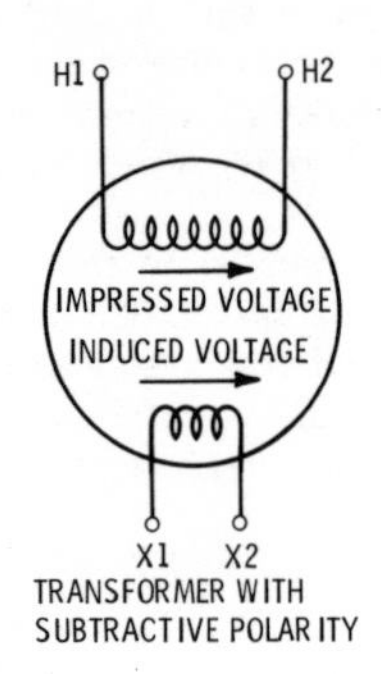

TRANSFORMER WITH SUBTRACTIVE POLARITY

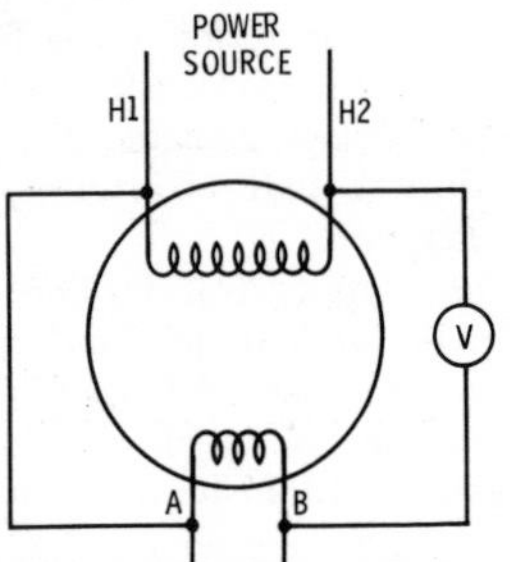

CHECKING TRANSFORMER POLARITY. IF ADDITIVE, THE VOLTAGE WILL READ THE SUM OF THE HIGH AND LOW VOLTAGE. IF SUBTRACTIVE POLARITY, THE VOLTAGE WILL READ THE DIFFERENCE BETWEEN THE HIGH AND LOW VOLTAGE.

SINGLE-PHASE TRANSFORMER CONNECTIONS

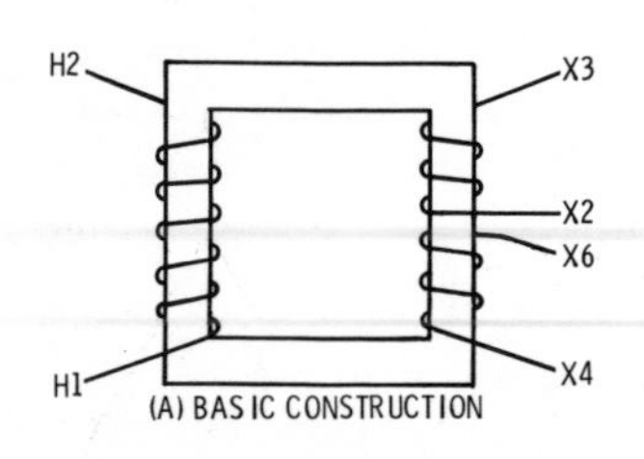

(A) BASIC CONSTRUCTION

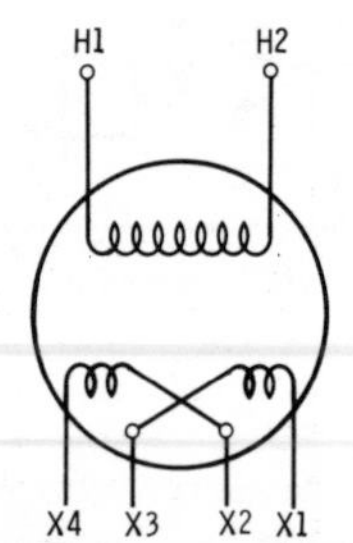

(B) LEADS X_2 AND X_3 CROSSED INTERNALLY

SINGLE-PHASE TRANSFORMERS WITH DUAL LOW-VOLTAGE WINDINGS

#1 #2

H1 H2

(B) WINDINGS (LOW-VOLTAGE) IN PARALLEL

(A) WINDINGS (LOW-VOLTAGE) IN SERIES

X4 X2 X3 X1

120 V 120 V

240 V

120 V

TWO-PHASE, FOUR-WIRE CONNECTION FOR TWO SINGLE-PHASE TRANSFORMERS

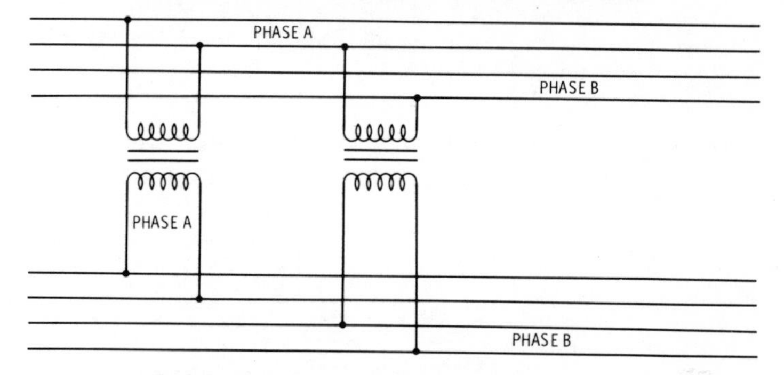

THREE SINGLE-PHASE TRANSFORMERS CONNECTED DELTA-DELTA AND VECTORS

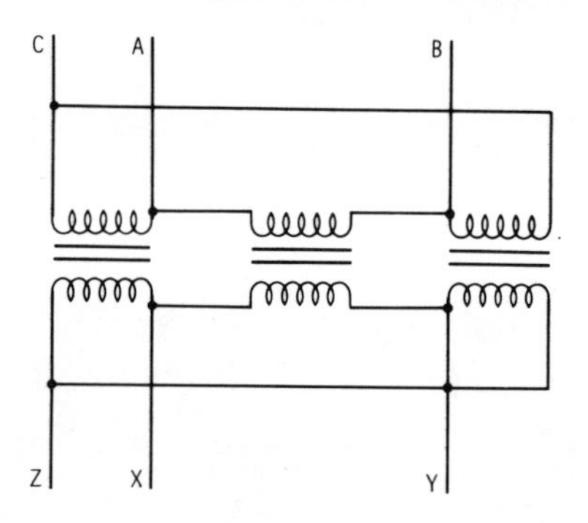

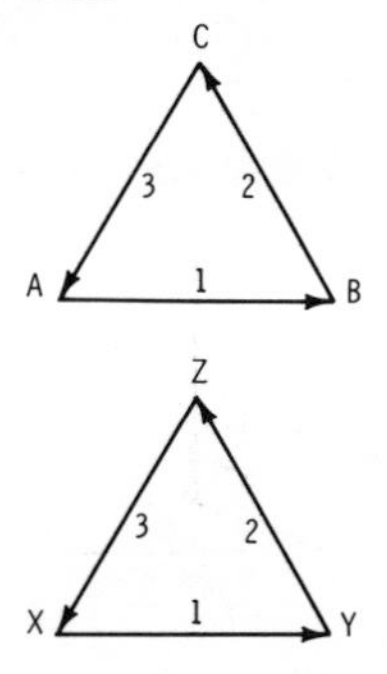

OPEN-DELTA CONNECTIONS USING TWO SINGLE-PHASE TRANSFORMERS

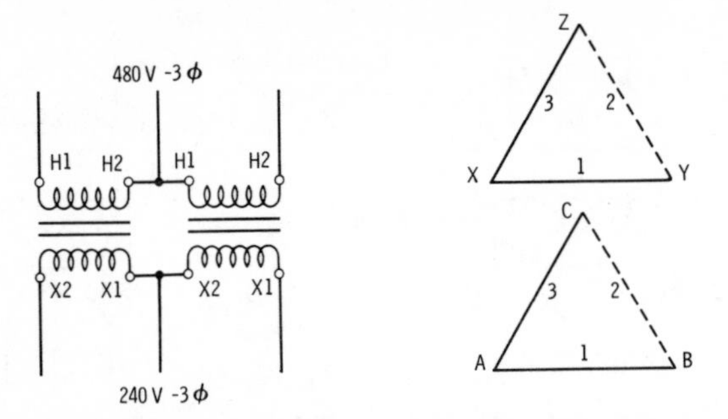

Both transformers are the same polarity—Looking at the vectors, the dashed line is termed the phantom-phase. Its voltage is just slightly higher than that across the two transformers. Also, some capacity is lost. For instance, say we had two 10 kVA transformers connected in open-delta, one would be led to think that 20 kVA could be obtained from them. This is not the case. You get 86.6% of the total capacity, so in this case, you actually have 86.6% of the 20 kVA as that capacity which is available, or 17.32 kVA.

OPEN-DELTA CONNECTIONS USING TWO SINGLE-PHASE TRANSFORMERS OF OPPOSITE POLARITIES

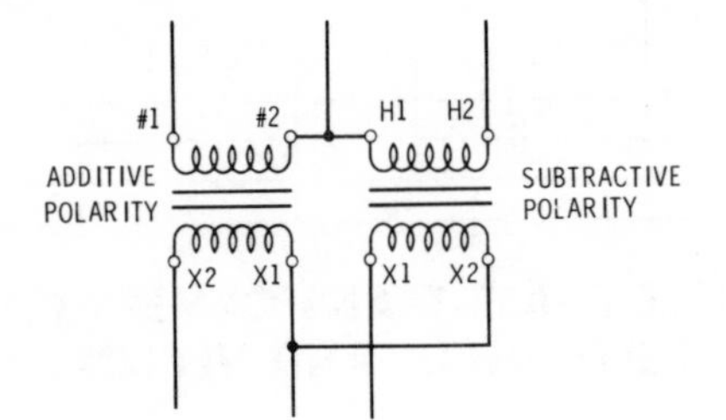

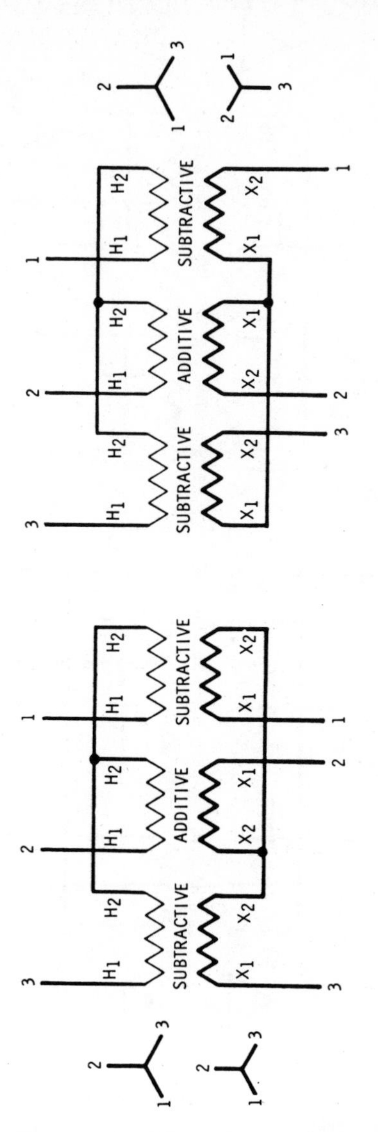

Connections for Star-Star Power Transformer Group to Obtain Additive or Subtractive Line Polarity.

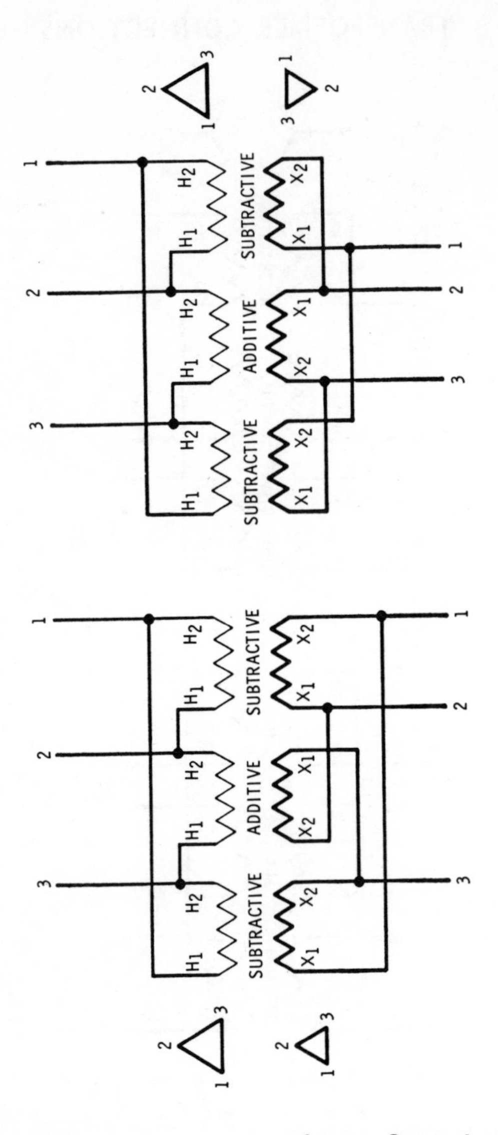

Connections for Delta-Delta Power Transformer Group to Obtain Additive or Subtractive Line Polarity.

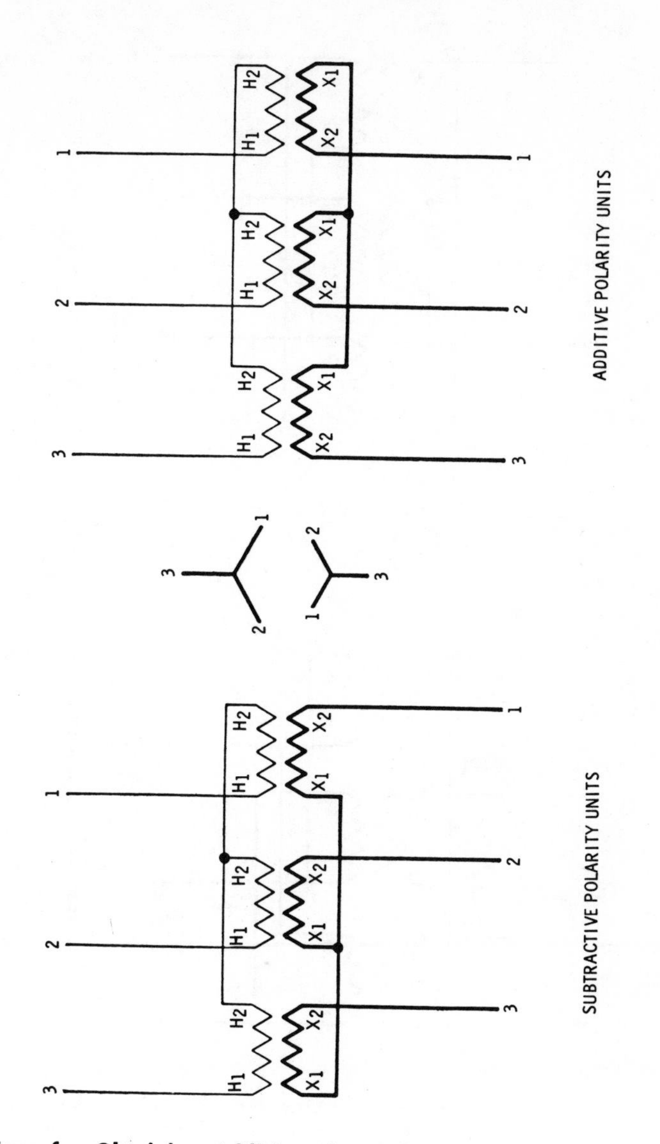

Connections for Obtaining Additive Line Polarity with Transformer Units of Either Additive or Subtractive Polarity in Star-Star Groups.

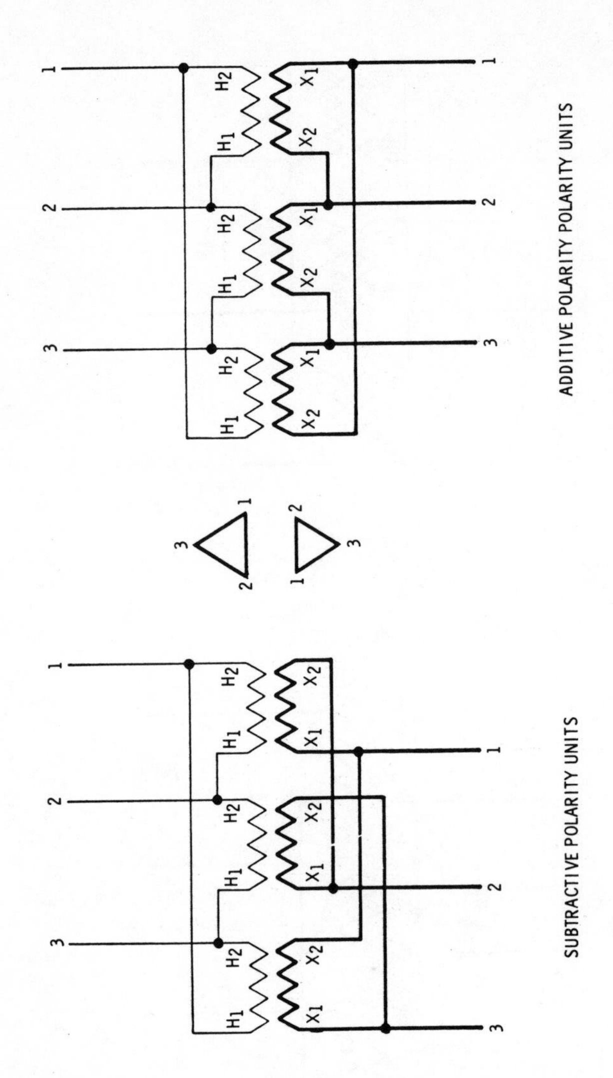

Connections for Obtaining Additive Line Polarity with Transformer Units of Either Additive or Subtractive Polarity in Delta-Delta Groups.

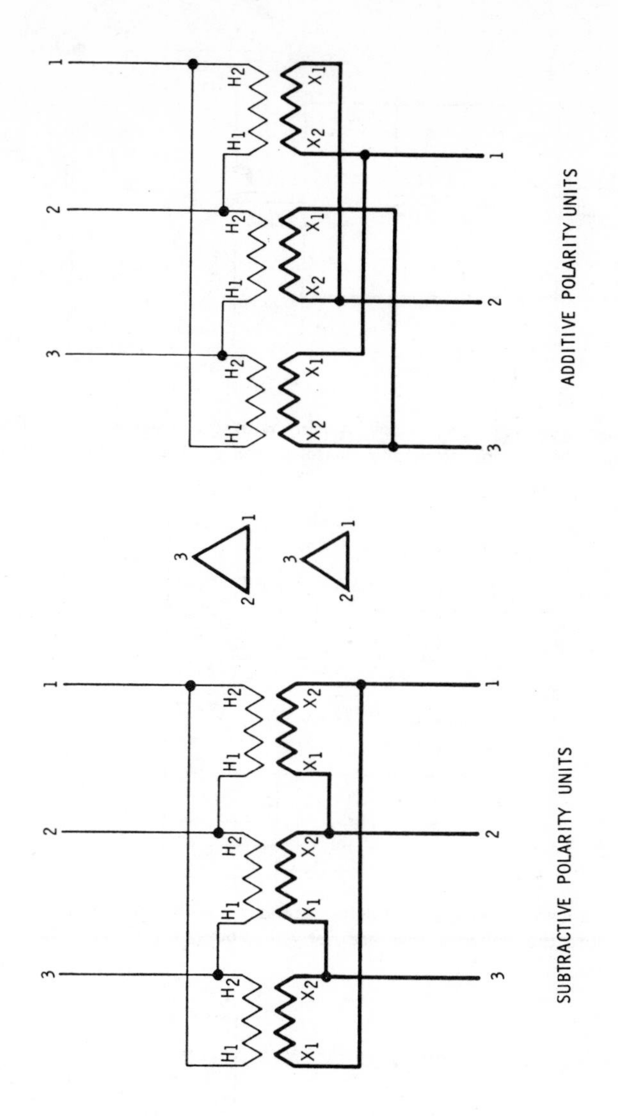

Connections for Obtaining Subtractive Line Polarity with Transformer Units of Either Additive or Subtractive Polarity in Delta-Delta Groups.

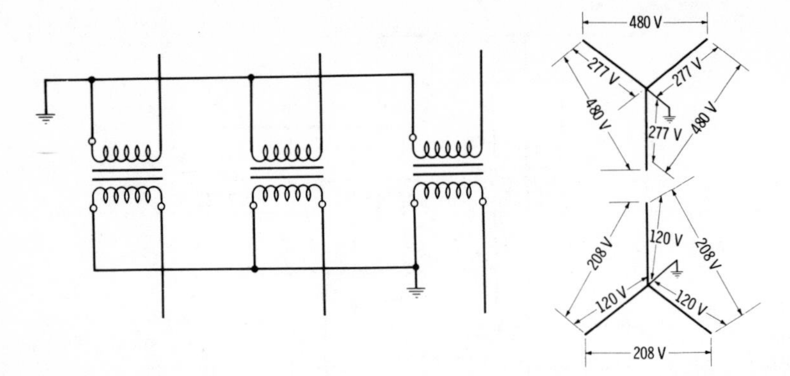

Three-phase wye-wye connections, 4-wire with grounded neutral, showing voltage relations. The voltage from any phase to the neutral is, phase-voltage divided by 1.73.

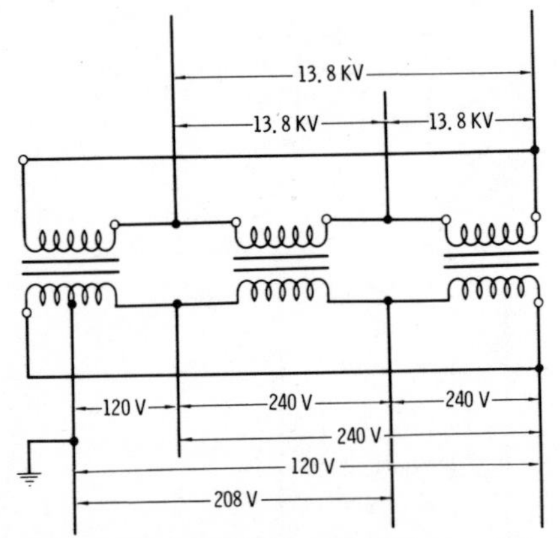

Wye primary and 4-wire closed-delta secondary, with mid point of one secondary grounded. Note the voltages that are derived. You get 208 volts from the neutral to the wild-leg.

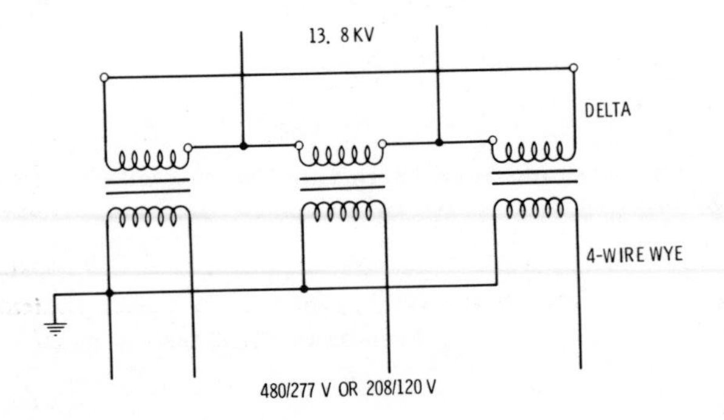

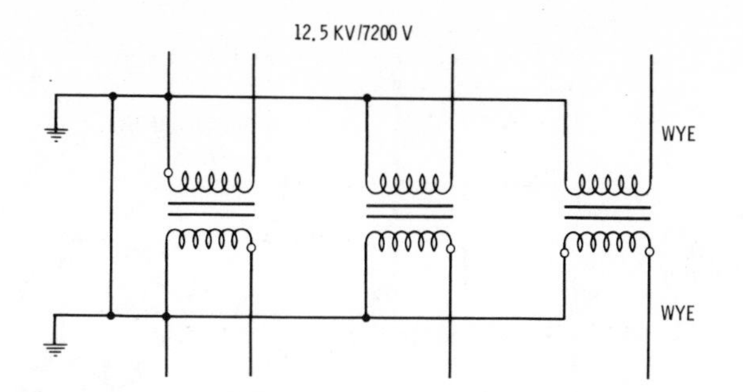

Zero-resonance can be a problem on utility company power lines. This is a resonance effect, which when a three-phase line, single-phases, will cause a build up of voltage and burn out of transformers. Many utilities use a delta primary and a wye secondary to eliminate this. Other utilities have not had trouble with wye wye installations, as long as the neutrals of the wyes are tied together and properly grounded.

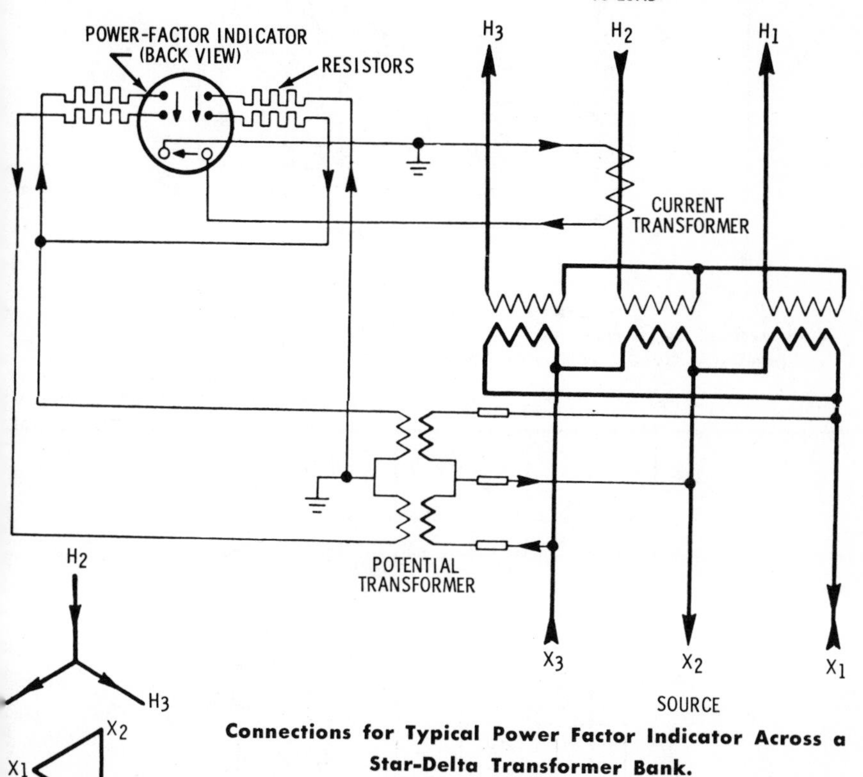

Connections for Typical Power Factor Indicator Across a Star-Delta Transformer Bank.

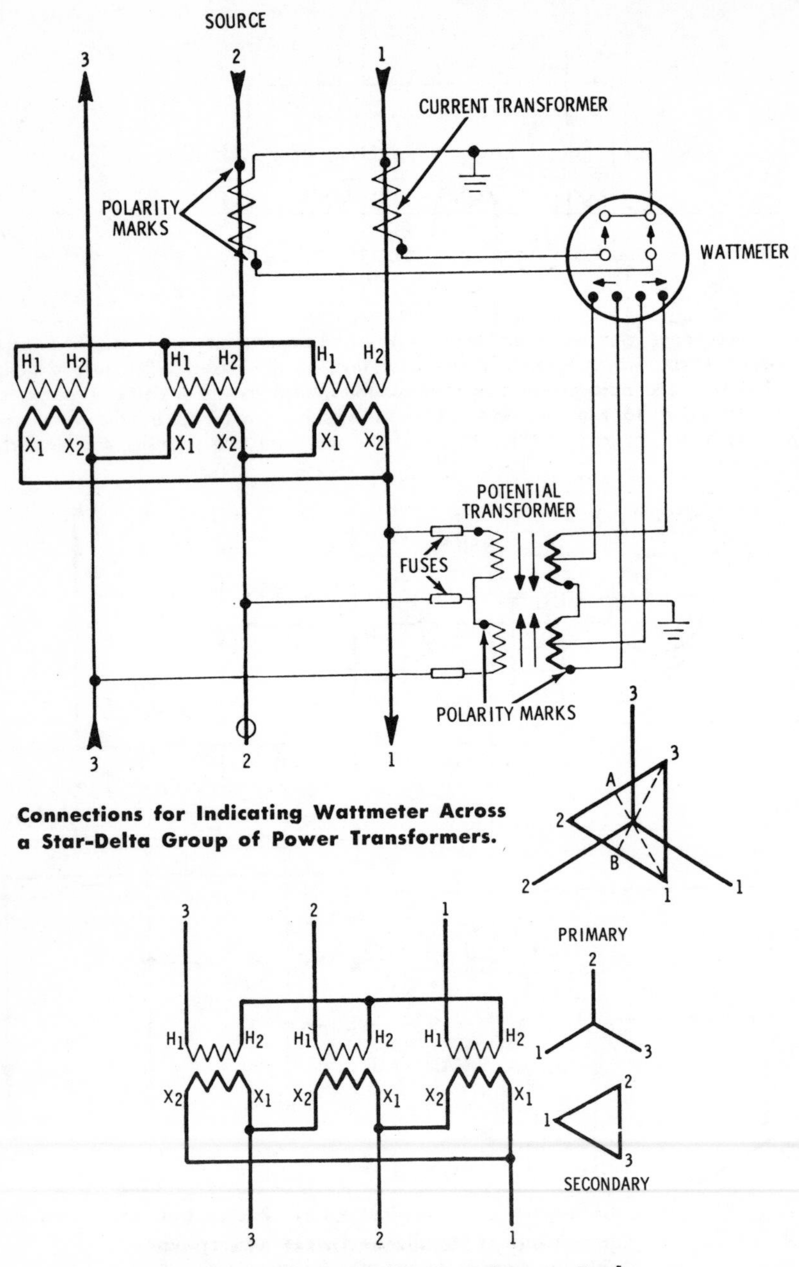

Connections for Indicating Wattmeter Across a Star-Delta Group of Power Transformers.

Connections of Star-Delta Transformer Bank, Additive Polarity Units.

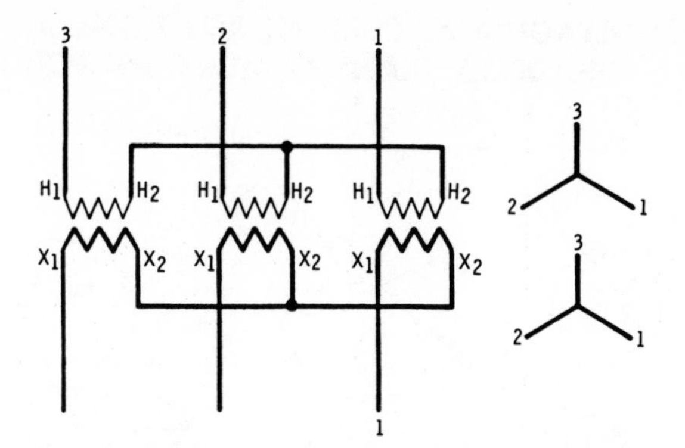

Connections of Star-Star Transformer Bank, Subtractive Polarity Units.

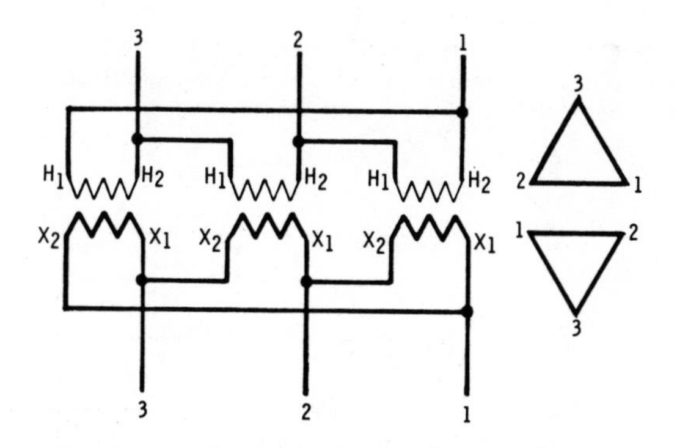

Connections of Delta-Delta Transformer Bank, Additive Polarity Units.

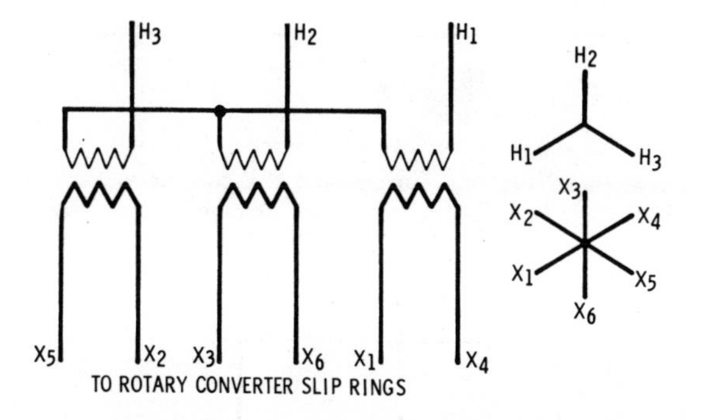

Connection of Star-Diametrical Transformer Bank for Rotary Converter Service.

VOLTAGE AND CURRENT RELATIONS IN WYE AND DELTA TRANSFORMER CONNECTIONS

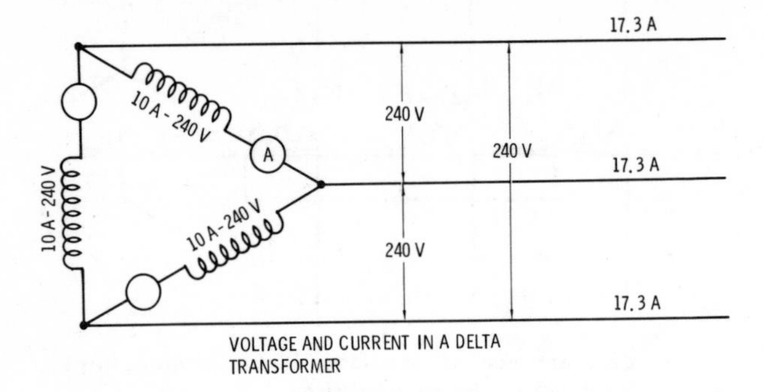

VOLTAGE AND CURRENT IN A DELTA TRANSFORMER

Note that in a delta bank, the current in each transformer is the line current divided by 1.73. This of course is assuming three-phase load. The voltage across each transformer is the same as the line phase voltages.

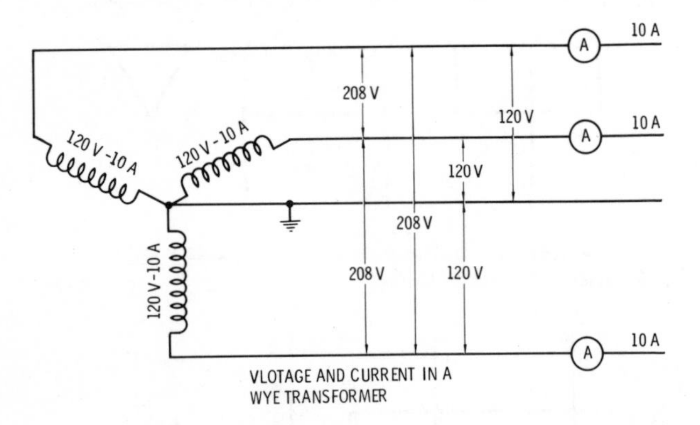

VLOTAGE AND CURRENT IN A WYE TRANSFORMER

Note that in a wye bank, the line current and the current in each transformer is the same, while the voltage between phases is the voltage of each transformer multiplied by 1.73. The voltage between each phase and the neutral is the same as the voltage of each transformer.

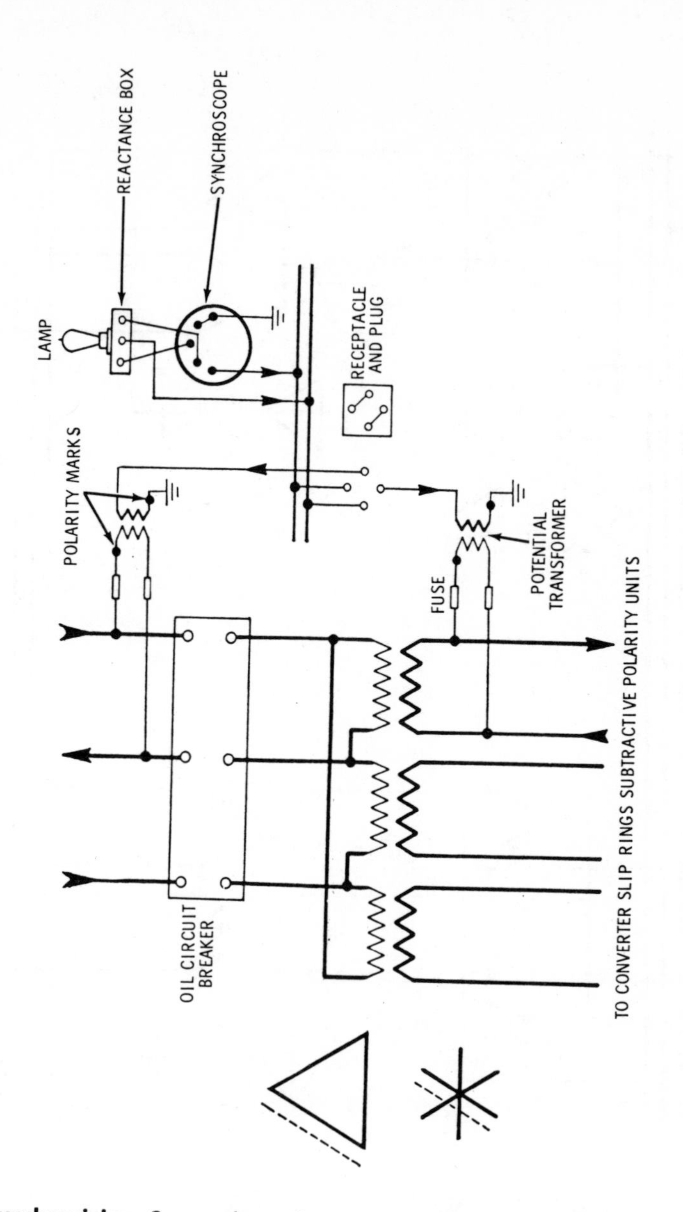

Synchronizing Connections Across a Delta-Diametrical Bank of Transformers Consisting of Subtractive Polarity Units.

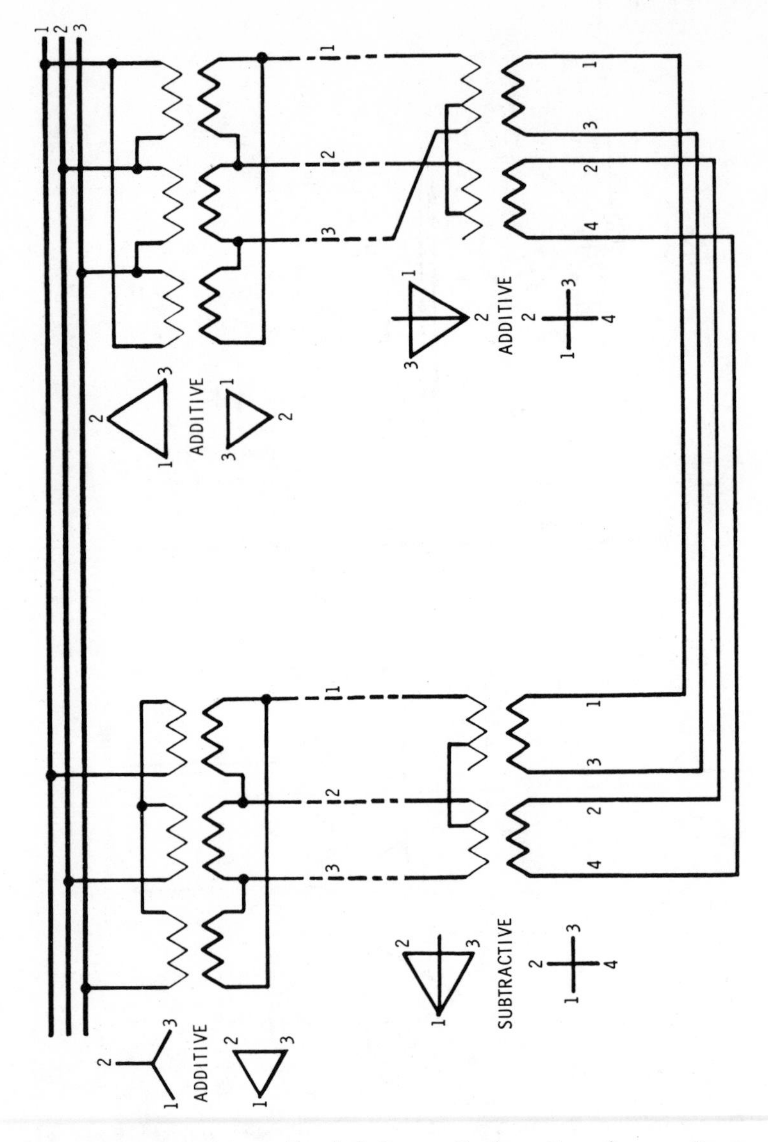

Three-Phase Voltages Carried Across Various Transformer Group Forms Resulting in In-Phase Voltages on the Quarter-Phase System.

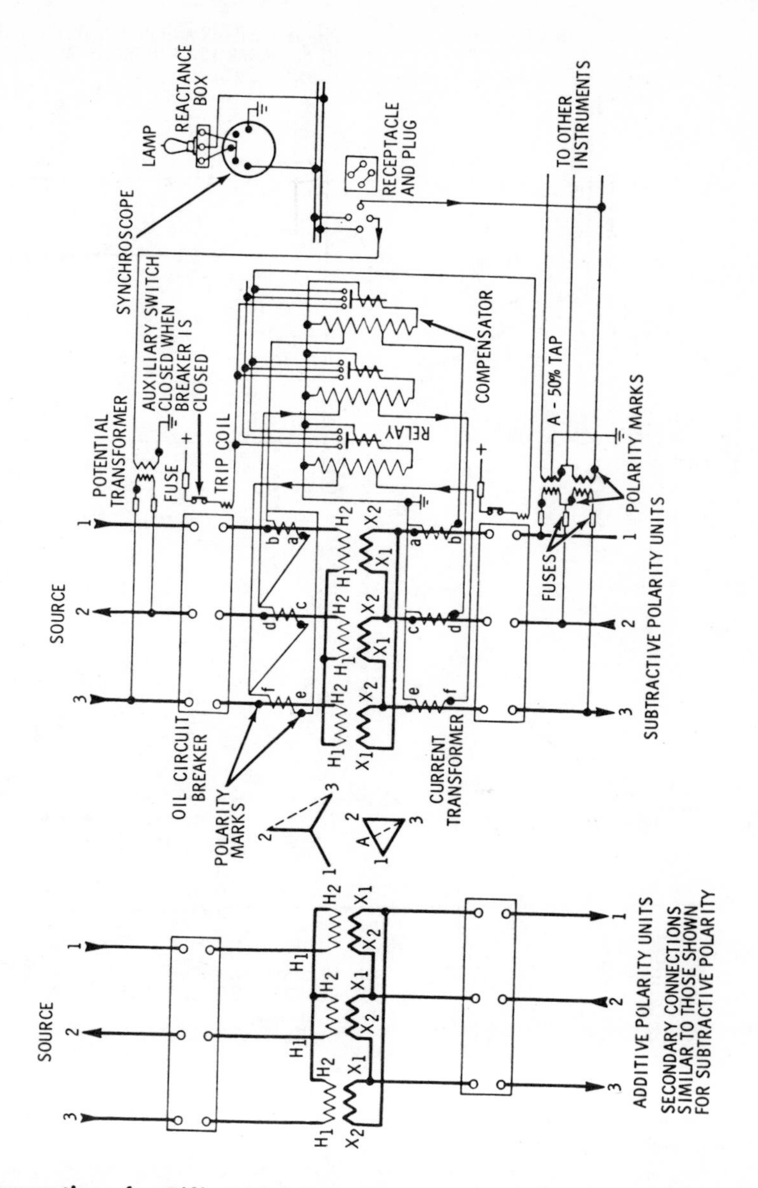

Connections for Differential Protection and Synchronizing Across a Star-Delta Bank.

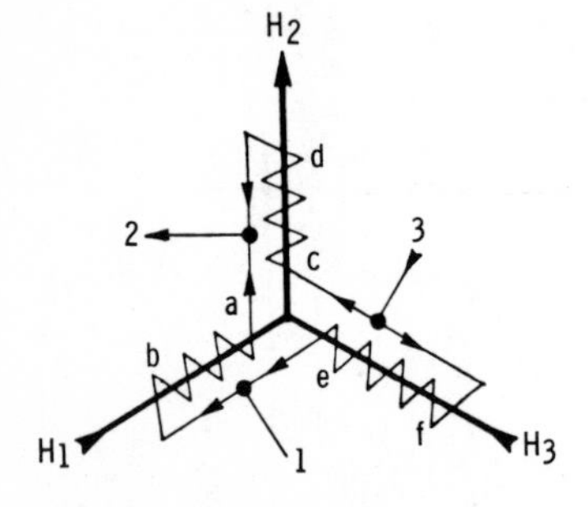

PRIMARIES AND SECONDARIES OF CURRENT TRANSFORMERS ON HIGH-VOLTAGE SIDE

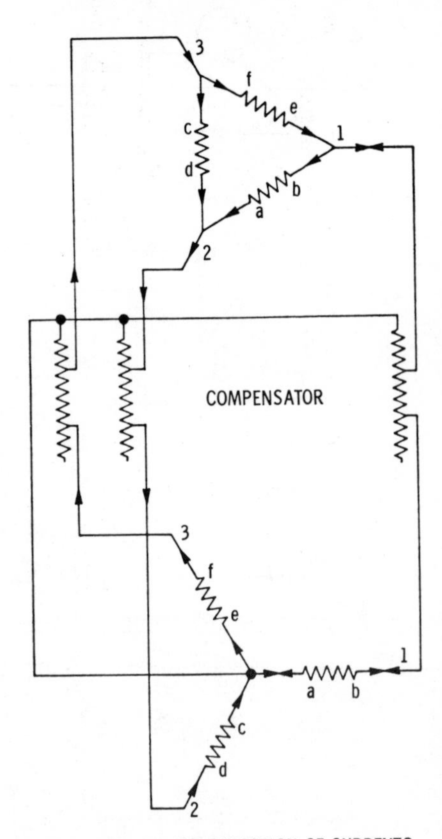

PHASE POSITION AND DIRECTION OF CURRENTS IN SECONDARIES ON LOW-VOLTAGE SIDE (5. 00 AMPERES)

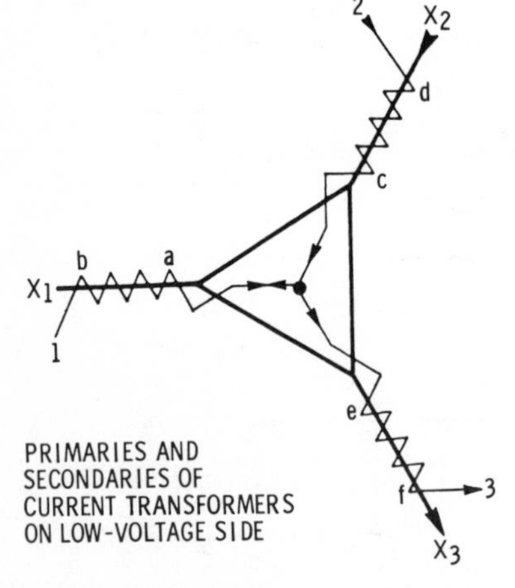

PRIMARIES AND SECONDARIES OF CURRENT TRANSFORMERS ON LOW-VOLTAGE SIDE

Vector Diagram Representing Current Relations at a Given Instant in Current Transformers Arranged for Differential Protection with Power Transformers Connected in Star-Delta.

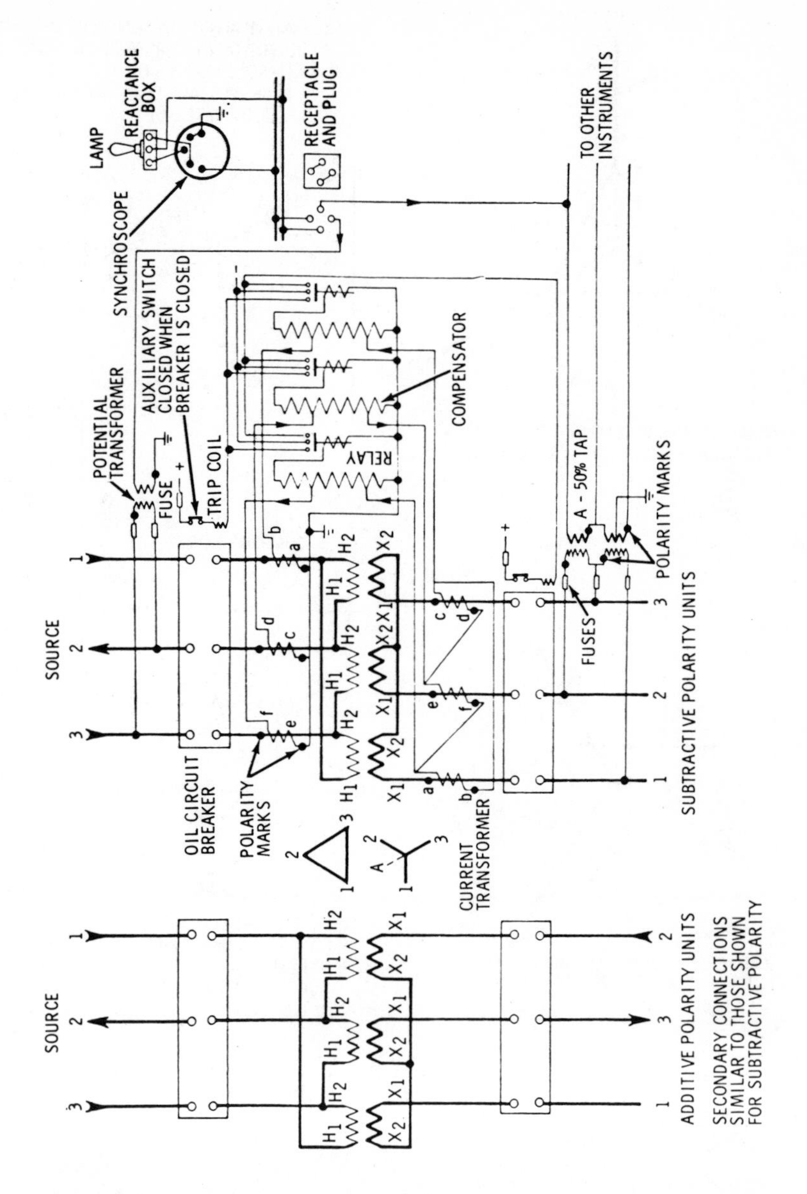

Typical Connections for Differential Protection and Synchronizing Across a Delta-Star Bank.

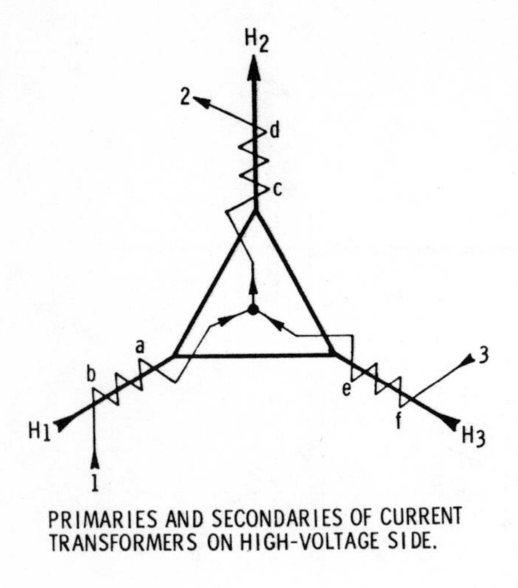

PRIMARIES AND SECONDARIES OF CURRENT TRANSFORMERS ON HIGH-VOLTAGE SIDE.

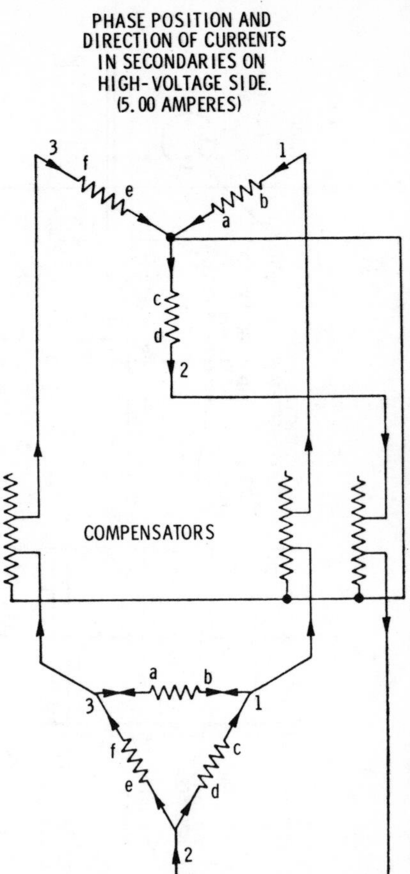

PHASE POSITION AND DIRECTION OF CURRENTS IN SECONDARIES ON LOW-VOLTAGE SIDE. (8.66 AMPERES)

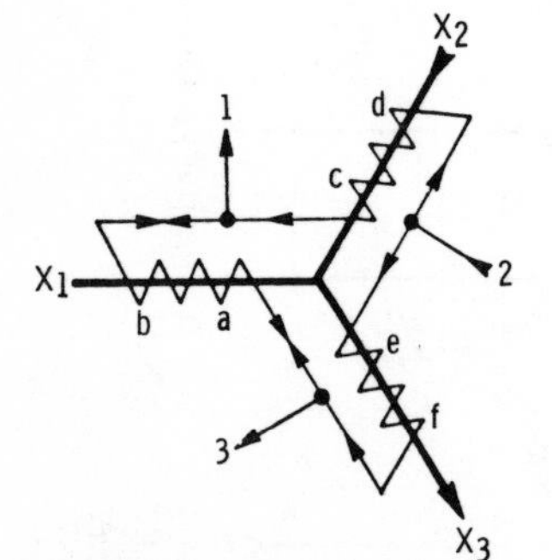

PRIMARIES AND SECONDARIES OF CURRENT TRANSFORMERS ON LOW-VOLTAGE SIDE.

Vector Diagram Representing Current Relations at a Given Instant in Current Transformers Arranged for Differential Protection with Power Transformers Connected in Delta-Star.

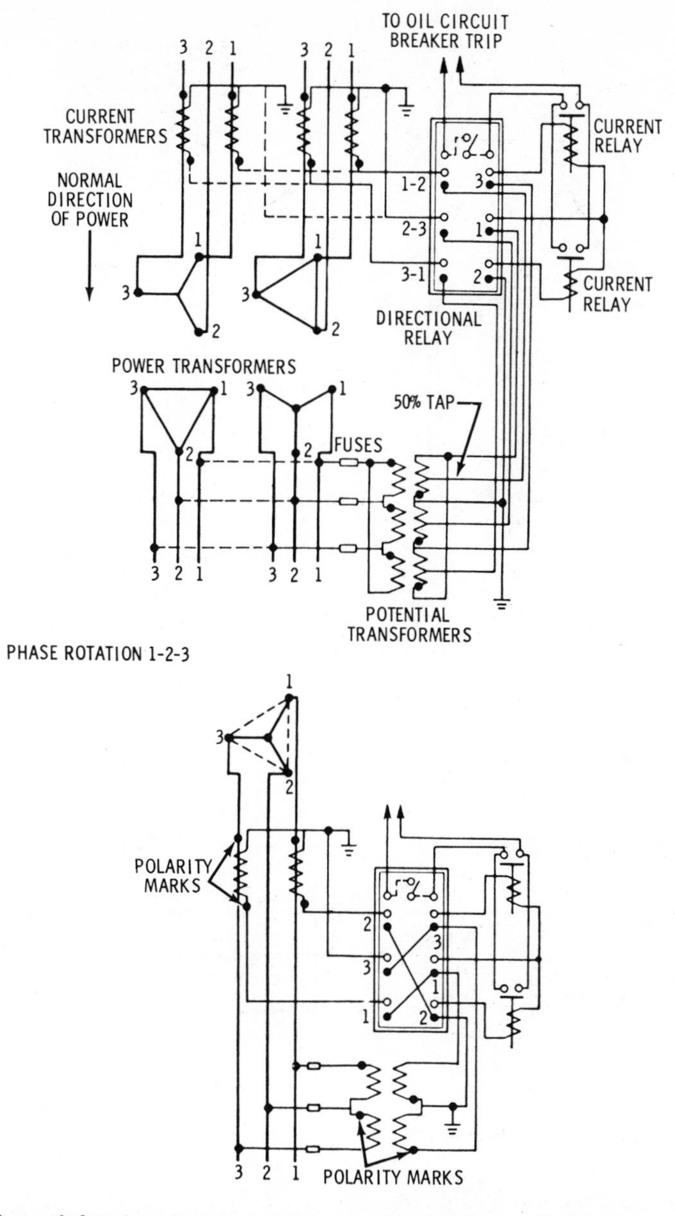

Differential Relay Connected Across a Star-Delta or Delta-Star Group of Power Transformers.

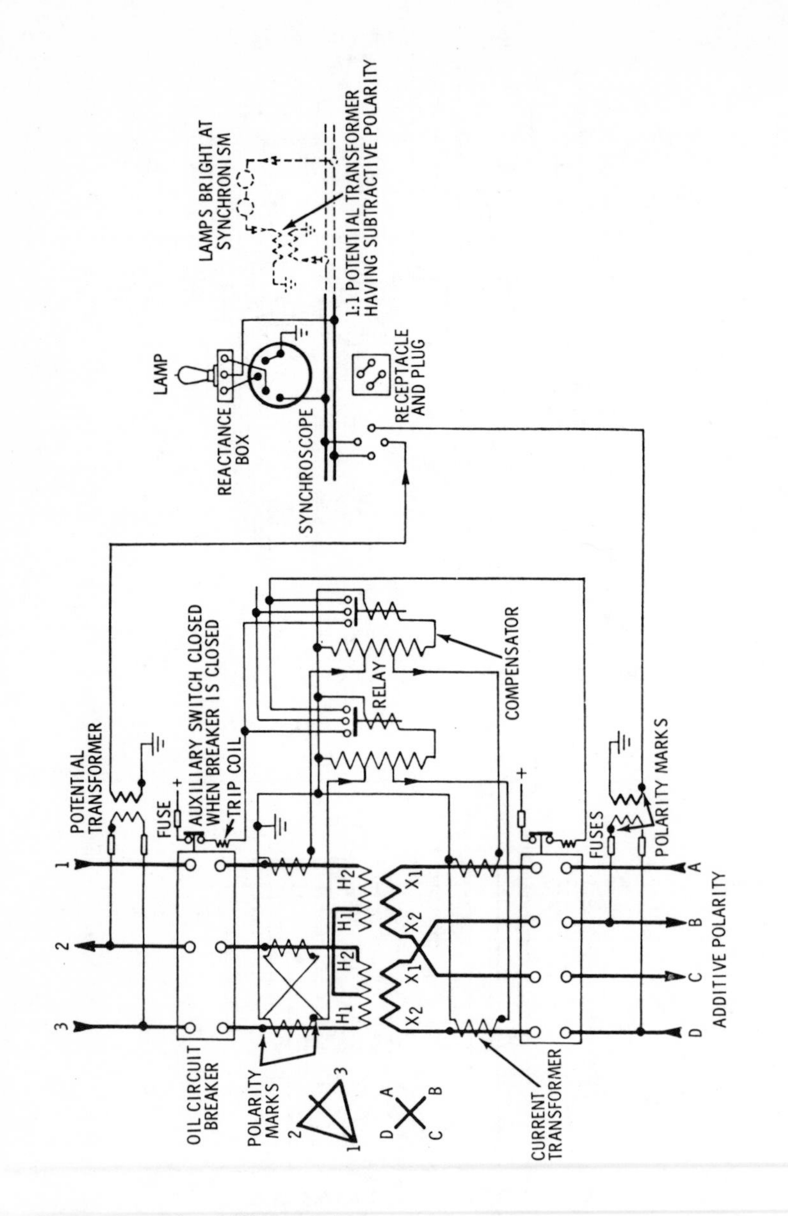

Connections for Differential Protection and Synchronizing Across a Scott-Connected Bank of Additive Polarity Transformers.

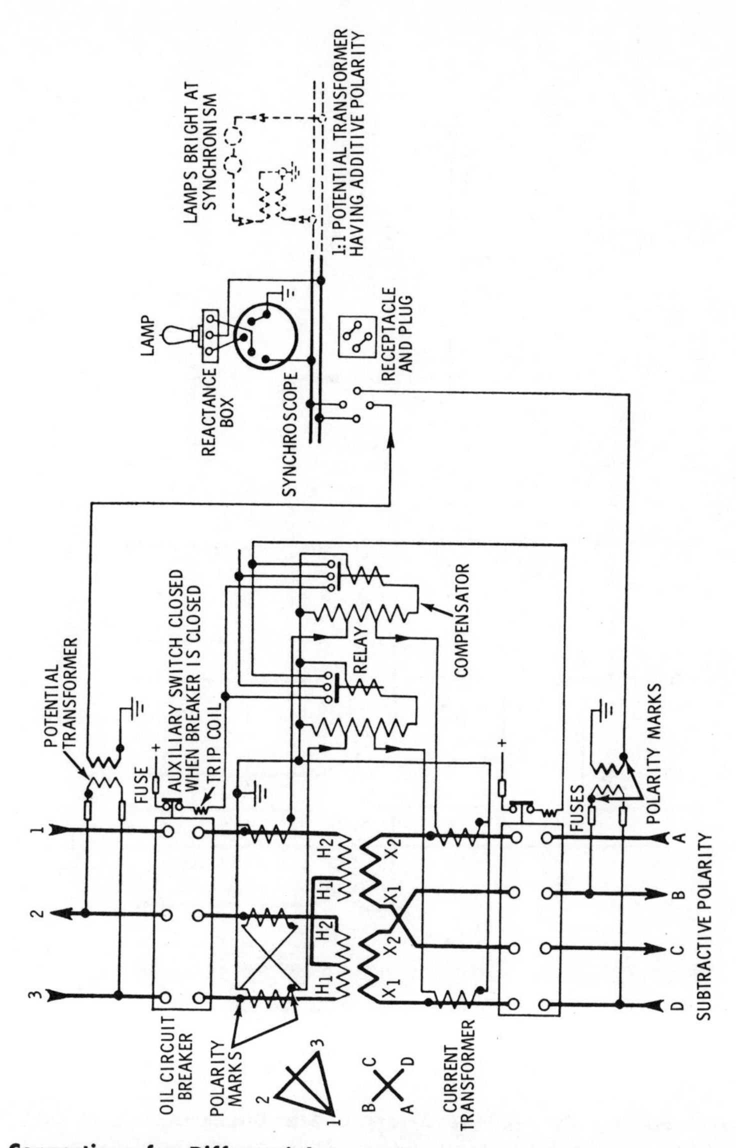

Connections for Differential Protection and Synchronizing Across a Scott-Connected Bank of Subtractive Polarity Transformers.

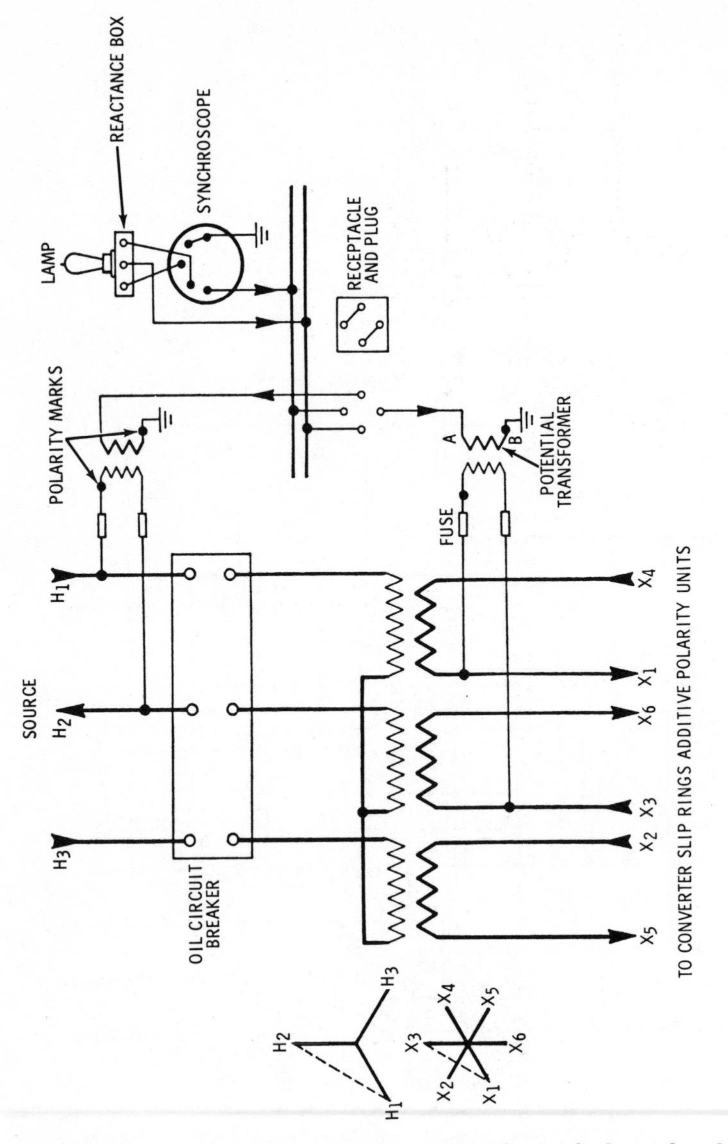

Synchronizing Connections Across a Star-Diametrical Bank of Transformers Consisting of Additive Polarity Units.

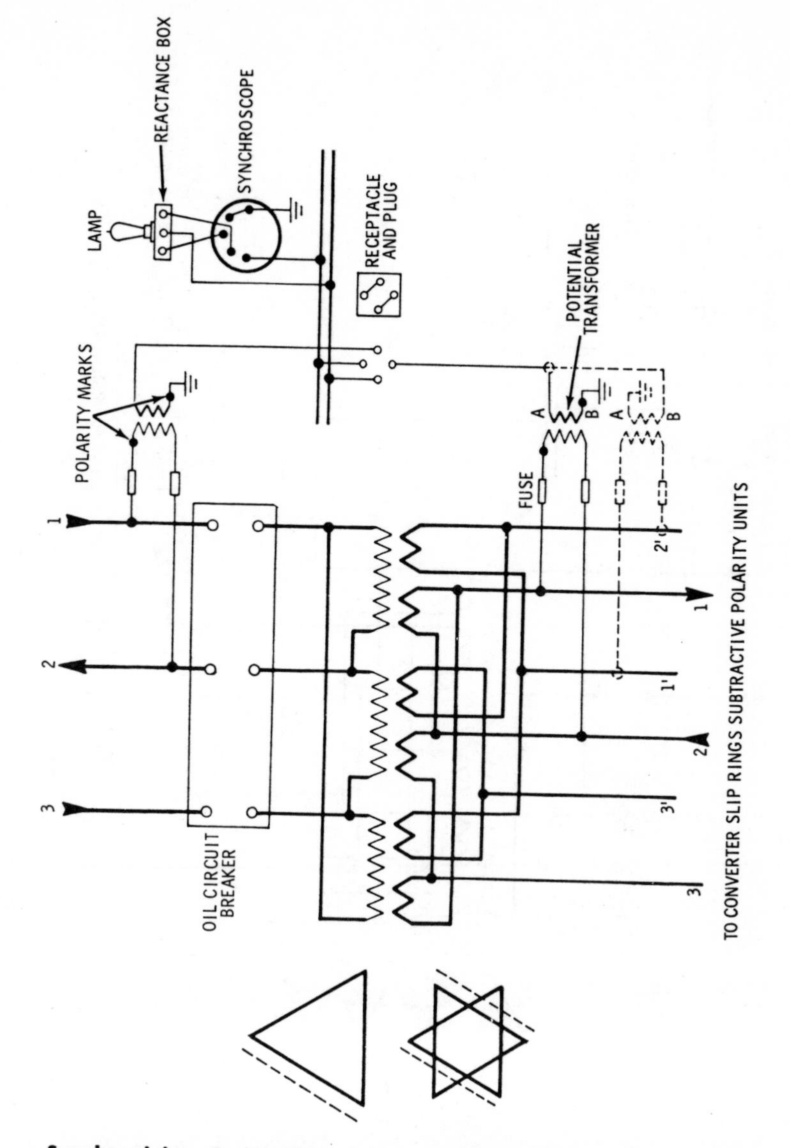

Synchronizing Connections Across a Delta-Double Delta Bank of Transformers Consisting of Subtractive Polarity Units.

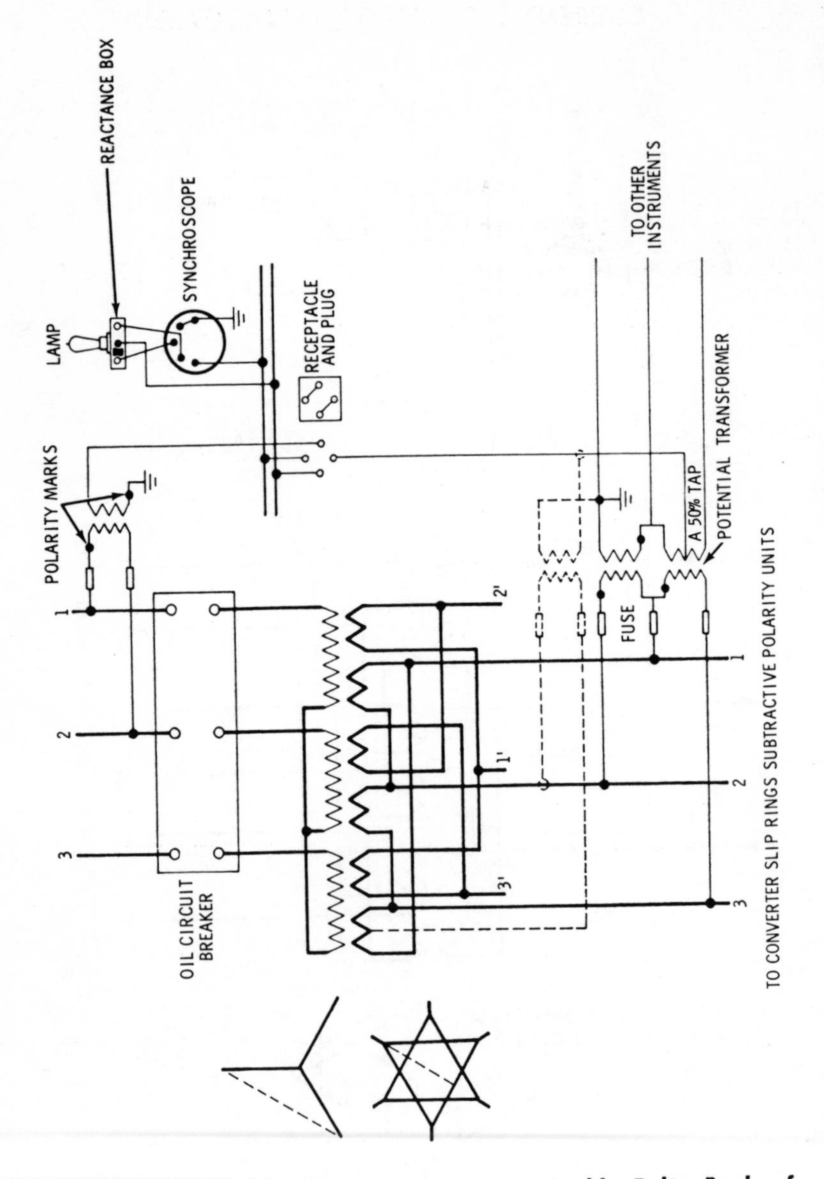

Synchronizing Connections Across a Star-Double Delta Bank of Transformers Consisting of Subtractive Polarity Units.

CURRENT-DIFFERENTIAL PROTECTION

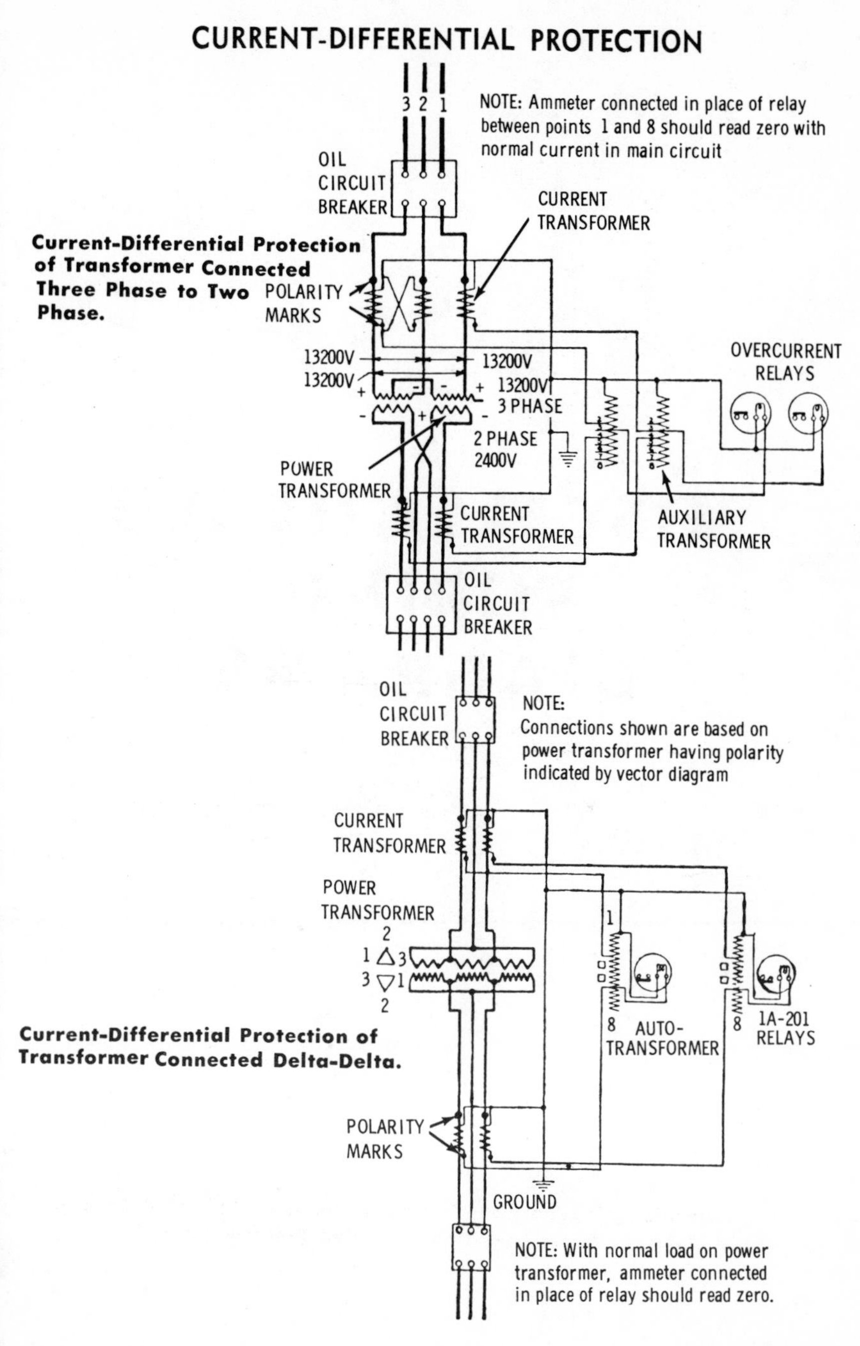

Current-Differential Protection of Transformer Connected Three Phase to Two Phase.

Current-Differential Protection of Transformer Connected Delta-Delta.

MISCELLANEOUS WIRING DIAGRAMS

GROUND-FAULT CIRCUIT-INTERRUPTERS

These are commonly called GFIC's and are intended as life protectors. They may also eliminate fires from electrical ground-faults, but their primary purpose is life protection.

Their use is a requirement of the National Electrical Code for many places; (1) receptacle outlets, lighting fixtures, and lighting outlets near swimming, wading, therapeutic, and decorative pools and fountains; (2) circuits for underwater lighting fixtures; (3) electrical equipment used with storable pools; (4) branch circuits supplying fountain electrical equipment; (5) residential outdoor and bathroom receptacle outlets, and (6) receptacles on 15 and 20 ampere circuits used on construction sites. It is also pointed out that they may be used with additional safety on any other circuits.

Class A, GFCI's are designed to open the circuit when the leakage to ground reaches 5 milliamperes, and to operate in a maximum of 5 milliseconds. Class A devices are to be used on all new installations.

Class B GFCI's operate at a maximum of 20 milliamperes, and are designed to be used on underwater lighting in existing swimming pools only.

These devices are available in many forms, to fit the application. Such as circuit breakers, 1 and 2 pole, to fit branch circuits or feeder circuits, plug-in assemblies to plug into existing receptacles, multiple outlet assemblies with cord to plug into an existing receptacle, single outlet receptacle or an outlet receptacle from which other receptacles may be added to the down side of this receptacle.

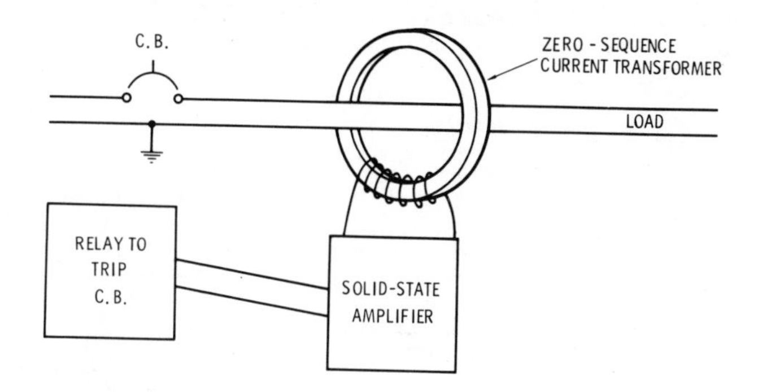

The zero-sequence current transformer covers all circuit conductors but not the equipment grounding conductors. As long as there is no leakage to ground on the load side, the current in the current carrying conductors balance out and no current flows in the secondary of the zero-sequence current transformer. If a ground-fault occurs on the load side, part of the current which would return through the neutral will bypass and return through the ground, thus an unbalanced current results in the current-carrying conductors and this induces current in the secondary of the transformer, which is amplified enough to actuate the circuit breaker trip mechanism relay, shutting off the circuit.

All GFCI's have a test button, which may be pressed occasionally and bypasses 5 milliamperes to ground, to check the operation of the GFCI. The circuit breaker mentioned above may be the circuit breaker in anyone of the types of GFCI's mentioned on the previous page.

Should trouble develop on a GFCI, do not attempt to repair it, replace it and send the defective one to the factory for repair and re-calibration.

Should a ground-fault occur on portable tools, etc., on a circuit with an equipment grounding conductor, part of the neutral current goes back over this grounding conductor, unbalancing the current through the zero-sequence transformer, actuates the relay and trips the breaker.

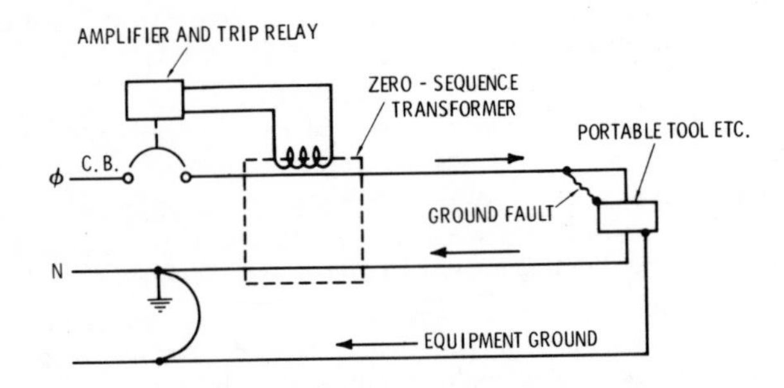

Should a ground-fault occur on portable tools, etc., on a circuit that has no equipment grounding conductor and a person touches the tool, the current passing through this person to earth and back to the service ground, will actuate the trip mechanism at 0.005 ampere, within 1/40th of a second.

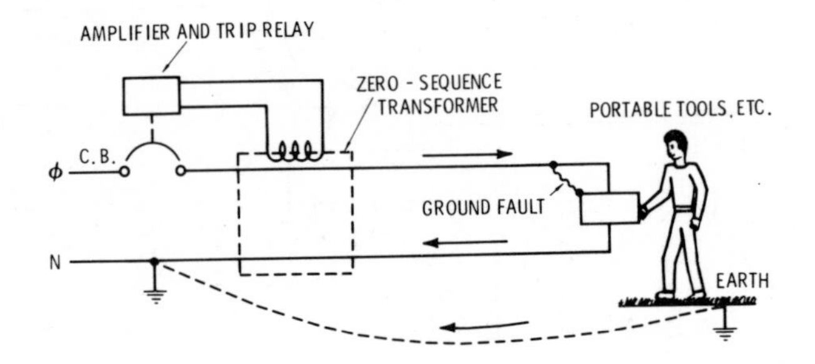

On services of 1000 amperes and larger, ground-fault interrupters are a requirement of the National Electrical Code. One item must be brought to your attention and that is, with GF protection at the service, should a GF occur anywhere on the system the entire load will be dropped. In designing systems, it is well to also put GFI's on feeder circuits down-stream, so they will act if a GF occurs on one feeder and just drop-out that feeder. The principle of operation is the same as with GFCI's, usually three-phase and neutral are involved.

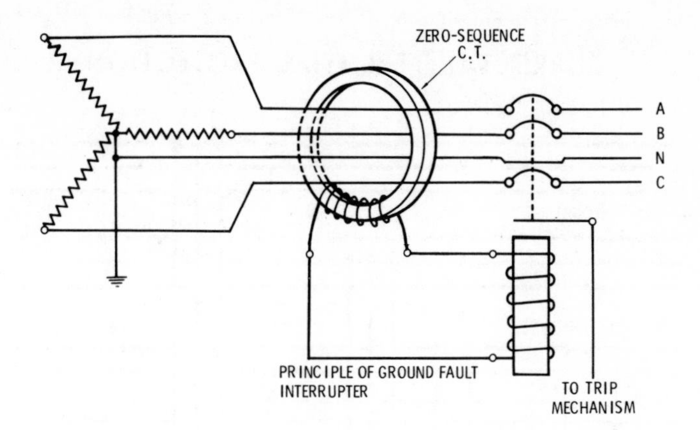

PRINCIPLE OF GROUND FAULT INTERRUPTER

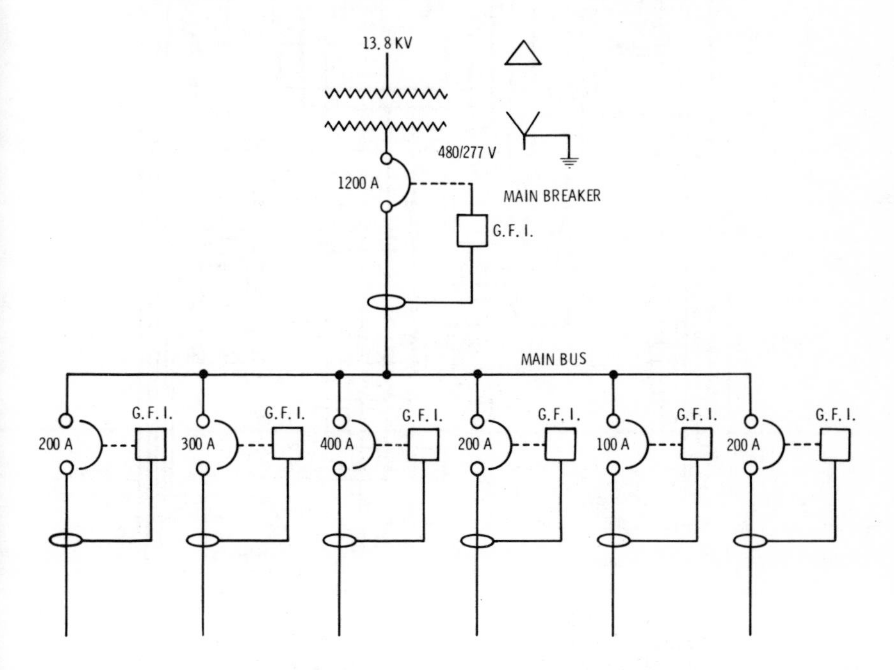

Typical one-line diagram of GFI's, main and feeder, coordinated to prevent total tripping of main breaker on a ground-fault.

CURRENT-DIFFERENTIAL PROTECTION

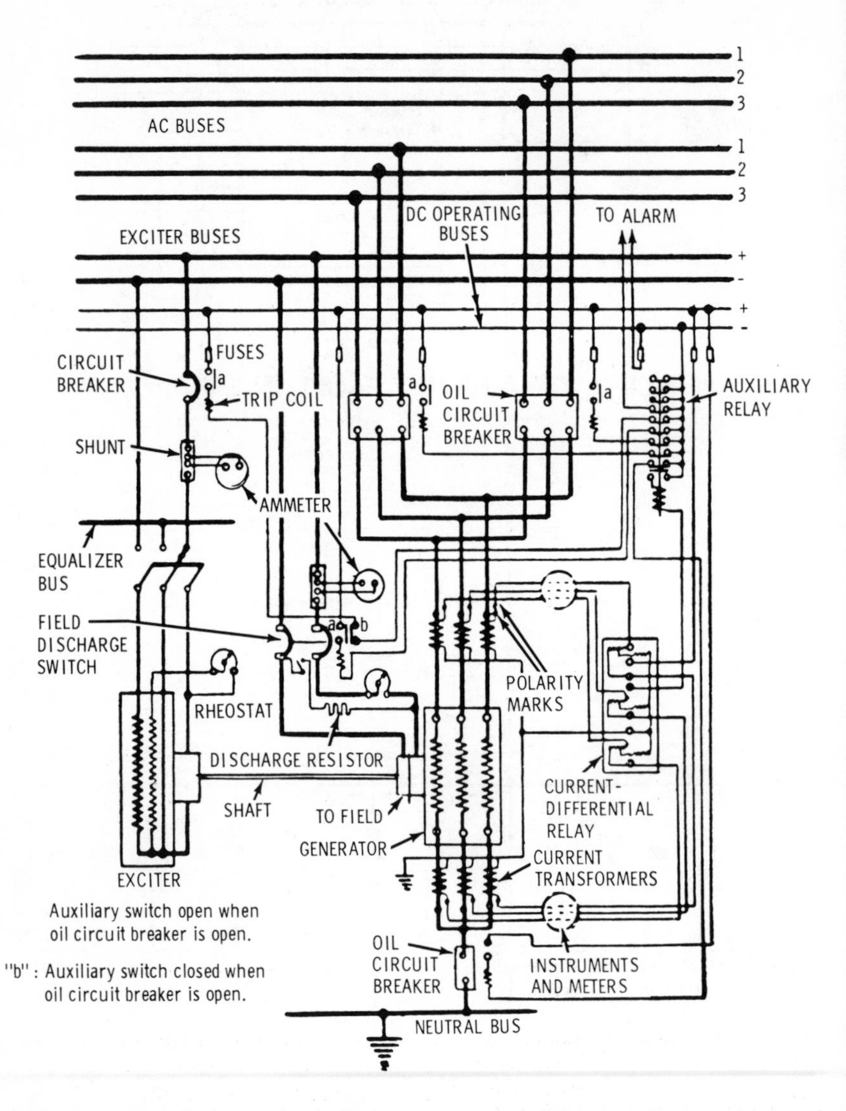

Complete Diagram of Connections for Current-Differential Protection of Three-Phase, Star-Connected, Alternating-Current Generator with Grounded Neutral and Direct-Current Exciter.

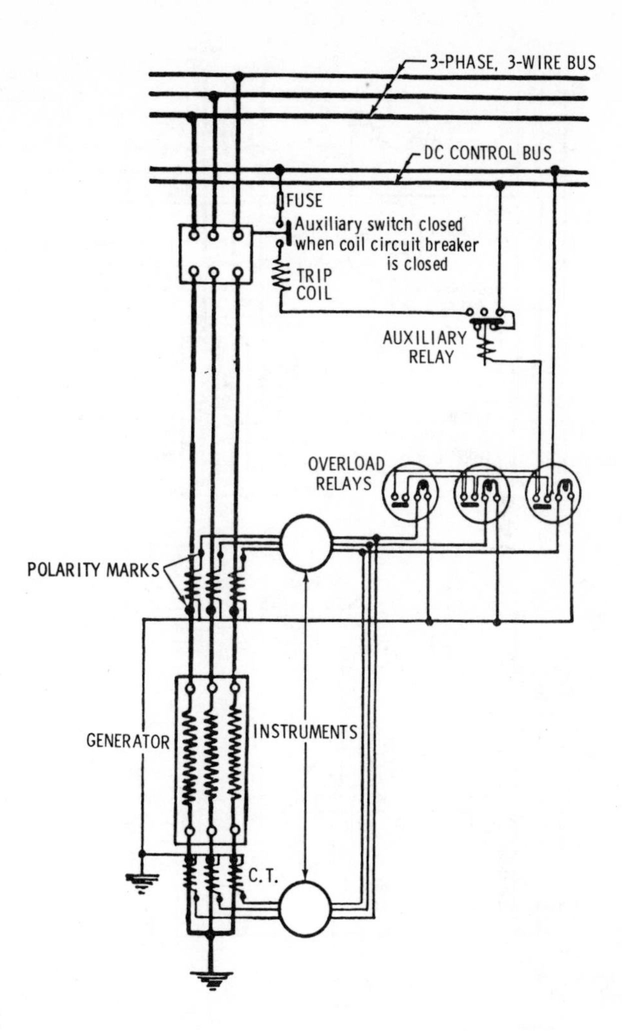

Current-Differential Protection of Station Auxiliary Feeder Using Induction-Type Overload Relays.

Current-differential protection consists essentially of current transformers installed at each end of the generator windings with their secondaries connected in series and relays connected differentially so that their functioning depends upon a difference of current flowing through the two sets of current transformers.

It is obvious that the differential protection disconnects the machine only in the case of electrical failure in the machine or its connecting leads.

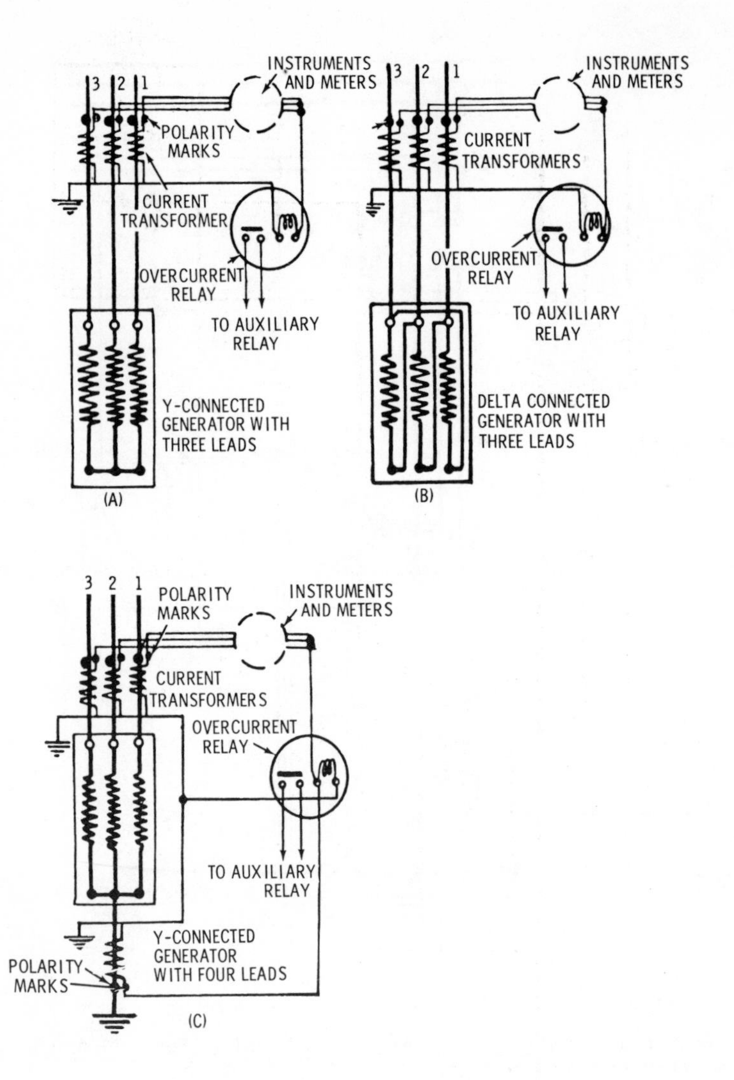

Method of obtaining differential protection when such provision was not made in the manufacture of the machine (where the phase leads were not brought out). Figs. A, B, and C show the connection used as expedients in such cases. Fig. A shows current differential protection of a three-phase, Y-connection, AC generator, neutral not brought out. Fig. B Current-differential protection of a three-phase, delta-connected, AC generator, phase leads not brought out. Fig. C Current-differential protection of a three-phase, Y-connected, AC generator, neutral only brought out.

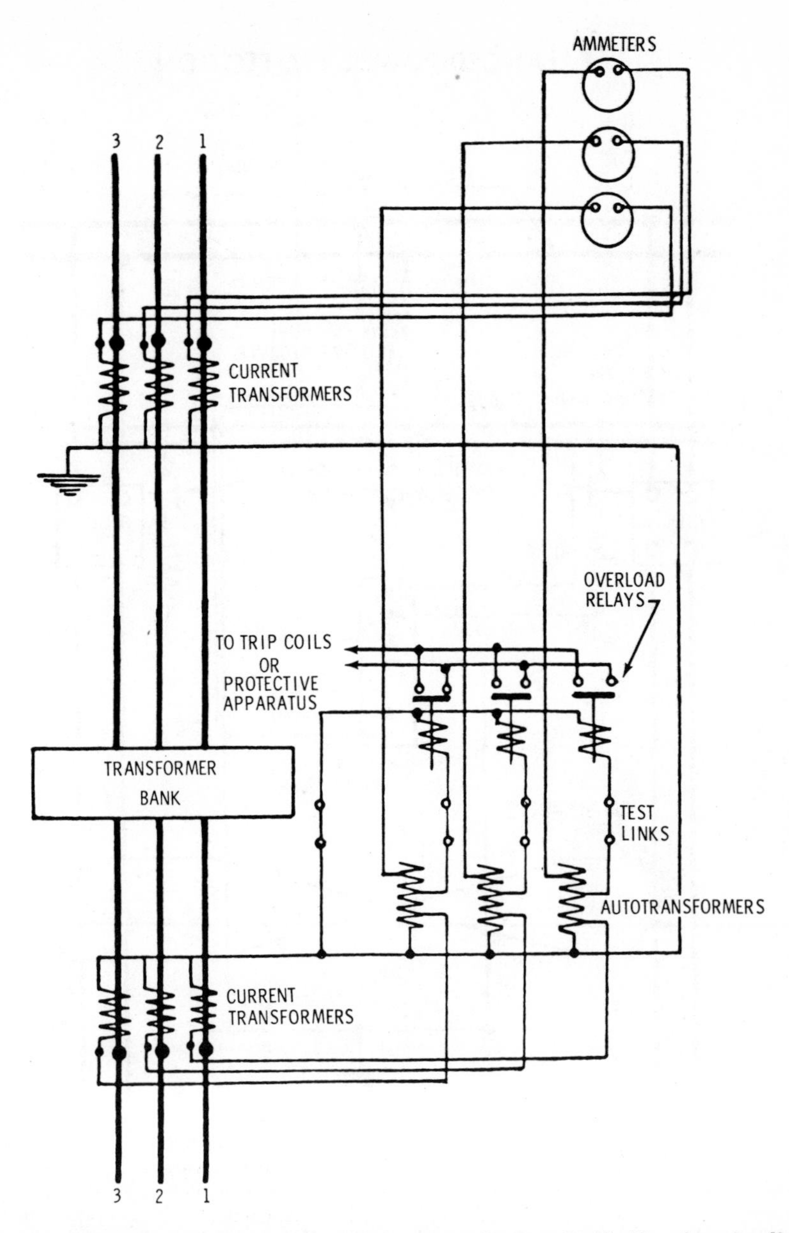

Current-Differential Protection Over a Transformer Bank Showing Application of Test-Links Facilitating Testing of Relay Coils.

BALANCED-POWER PROTECTION

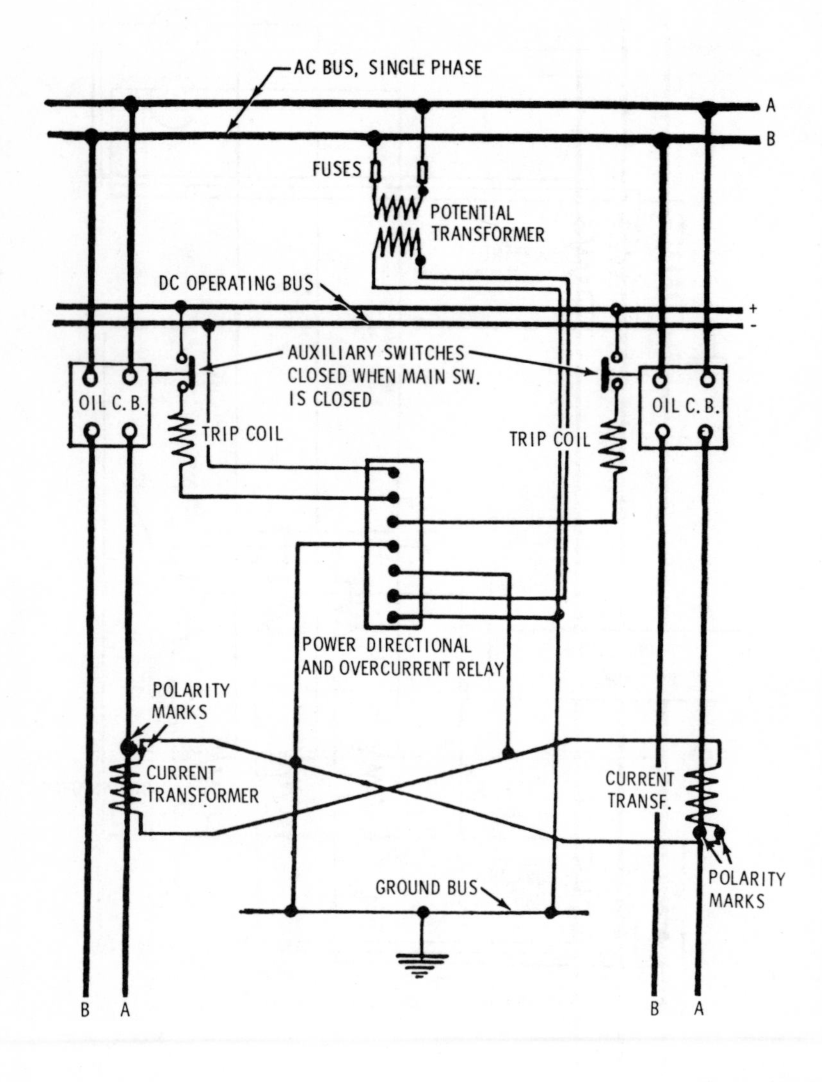

Balanced Power Connection for Two Parallel Incoming, Outgoing or Tie Lines, Single-Phase DC Current Used for Tripping Oil Circuit Breakers.

BALANCED AND OVERCURRENT PROTECTION

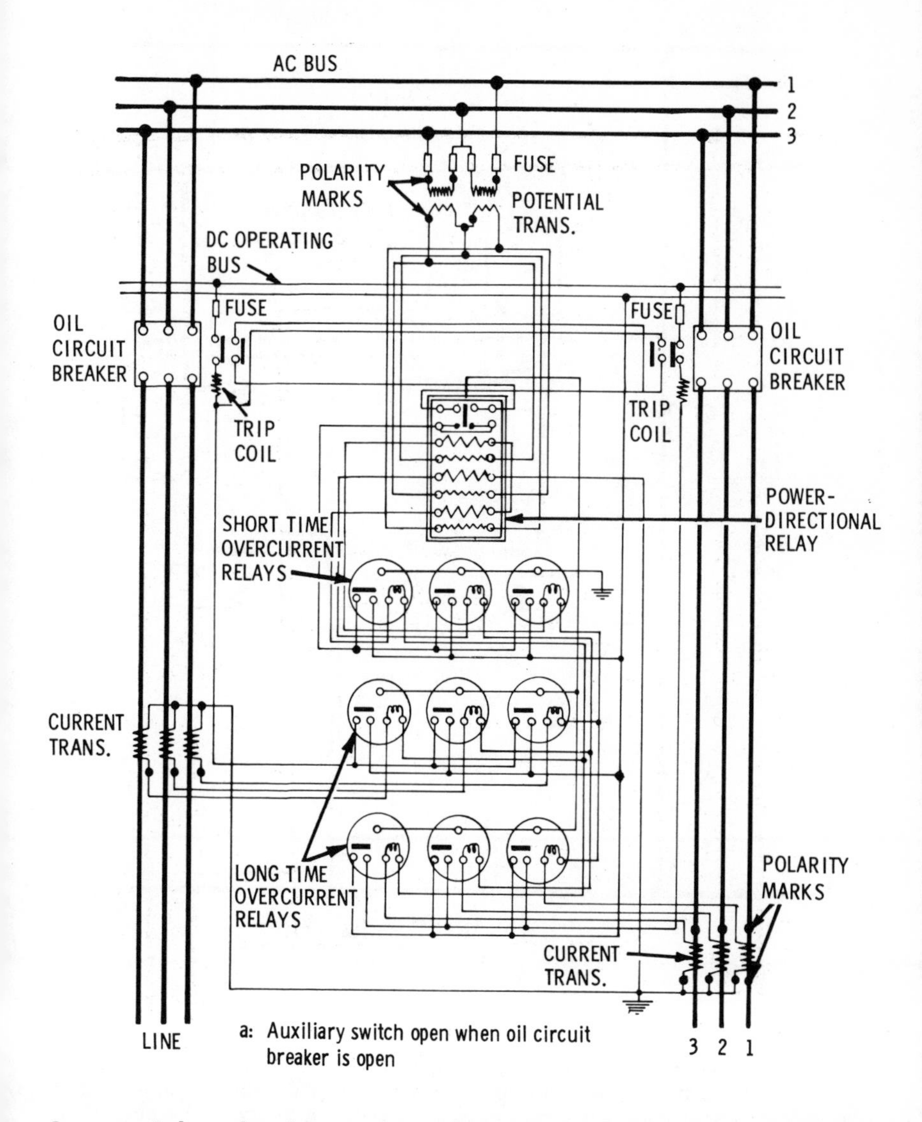

Separate Balanced and Overcurrent Protection for Two Parallel Three-Phase Incoming, Outgoing or Tie Lines with Grounded Neutral.

BELL-ALARM CONNECTIONS

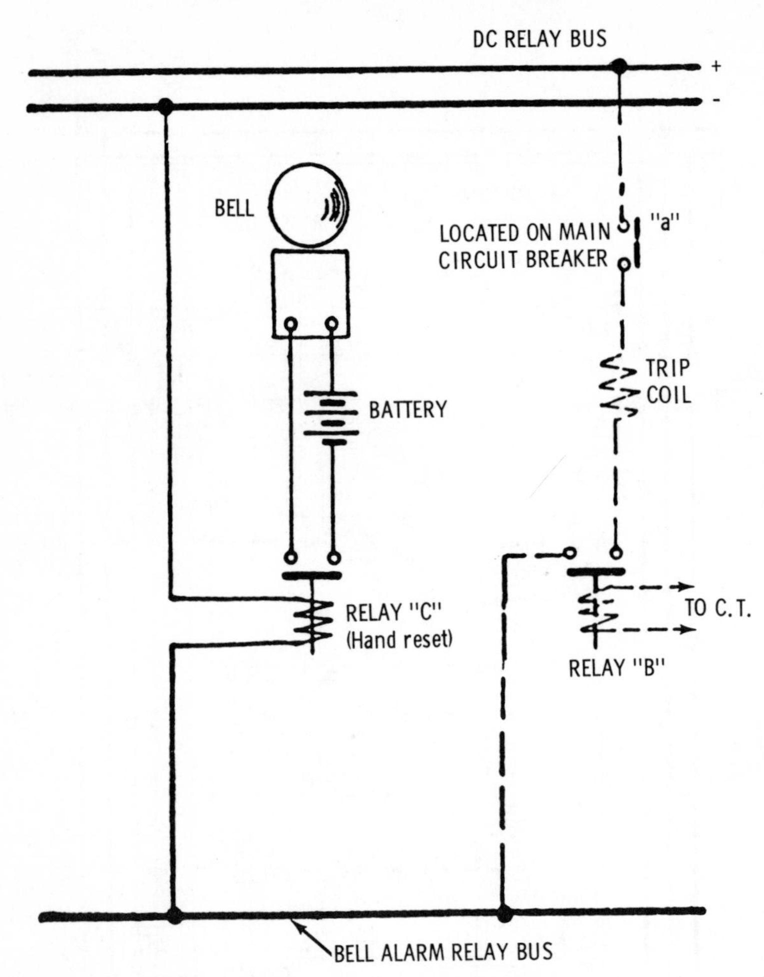

Operation—"A" is an auxiliary switch closed when main breaker is closed.

When current from transformer exceeds current setting of relay "B," the relay closes its contacts, energizing the trip coil and relay "C" which in turn closes the battery circuit, sending a current through the bell, and notifying the operator that the oil circuit breaker is out of service.

Note, that relay "C" is usually hand reset, and is reset for service by the attendant.

EMERGENCY SWITCHING METHODS

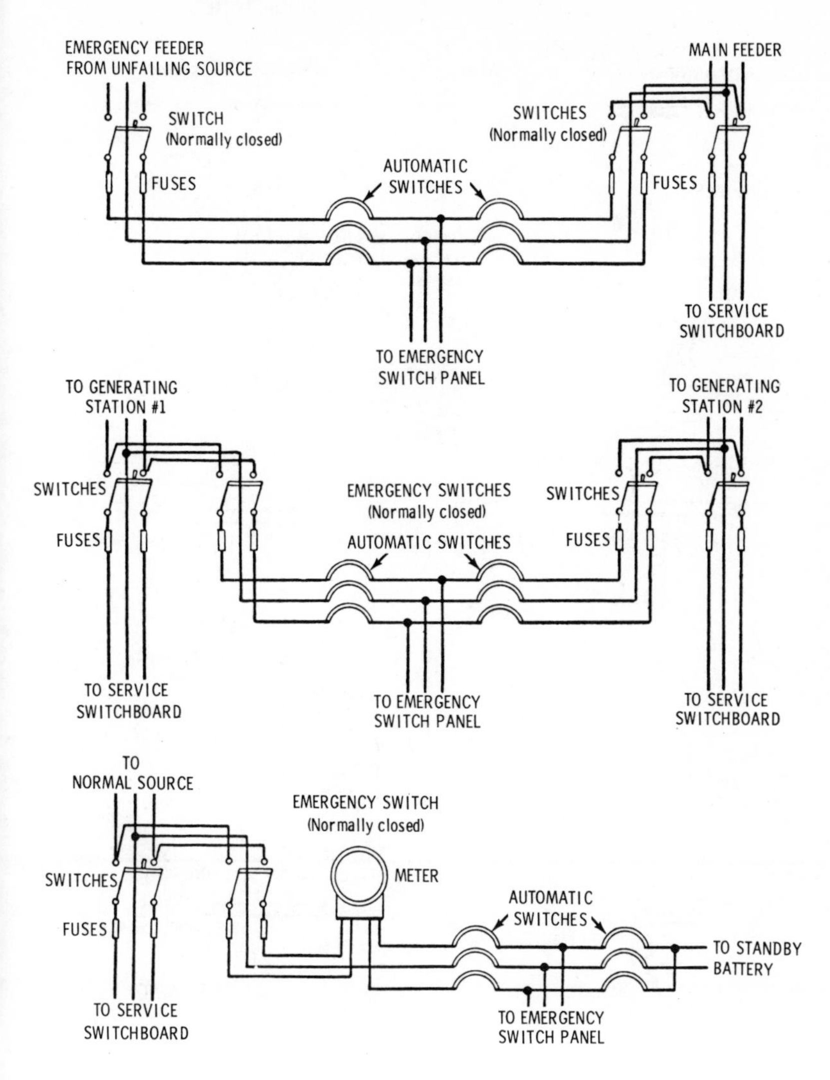

Operation—The above diagrams represent emergency switching methods generally utilized in public institutions such as hospitals, schools, etc. where it is imperative that an unfailing lighting source be available. The automatic switches shown will automatically close themselves when the normal supply fails, causing the service to remain uninterrupted.

MOTOR DISCONNECTING AND CONTROL METHODS

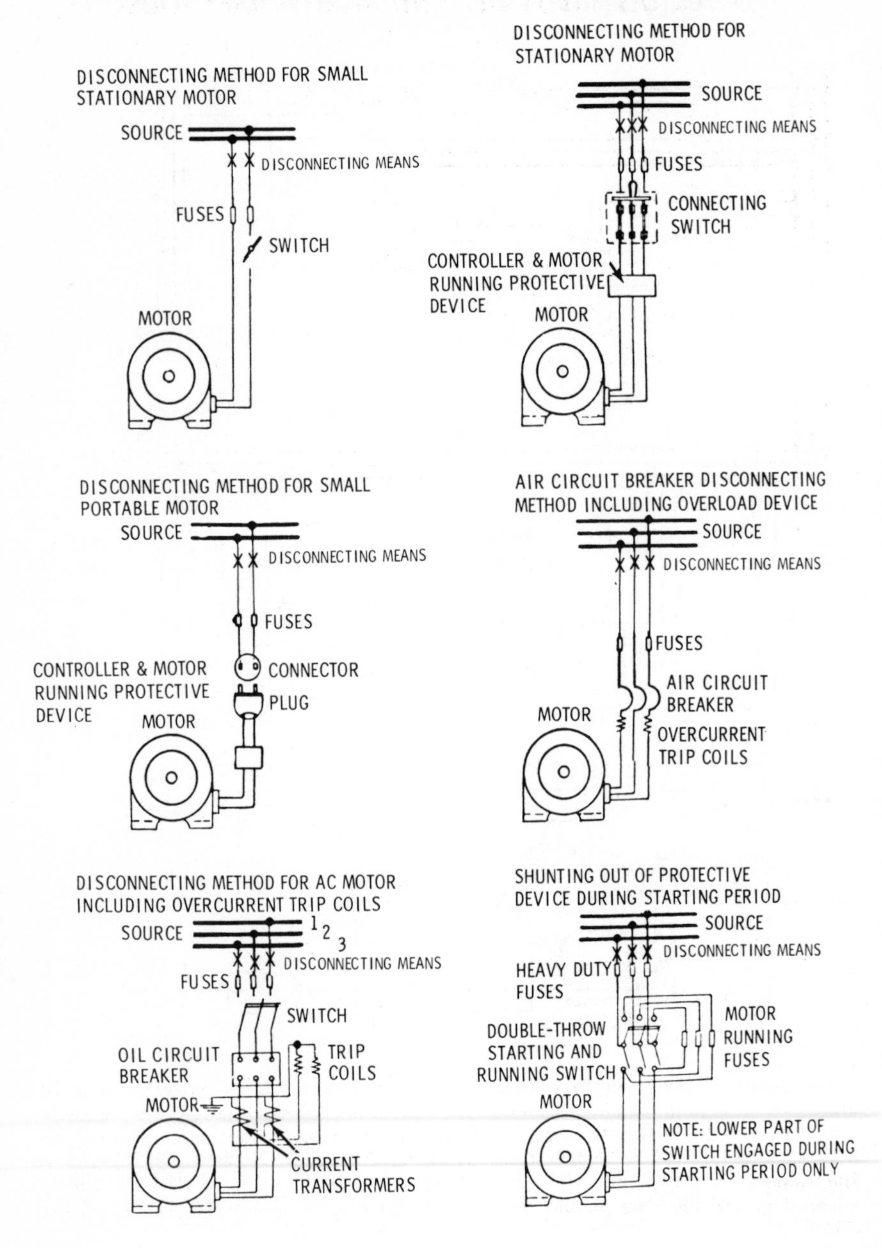

POWER DISTRIBUTION AND WIRING METHODS

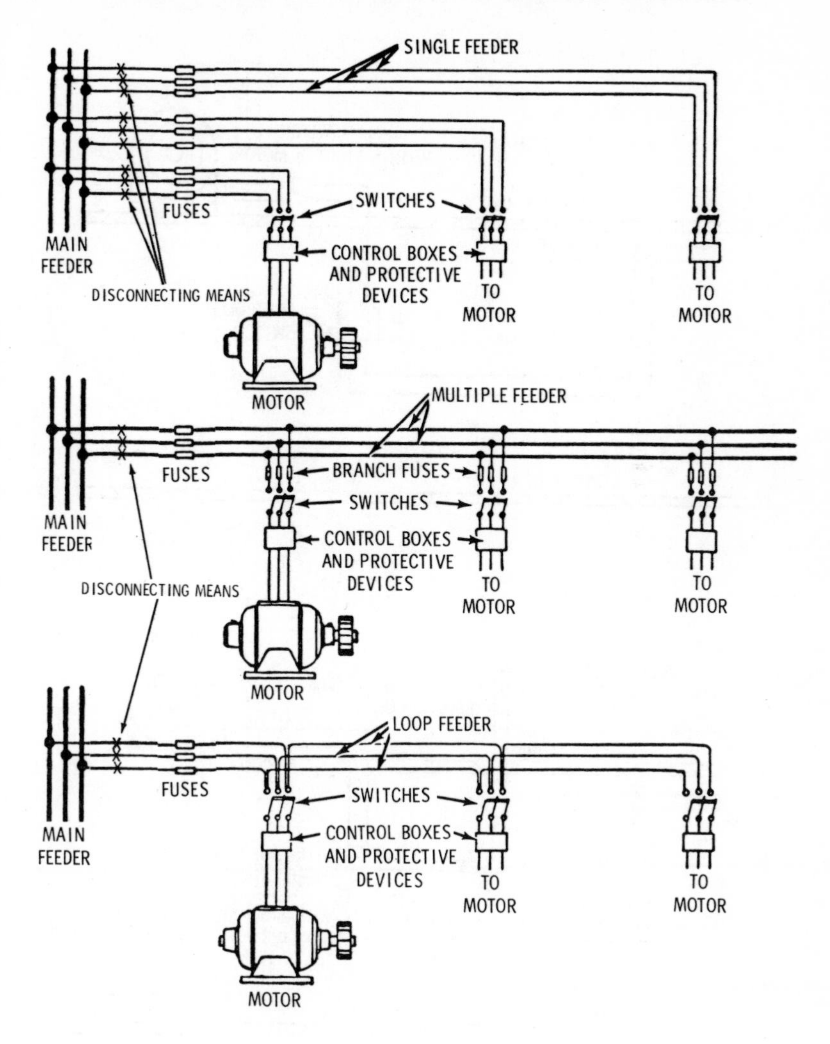

The above diagrams suggest three methods of motor wiring. However, due to individual operating conditions, the type of wiring adopted should be carefully analyzed for the particular case involved. Certain applications, for example, by their very nature will prove themselves better suited for one scheme than the other. The National Electric Code as well as any local requirements should be strictly adhered to and the wire should be of ample capacity to prevent excessive voltage drops.

REMOTE-CONTROL WIRING FOR SMALL MOTORS

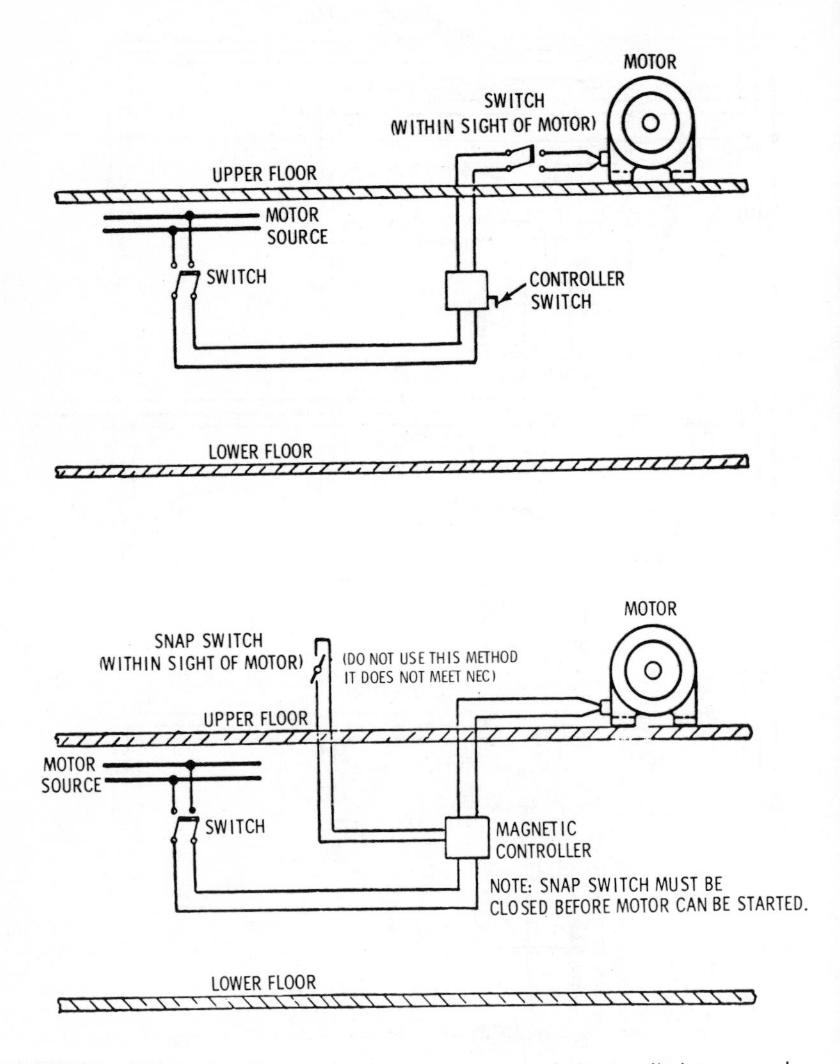

Note—All motors and control wiring should be carefully installed in accordance with the National Electric Code and any local requirement. The wire should be of ample capacity based on a maximum drop of 2 per cent of line voltage at full load current.

SWITCHING PANELS AND FEEDER DISTRIBUTION METHODS

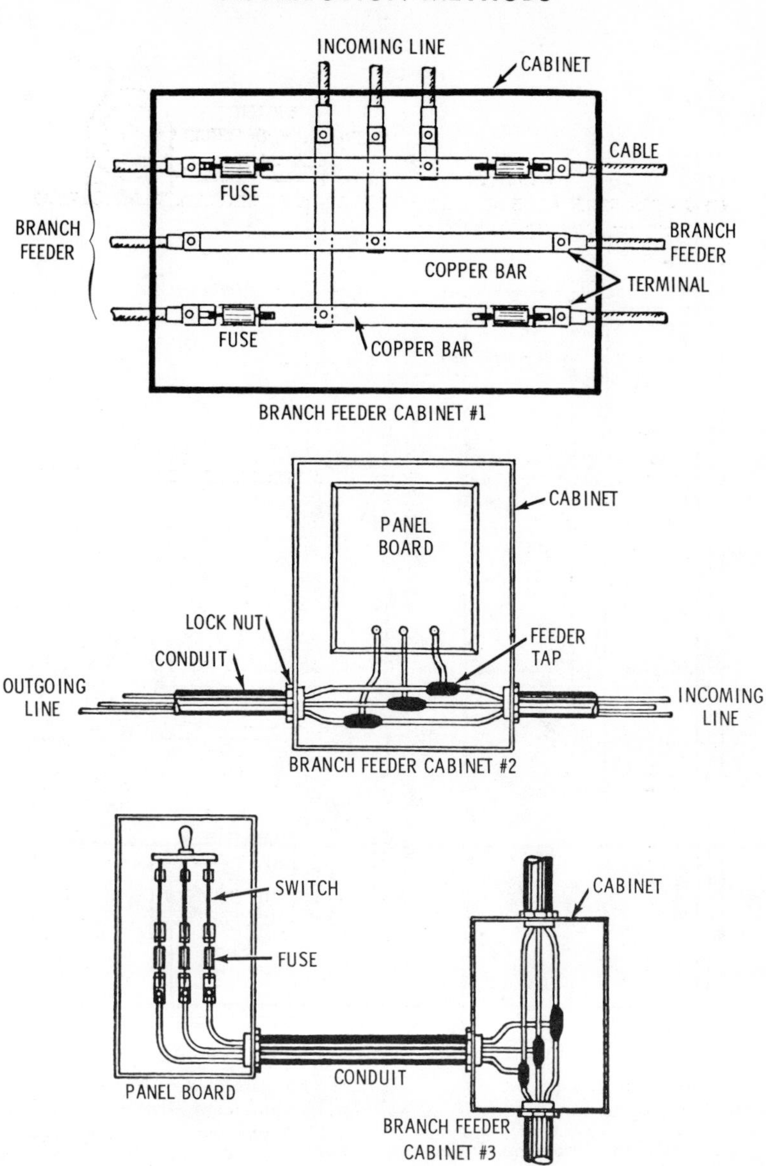

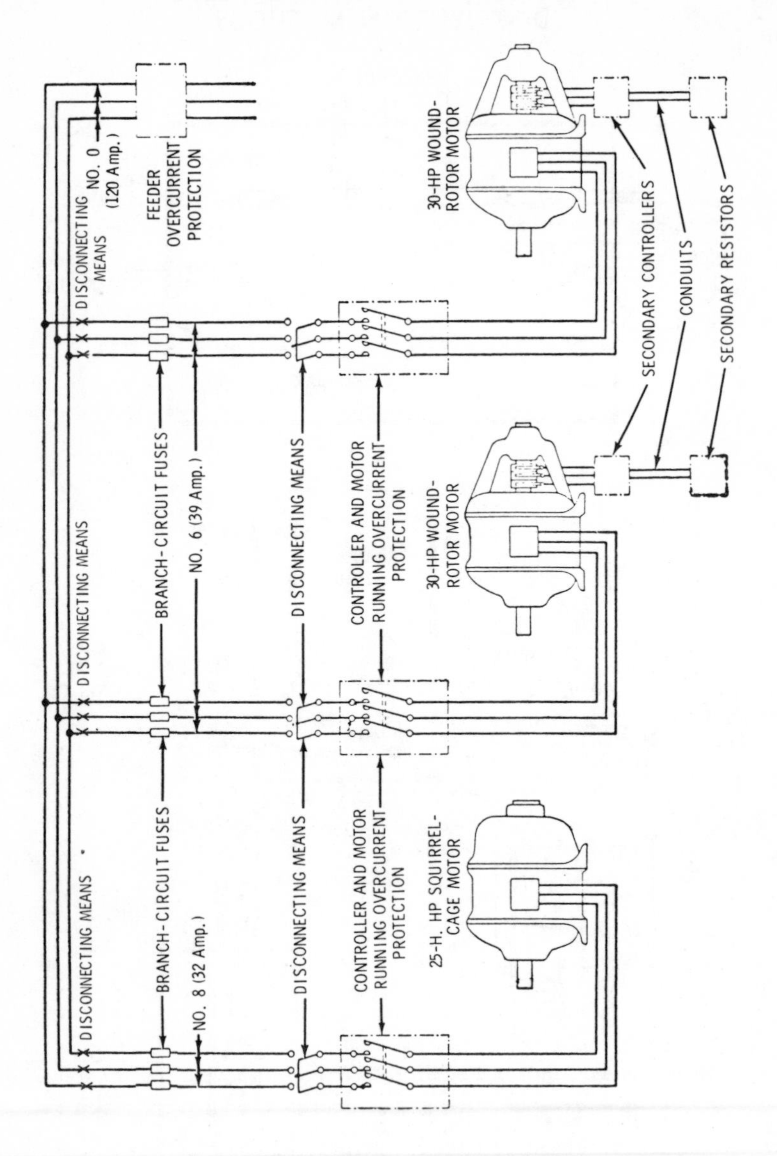

Wiring Diagram Showing Individual Branch Circuits, Feeder Circuit and Protective Devices in a Typical Motor Layout.

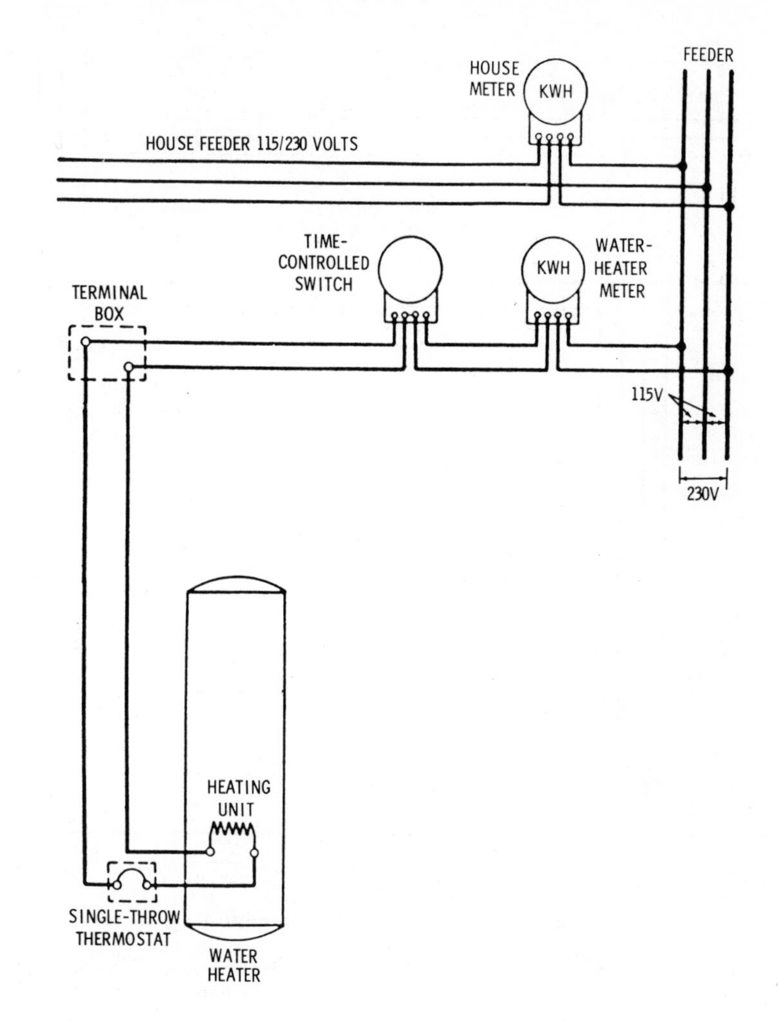

Wiring Diagram for Limited-Demand Service on a Single-Unit Water Heater —In a circuit of this type, the single-throw, single-pole thermostat functions to close and open the circuit to the heating unit at specified temperatures according to its setting. The time-controlled switch will determine the hours of the day when the circuit will be opened or closed, thus preventing unlimited use of hot water during the hours of greatest power load.

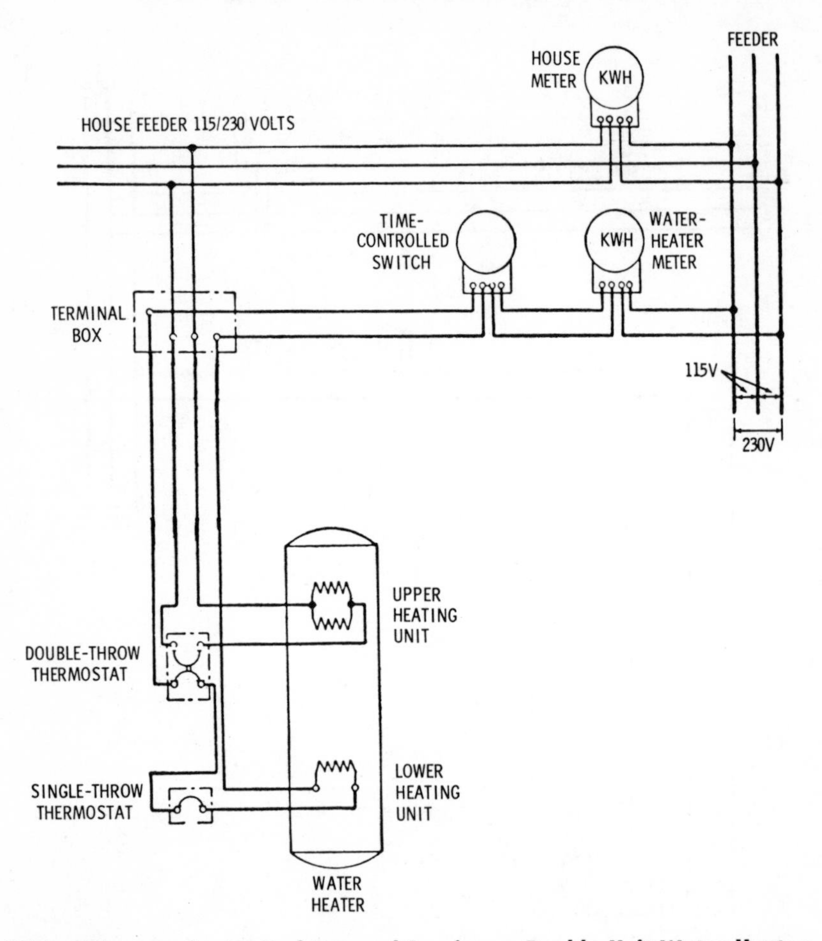

Wiring Diagram for Limited-Demand Service on Double-Unit Water Heater—In a wiring method of this type, the upper heating unit is controlled by a double-throw, single-pole, thermostatic switch. This switch has two sets of contacts, one of which controls the flow of current to the upper heating unit and the other controls the flow of current to the lower thermostat. The lower heating unit is controlled by a single-throw, single-pole, thermostatic switch. This switch has only one set of contacts and opens and closes in response to the temperature of the water in the lower tank. The function of the time switch is to prevent unlimited use of hot water during a predetermined time (or times), usually during that period of the day when the general demand for power is the greatest.

INTERNAL AND EXTERNAL CONNECTIONS FOR DUAL-VOLTAGE THREE-PHASE SQUIRREL CAGE INDUCTION MOTORS

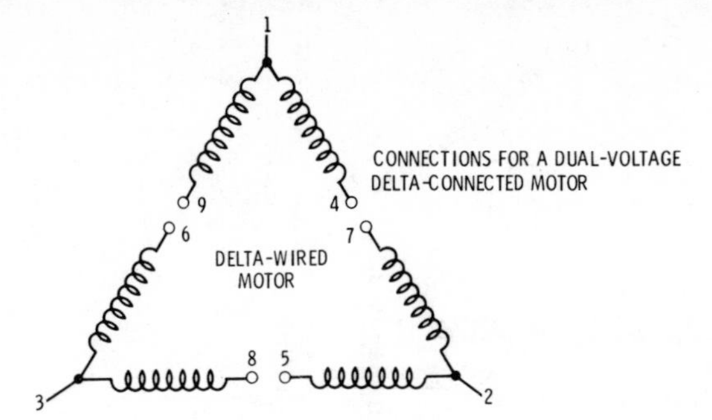

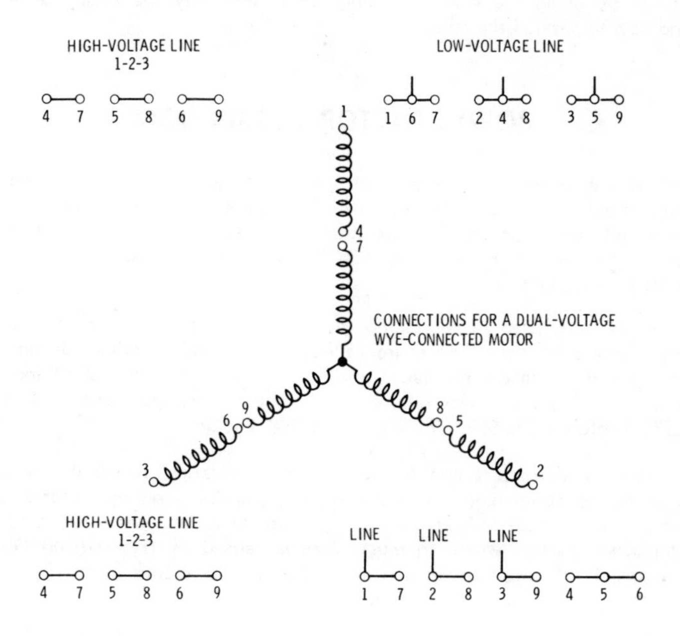

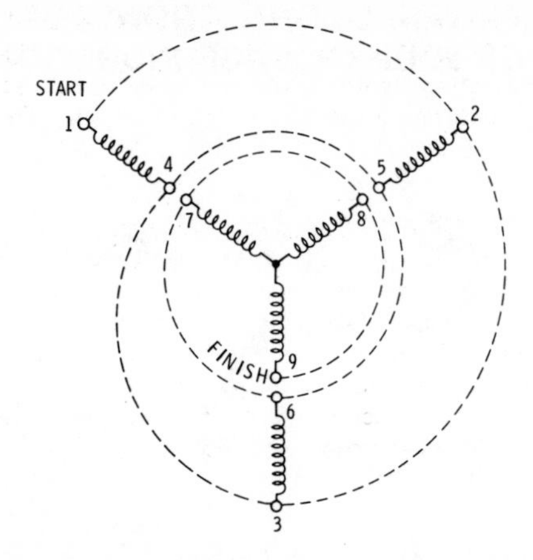

An easy method of remembering how to connect a wye-wound motor for high and low voltage. By using the ever-expanding circle, one may see at a glance how to series and how to parallel the coils.

POWER FACTOR CORRECTION

On AC where inductive loads such as motors are used, there is a lagging power factor developed from the inductive load. This sometimes has to be corrected, as low power factors in many cases cause rate increases. Correctition is done by capacitors. As will be seen later, they may be installed on individual motors, or at the load center for the plant.

The term Kvar is used in expressing the values of capacitative reactance involved in power factor correction. These are called kilovars and reactive kilovolt amperes. Most of the reactive loads that cause power factor problems consist of inductive reactance, causing lagging power failure. Thus, when Kvars are mentioned, it may be assumed that reference is being made to capacitive reactance.

You will find Varmeters and Kilovar meters (indicating types) on switchboards. Zero is in the top center and one side is marked plus (+) and the other side marked minus (−). The plus side indicates a lagging power factor and the minus side indicates a leading power factor, which in most cases is caused by transmission line capacitance, especially if the transmission lines are a grid system and of considerable length, with a comparative light loading.

To read power factor, one must either use a special power factor meter, or you may use wattmeters, voltmeters, and ammeters. The volts times the amperes will give you apparent power and the wattmeter the true power. The answer will be the cosine of the angle of lag or lead and is obtained by the formula; PF = Watts × 100/Volts × Amperes. The 100 is used so that instead of getting a decimal answer you get a % answer.

If Varmeters, or kilovarmeters or mega varmeters are used and an indicating wattmeter is present, its reading may be used and from these figures the power factor may be obtained. Be certain that you use the same quantities in vars and watts, that is, if Kilo is used in one figure use Kilo in the other, etc. The formula for power factor is:

$$\text{Power Factor} = \frac{\text{Watts}}{\sqrt{\text{Watts}^2 + \text{Vars}^2}}$$

Tables may be used to arrive at the power factor correction, that is to obtain the necessary Kvars to bring the power factor up to what ever percent you wish it to be; In the left-hand column find the present power factor, say 75%, if you wish to correct to 95% PF, lay a straight-line from 75 in the left-hand column and under 95 in the top column you find the factor 0.553. If the Kilowatt reading was 38.795 KW, multiply the 38,795KW by the constant 0.553 and you will arrive at 21.45 kvars needed to correct to 95% PF.

Table 1. Capacitor Kvar Table For Improving Power Factor

Desired Power Factor in Percent

Original Power Factor in Percent	80%	81	82	83	84	85	86	87	88	89	90	91	92	93	94	95
50%	.982	1.008	1.034	1.060	1.086	1.112	1.139	1.165	1.192	1.220	1.248	1.276	1.303	1.337	1.369	1.402
51	.936	.962	.988	1.014	1.040	1.066	1.093	1.119	1.146	1.174	1.202	1.230	1.257	1.291	1.320	1.357
52	.894	.920	.946	.972	.998	1.024	1.051	1.077	1.104	1.132	1.160	1.188	1.215	1.249	1.281	1.135
53	.850	.876	.902	.928	.954	.980	1.007	1.033	1.060	1.088	1.116	1.144	1.171	1.205	1.237	1.271
54	.809	.835	.861	.887	.913	.939	.966	.992	1.019	1.047	1.075	1.103	1.130	1.164	1.196	1.230
55	.769	.795	.821	.847	.873	.899	.926	.952	.979	1.007	1.035	1.063	1.090	1.124	1.156	1.190
56	.730	.756	.782	.808	.834	.860	.887	.913	.940	.968	.996	1.024	1.051	1.085	1.117	1.151
57	.692	.718	.744	.770	.796	.822	.849	.875	.902	.930	.958	.986	1.013	1.047	1.079	1.113
58	.655	.681	.707	.733	.759	.785	.812	.838	.865	.893	.921	.949	.976	1.010	1.042	1.076
59	.618	.644	.670	.696	.722	.748	.775	.801	.828	.856	.884	.912	.939	.973	1.005	1.039
60	.584	.610	.636	.662	.688	.714	.741	.767	.794	.822	.849	.878	.905	.939	.971	1.005
61	.549	.575	.601	.627	.653	.679	.706	.732	.759	.787	.815	.843	.870	.904	.936	.970
62	.515	.541	.567	.593	.619	.645	.672	.698	.725	.753	.781	.809	.836	.870	.902	.936
63	.483	.509	.535	.561	.587	.613	.640	.666	.693	.721	.749	.777	.804	.838	.870	.904
64	.450	.476	.502	.528	.554	.580	.607	.633	.660	.688	.716	.744	.771	.805	.837	.871
65	.419	.445	.471	.497	.523	.549	.576	.602	.629	.657	.685	.713	.740	.774	.806	.840
66	.388	.414	.440	.466	.492	.518	.545	.571	.598	.626	.654	.682	.709	.743	.775	.809
67	.358	.384	.410	.436	.462	.488	.515	.541	.568	.596	.624	.652	.679	.713	.745	.779
68	.329	.355	.381	.407	.433	.459	.486	.512	.539	.567	.595	.623	.650	.684	.716	.750
69	.299	.325	.351	.377	.403	.429	.456	.482	.509	.537	.565	.593	.620	.654	.686	.720
70	.270	.296	.322	.348	.374	.400	.427	.453	.480	.508	.536	.564	.591	.625	.657	.691
71	.242	.268	.294	.320	.346	.372	.399	.425	.452	.480	.508	.536	.563	.597	.629	.663
72	.213	.239	.265	.291	.317	.343	.370	.396	.423	.451	.479	.507	.534	.568	.600	.634
73	.186	.212	.238	.264	.290	.316	.343	.369	.396	.424	.452	.480	.507	.541	.573	.607
74	.159	.185	.211	.237	.263	.289	.316	.342	.369	.397	.425	.453	.480	.514	.546	.580
75	.132	.158	.184	.210	.236	.262	.289	.315	.342	.370	.398	.426	.453	.487	.519	.553

A Monograph and ruler may also be used. The dotted line illustrates the ruler on the chart, for the problem just worked, the middle column gives the percent reactive

Kva, change it to a decimal and use as you did the correction factor in the previous chart.

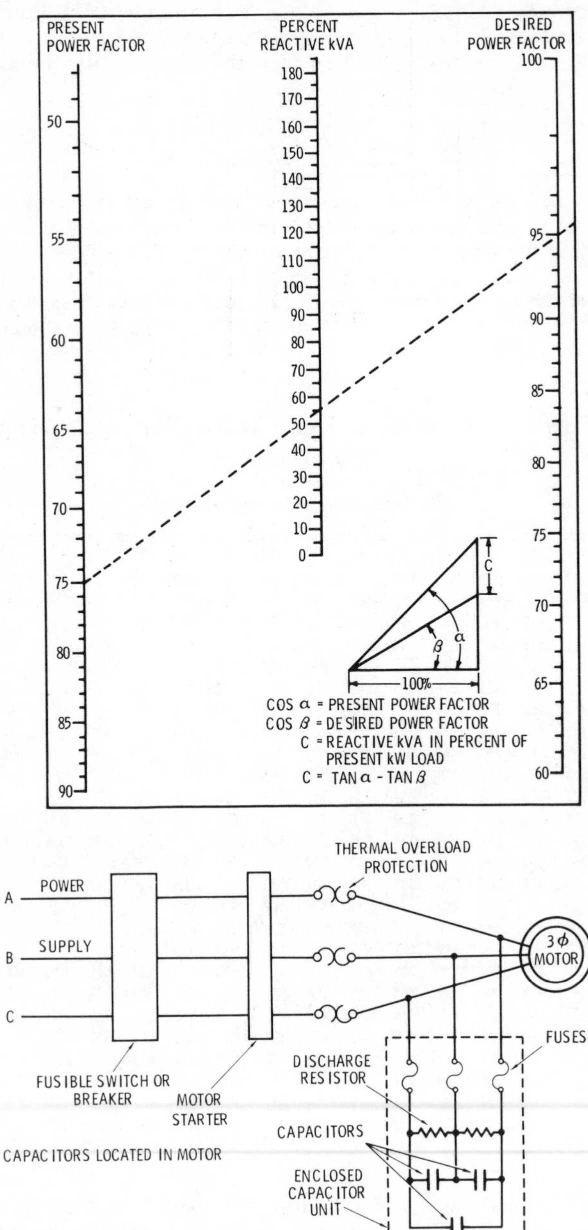

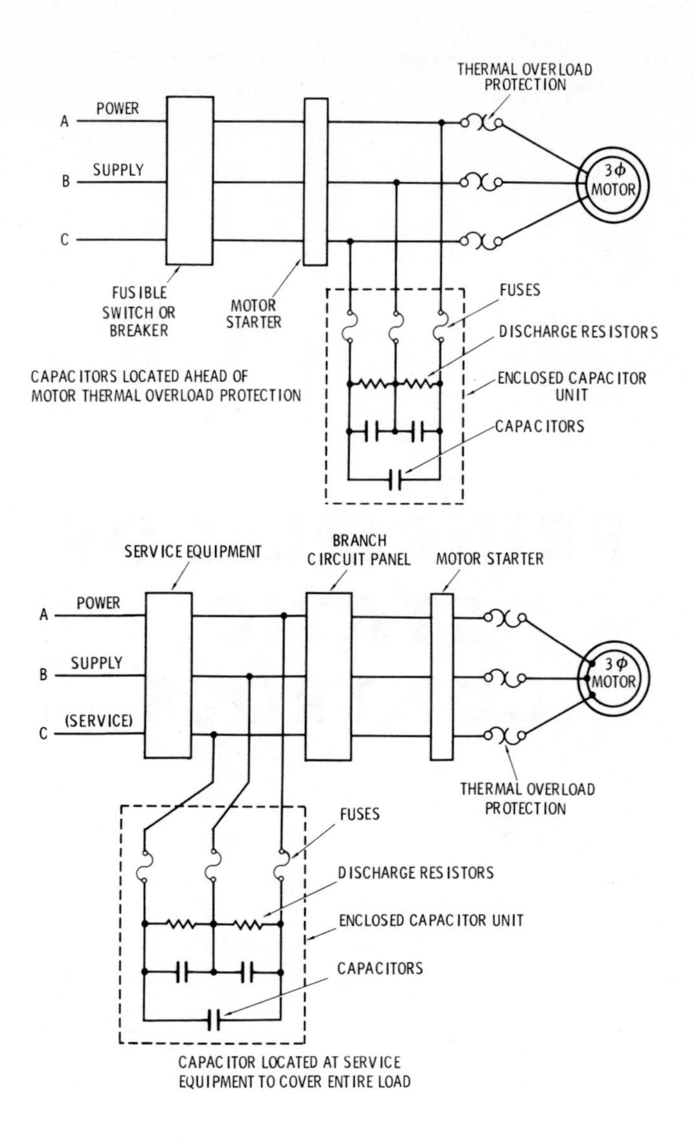
THERMAL OVERLOAD PROTECTION
POWER
SUPPLY
A
B
C
3φ MOTOR
FUSIBLE SWITCH OR BREAKER
MOTOR STARTER
FUSES
DISCHARGE RESISTORS
CAPACITORS LOCATED AHEAD OF MOTOR THERMAL OVERLOAD PROTECTION
ENCLOSED CAPACITOR UNIT
CAPACITORS
SERVICE EQUIPMENT
BRANCH CIRCUIT PANEL
MOTOR STARTER
POWER
SUPPLY
(SERVICE)
A
B
C
3φ MOTOR
THERMAL OVERLOAD PROTECTION
FUSES
DISCHARGE RESISTORS
ENCLOSED CAPACITOR UNIT
CAPACITORS
CAPACITOR LOCATED AT SERVICE EQUIPMENT TO COVER ENTIRE LOAD

PRINCIPLES OF STATIC CONTROLS

Static control is a system of machine or process control in which there are no moving parts or electric contacts. By using solid-state devices instead of relays results in a material extension of the life of control systems.

Logic function terms are used in the language of static control and its circuit requirements. There are four logic elements and a time control, used to fullfill these. The logic elements are AND, NOT, OR, and MEMORY.

If the output from an AND element are required, it is necessary to have A and B and C and D all present. With OR elements, A or B or C or any combination of inputs will produce an output. With a NOT element an A input will cause an output not to occur, and when the A input is not present an output will occur.

The MEMORY element provides a maintained output with a momentary A input. The output will continue after the A input is removed. When a memory B input is received, the output is turned off, thus, it is necessary that another A input is received to turn the element on again. If a power failure occurs followed by a return of power, the Memory element will recall its last output condition (on or off) and résume operation.

In the time delay element, input A will result in an output after a preset period of time. The output will continue until input A is removed.

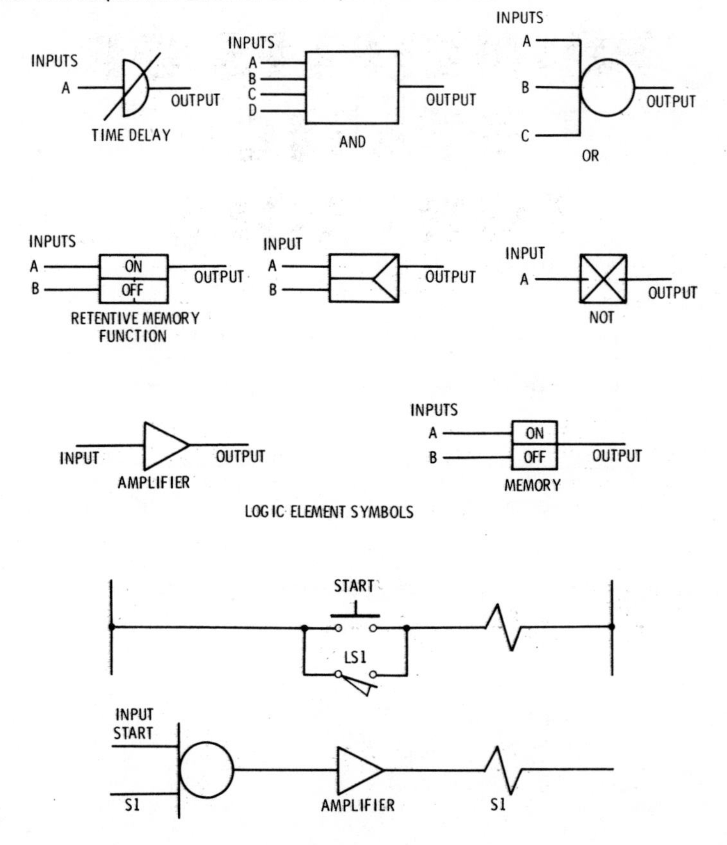

LOGIC ELEMENT SYMBOLS

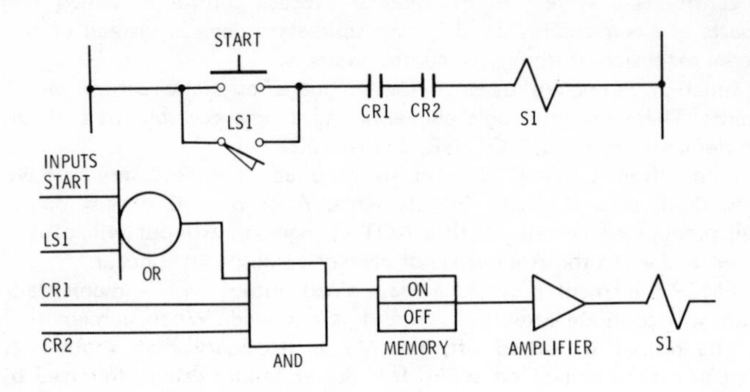
START
LS1
CR1
CR2
S1
INPUTS
START
LS1
OR
CR1
CR2
AND
ON
OFF
MEMORY
AMPLIFIER
S1

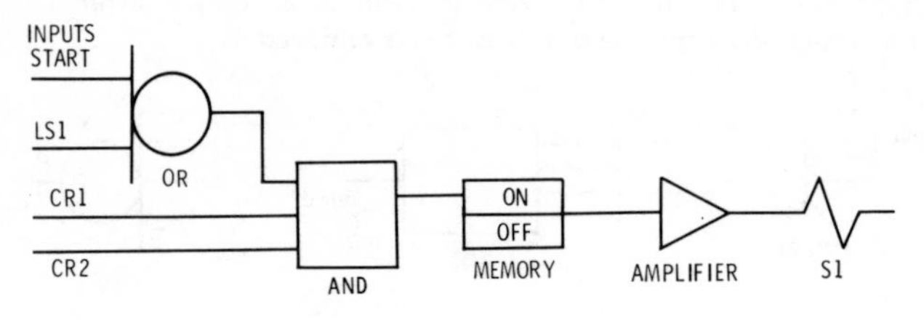
INPUTS
START
LS1
OR
CR1
CR2
AND
ON
OFF
MEMORY
AMPLIFIER
S1

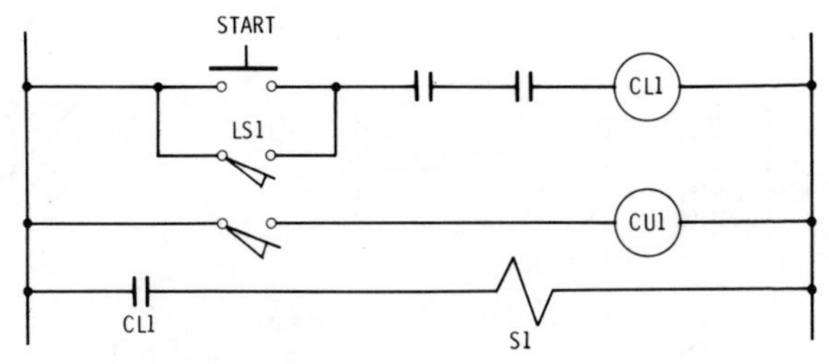
START
LS1
CL1
CU1
CL1
S1

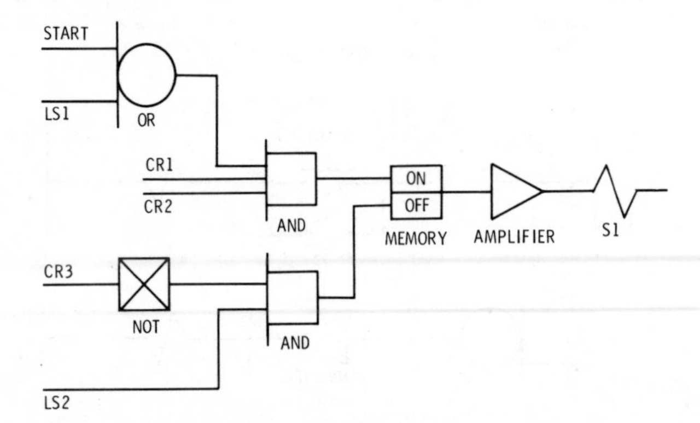
START
LS1
OR
CR1
CR2
AND
ON
OFF
MEMORY
AMPLIFIER
S1
CR3
NOT
AND
LS2

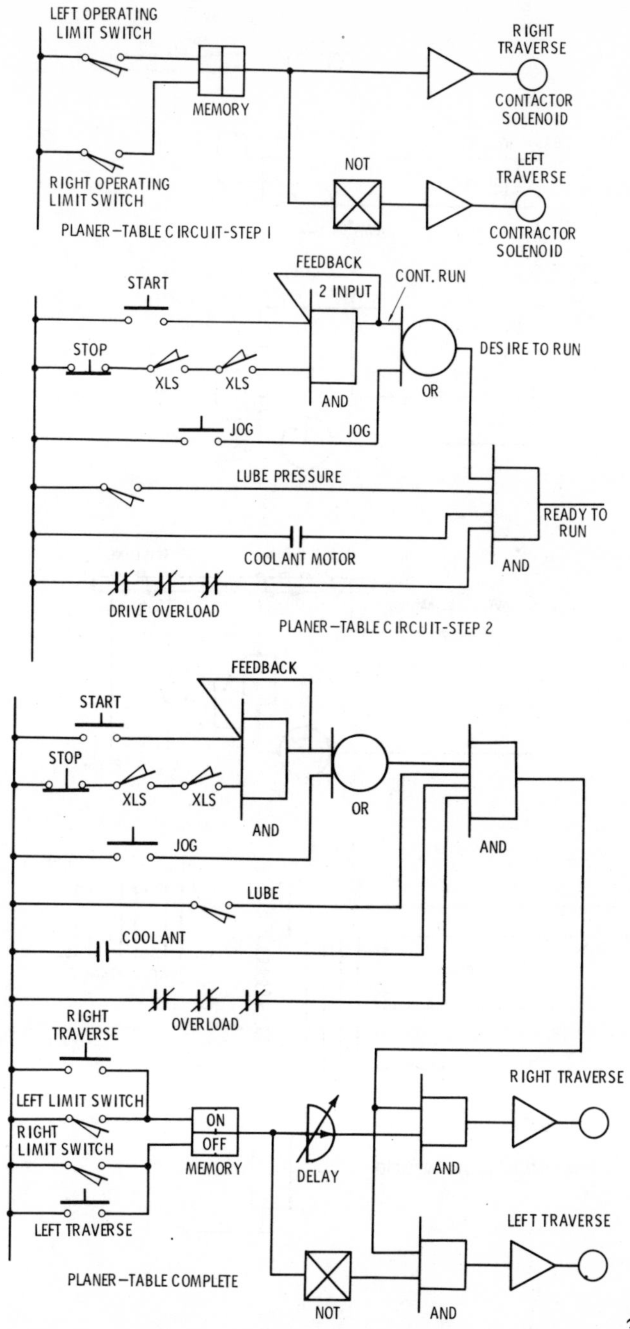
LEFT OPERATING
LIMIT SWITCH
MEMORY
RIGHT
TRAVERSE
CONTACTOR
SOLENOID
NOT
LEFT
TRAVERSE
RIGHT OPERATING
LIMIT SWITCH
CONTRACTOR
SOLENOID
PLANER—TABLE CIRCUIT-STEP I
FEEDBACK
START
2 INPUT
CONT. RUN
STOP
XLS
XLS
DESIRE TO RUN
AND
OR
JOG
JOG
LUBE PRESSURE
READY TO
RUN
COOLANT MOTOR
AND
DRIVE OVERLOAD
PLANER—TABLE CIRCUIT-STEP 2
FEEDBACK
START
STOP
XLS
XLS
OR
AND
AND
JOG
LUBE
COOLANT
OVERLOAD
RIGHT
TRAVERSE
RIGHT TRAVERSE
LEFT LIMIT SWITCH
ON
OFF
RIGHT
LIMIT SWITCH
MEMORY
DELAY
AND
LEFT TRAVERSE
LEFT TRAVERSE
PLANER—TABLE COMPLETE
NOT
AND

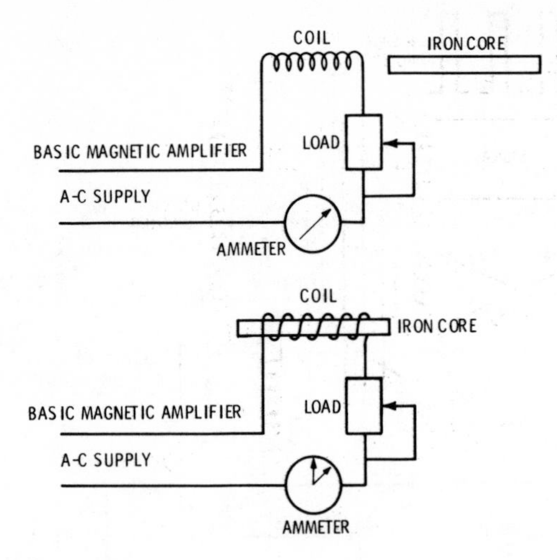
COIL
IRON CORE
BASIC MAGNETIC AMPLIFIER
LOAD
A-C SUPPLY
AMMETER
COIL
IRON CORE
BASIC MAGNETIC AMPLIFIER
LOAD
A-C SUPPLY
AMMETER

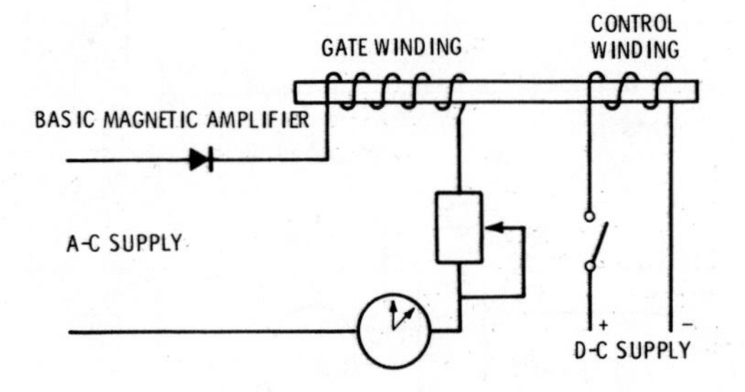
GATE WINDING
CONTROL WINDING
BASIC MAGNETIC AMPLIFIER
A-C SUPPLY
+
–
D-C SUPPLY

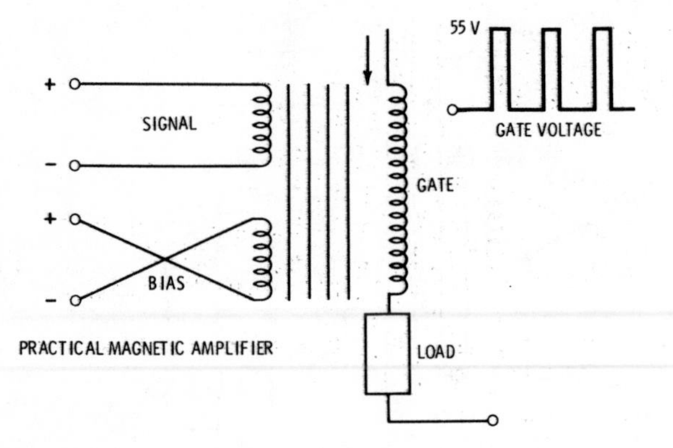
55 V
GATE VOLTAGE
+
SIGNAL
–
GATE
+
BIAS
–
PRACTICAL MAGNETIC AMPLIFIER
LOAD

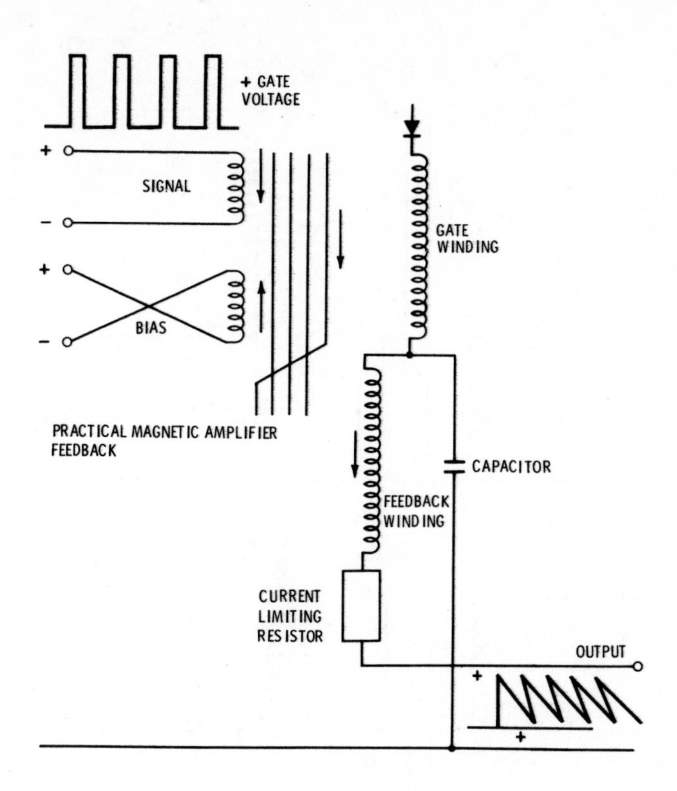

PRACTICAL MAGNETIC AMPLIFIER FEEDBACK

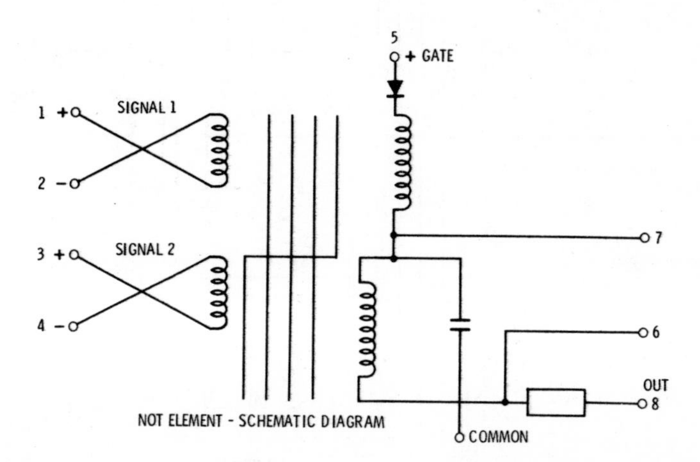

NOT ELEMENT - SCHEMATIC DIAGRAM

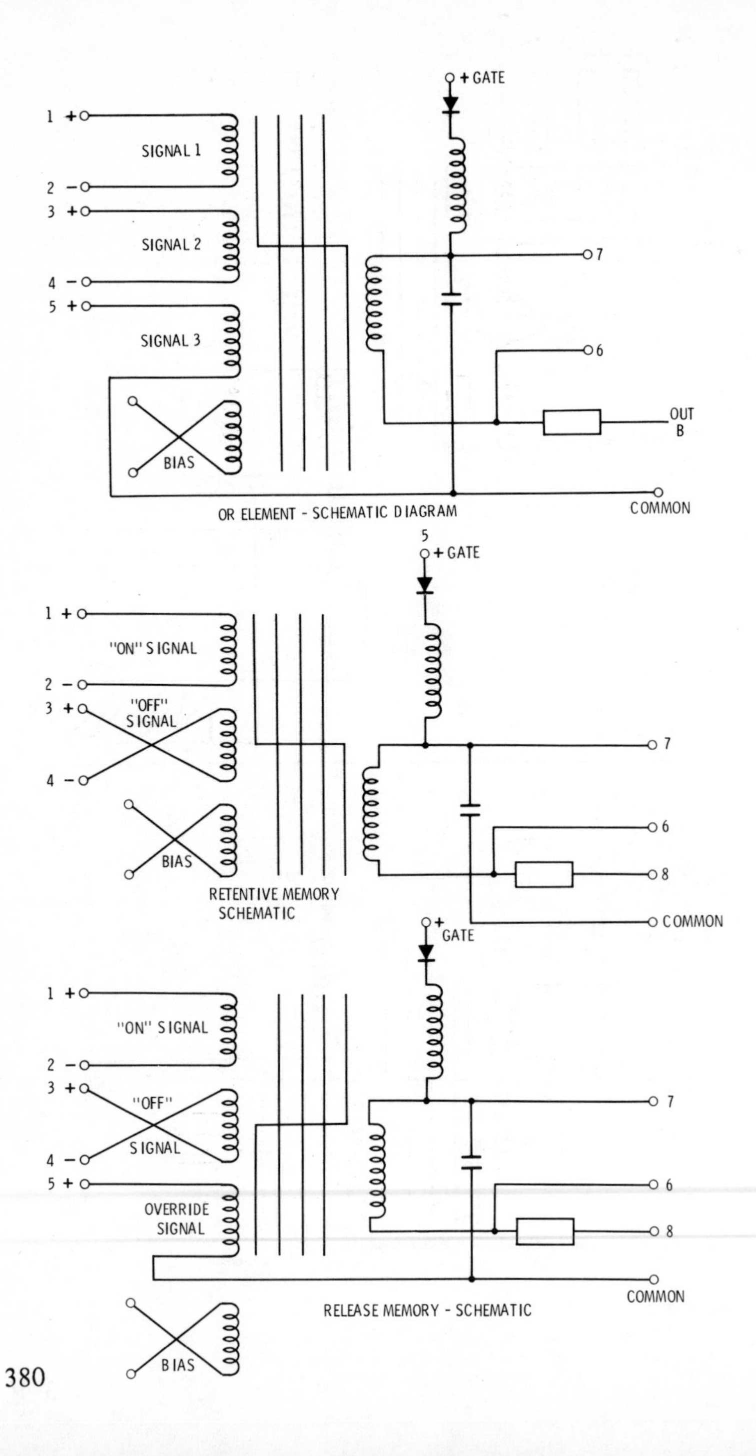
+ GATE
1 +
SIGNAL 1
2 −
3 +
SIGNAL 2
4 −
5 +
SIGNAL 3
BIAS
7
6
OUT B
COMMON
OR ELEMENT - SCHEMATIC DIAGRAM
5
+ GATE
1 +
"ON" SIGNAL
2 −
3 +
"OFF" SIGNAL
4 −
BIAS
7
6
8
COMMON
RETENTIVE MEMORY SCHEMATIC
+ GATE
1 +
"ON" SIGNAL
2 −
3 +
"OFF" SIGNAL
4 −
5 +
OVERRIDE SIGNAL
7
6
8
COMMON
RELEASE MEMORY - SCHEMATIC
BIAS

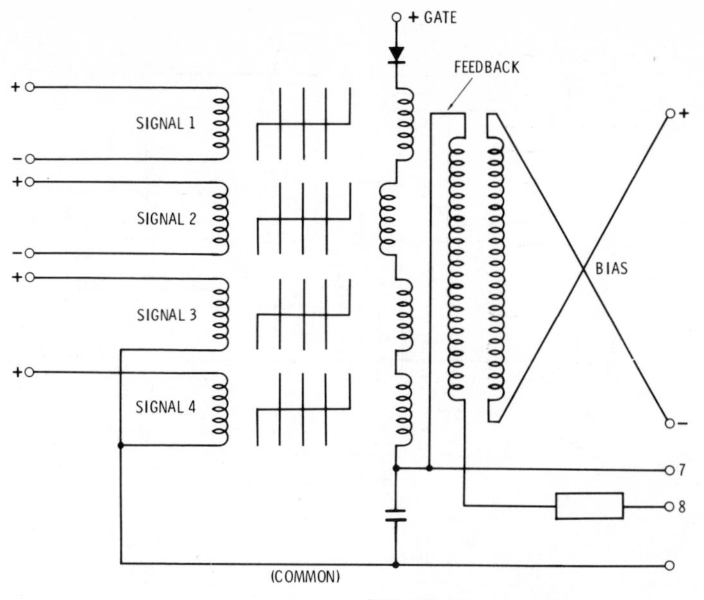

FOUR-INPUT "AND" SCHEMATIC

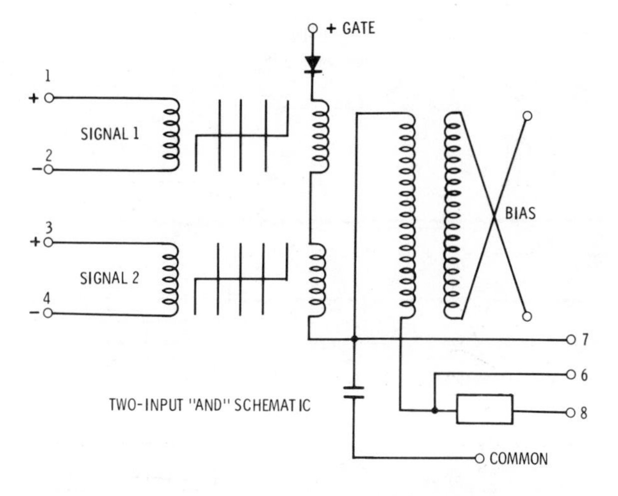

TWO-INPUT "AND" SCHEMATIC

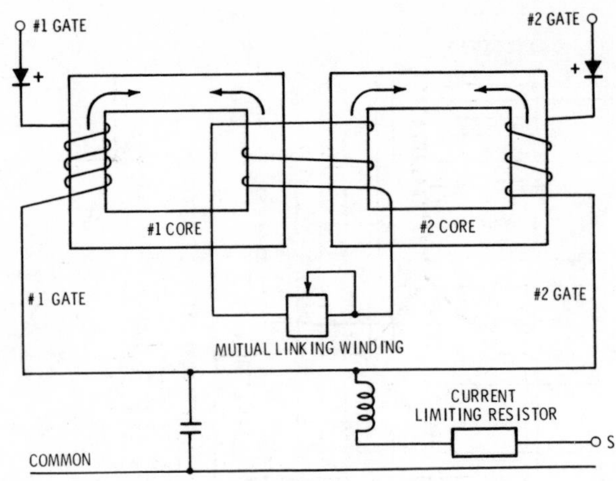

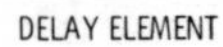

DELAY ELEMENT

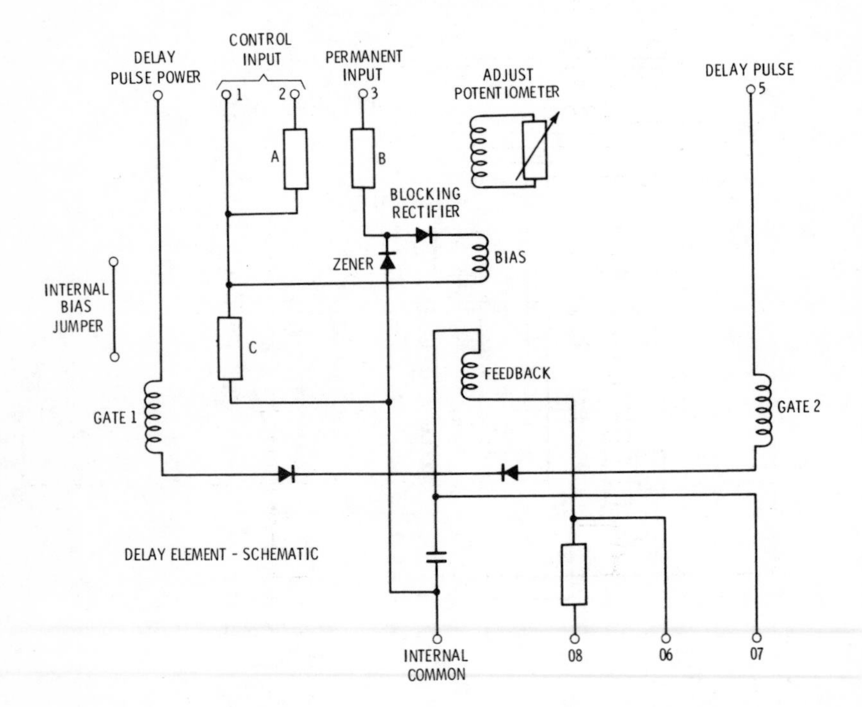

DELAY ELEMENT - SCHEMATIC

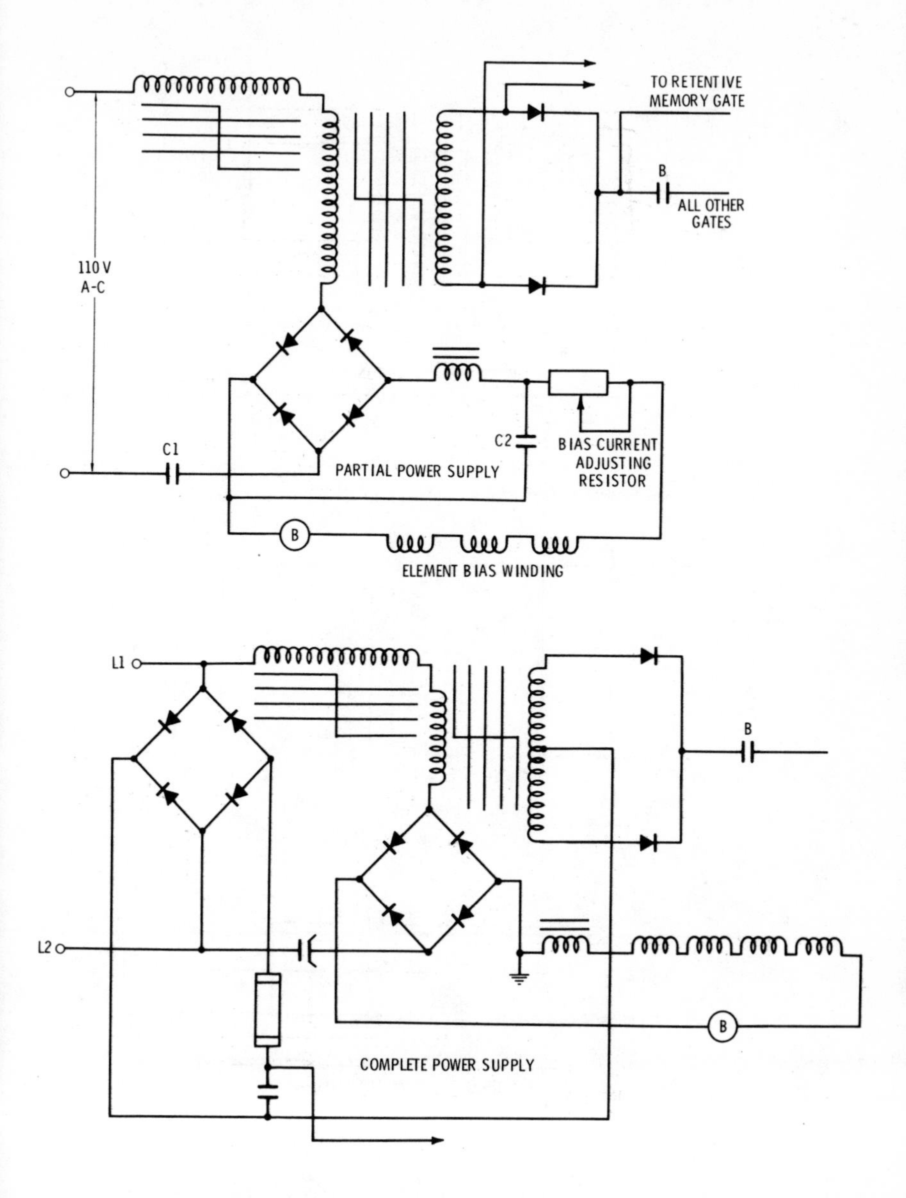
TO RETENTIVE MEMORY GATE
B
ALL OTHER GATES
110 V A-C
C1
C2
BIAS CURRENT ADJUSTING RESISTOR
PARTIAL POWER SUPPLY
B
ELEMENT BIAS WINDING
L1
L2
B
B
COMPLETE POWER SUPPLY

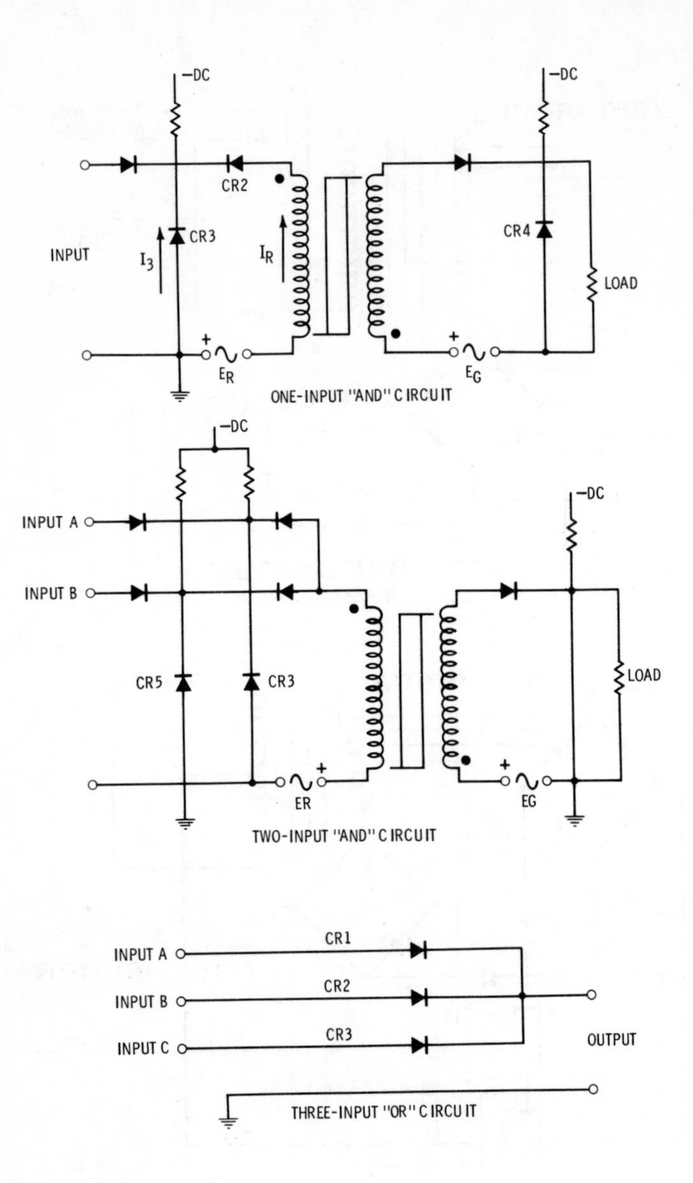

−DC
−DC
CR2
CR3
INPUT
I3
IR
CR4
LOAD
+
ER
+
EG
ONE-INPUT "AND" CIRCUIT
−DC
−DC
INPUT A
INPUT B
CR5
CR3
LOAD
+
ER
+
EG
TWO-INPUT "AND" CIRCUIT
CR1
INPUT A
CR2
INPUT B
CR3
INPUT C
OUTPUT
THREE-INPUT "OR" CIRCUIT

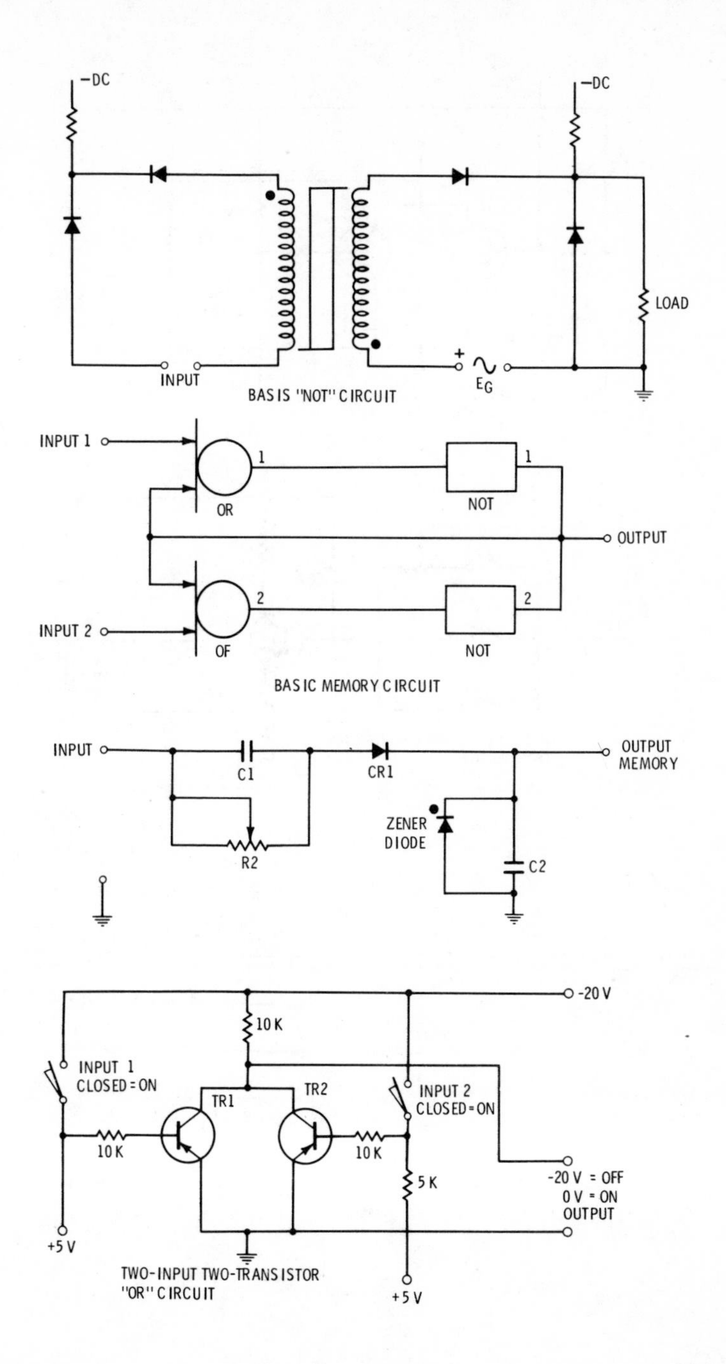
-DC
-DC
LOAD
INPUT
+
E_G
BASIS "NOT" CIRCUIT
INPUT 1
1
OR
1
NOT
OUTPUT
2
2
INPUT 2
OF
NOT
BASIC MEMORY CIRCUIT
INPUT
C1
CR1
OUTPUT
MEMORY
R2
ZENER
DIODE
C2
-20 V
10 K
INPUT 1
CLOSED = ON
TR1
TR2
INPUT 2
CLOSED = ON
10 K
10 K
5 K
-20 V = OFF
0 V = ON
OUTPUT
+5 V
TWO-INPUT TWO-TRANSISTOR
"OR" CIRCUIT
+5 V

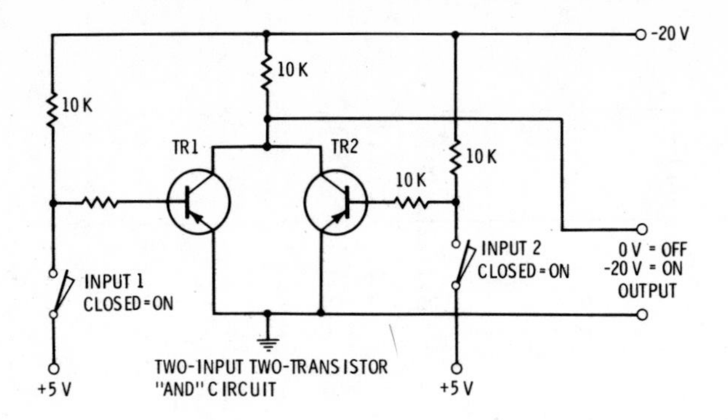

TWO-INPUT TWO-TRANSISTOR "AND" CIRCUIT

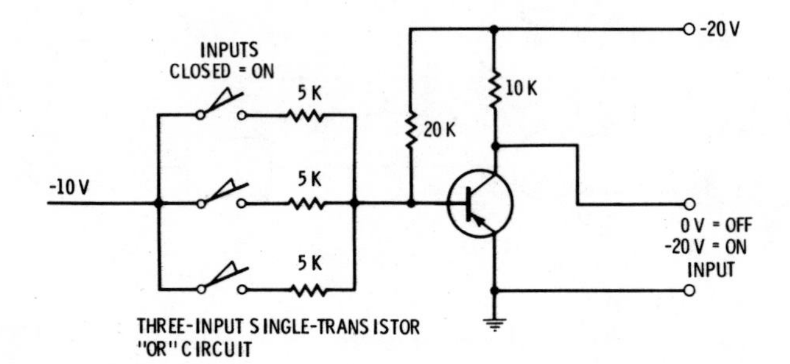

THREE-INPUT SINGLE-TRANSISTOR "OR" CIRCUIT

Notes

Notes

Notes

Notes